ELECTRICAL SYSTEMS
for FACILITIES MAINTENANCE PERSONNEL

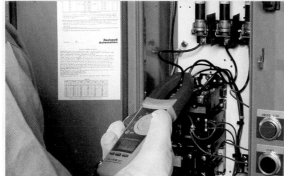

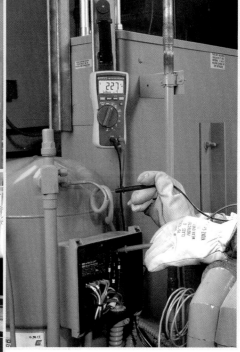

AMERICAN TECHNICAL PUBLISHERS
ORLAND PARK, ILLINOIS 60467-5756

Glen A. Mazur

Electrical Systems for Facilities Maintenance Personnel contains procedures commonly practiced in industry and the trade. Specific procedures vary with each task and must be performed by a qualified person. For maximum safety, always refer to specific manufacturer recommendations, insurance regulations, specific job site and plant procedures, applicable federal, state, and local regulations, and any authority having jurisdiction. The material contained is intended to be an educational resource for the user. American Technical Publishers, Inc. assumes no responsibility or liability in connection with this material or its use by any individual or organization.

American Technical Publishers, Inc., Editorial Staff

Editor in Chief:
    Jonathan F. Gosse
Vice President—Production:
    Peter A. Zurlis
Art Manager:
    James M. Clarke
Multimedia Manager:
    Carl R. Hansen
Technical Editor:
    Peter A. Zurlis
    James T. Gresens
Copy Editor:
    Jeana M. Platz
Cover Design:
    Jennifer M. Hines
Illustration/Layout:
    Jennifer M. Hines
    Samuel T. Tucker
    Melanie G. Doornbos
CD-ROM Development:
    Gretje Dahl
    Daniel Kundrat
    Adam T. Schuldt
    Nicole S. Polak
    Kathryn C. Deisinger
    Amanda N. Sidorowicz

Microsoft, Windows, Windows Vista, PowerPoint, and Internet Explorer are either registered trademarks or trademarks of Microsoft Corporation in the United States and/or other countries. Adobe, Acrobat, and Reader are registered trademarks of Adobe Systems Incorporated in the United States and/or other countries. Intel is a registered trademark of Intel Corporation in the United States and/or other countries. Firefox is a registered trademark of the Mozilla Foundation. National Electrical Code and NEC are registered trademarks of the National Fire Protection Association, Inc. Quick Quiz, Quick Quizzes, and Master Math are either registered trademarks or trademarks of American Technical Publishers, Inc.

© 2011 by American Technical Publishers, Inc.
All rights reserved

1 2 3 4 5 6 7 8 9 – 11 – 9 8 7 6 5 4 3 2 1

Printed in the United States of America

    ISBN 978-0-8269-1594-8

This book is printed on recycled paper.

ACKNOWLEDGMENTS

The authors and publisher are grateful to the following companies and organizations for providing technical information and assistance.

ASI Robicon
Baldor
Cleaver-Brooks
Electrical Apparatus Service Association, Inc.
Fluke Corporation
Furnas Electric Co.
Greenheck
Henny Penny Corporation
McQuay International
Megger Group Limited
Panduit Corp
Saftronics Inc.
Siemens
The Lincoln Electric Company
The Snell Group
UE Systems, Inc.
Werner Ladder Co.

CONTENTS

1. POWER TRANSMISSION AND DISTRIBUTION SYSTEMS — 1
Generators • Transmission Substations • Transmission Lines • Distribution Substations • Distribution Lines • Electrical Services • Building Power Requirements • Building Voltage-Level Requirements • Switchgear/Switchboards • Transformers • Power Protection and Monitoring • Automatic Transfer Switches • Feeder Panels • Conduit and Busway Systems • Motor Control Centers • Panelboards and Branch Circuits • Receptacles • Building Distribution System Drawings

2. PRACTICING ELECTRICAL SAFETY — 43
Safety Labels • Electrical Shock • Arc Flash and Arc Blast • Personal Protective Equipment • Lockout/Tagout • Basic First Aid • Fire Extinguishers • Grounding for Property and Equipment Safety • Double-Insulated (1000 V) Electric Safety Tools • Fuses, Circuit Breakers, and Overload Protection

3. ELECTRICAL QUANTITIES AND CIRCUITS — 67
DC Voltage • AC Voltage • Voltage Problems • Direct Current • Alternating Current • Current Problems • Resistive Circuits • Ohm's Law and Impedance • True Power • Reactive Power • Apparent Power • Phase Shift • Power Formula • Series Circuits • Parallel Circuits • Series-Parallel Circuits • Magnetism • Electromagnetism • Electric Circuit Sections

4. USING ELECTRICAL TEST INSTRUMENTS — 99
Test Instrument Usage • Test Instrument Terminology • Test Instrument Abbreviations • Test Instrument Symbols • Voltage Testers • Digital Multimeters • Multimeter Resistance Measurement • Multimeter Current Measurement • Multimeter Voltage Measurement • Megohmmeters • Megohmmeter Measurement Procedure

5. ELECTRICAL DISTRIBUTION SYSTEMS — 129
Testing Standard and Isolated-Ground Receptacles • Testing GFCI Receptacles • Testing 120 V, Single-Phase Locking and Nonlocking Receptacles • Special-Use Receptacles • Testing 240 V, Single-Phase Receptacles • Testing 120/240 V, Single-Phase Receptacles • Testing Three-Phase Receptacles • Neutral-to-Ground Voltage Troubleshooting • Measuring Circuit Loading • Testing Fuses • Testing Circuit Breakers • Voltage Drop • Voltage Unbalance • Current Unbalance • Improper Phase Sequence • Harmonic Distortion • K-Rated Transformers • Transformer Test Measurements

6. LIGHTING SYSTEMS — 171
Incandescent Lamps • Tungsten-Halogen Lamps • Light-Emitting Diode (LED) Lamps • Fluorescent Lamps • High-Intensity Discharge Lamps • Exit Lighting Systems • Emergency Exit Lighting Systems • Manual Lighting Switches • Troubleshooting Manually Operated Lighting Switches • Motion Sensors • Light Level Sensors • Contactors

7 HVAC SYSTEMS _____ 209
Electric Motor Malfunctions • Motor Environmental and Mechanical Problems • Single-Phase Motors • Three-Phase Motors • Troubleshooting Three-Phase Motors • Heating Elements • Electric Heating Circuit Devices • Normally Closed and Normally Open Valves • Troubleshooting Solenoid-Operated Directional Control Valves

8 MOTOR CONTROL _____ 255
Motor Power Circuits • Disconnect Switches • Power Circuit Terminal Identification • Troubleshooting Motor Power Circuits • Motor Control Circuits • Control Switches • Mechanical and Solid-State Relays • Timers • Troubleshooting Control Circuits • Magnetic Motor Starters • Electric Motor Drives • Motor Circuit Connections • Troubleshooting Motor Drives

APPENDIX _____ 295

GLOSSARY _____ 311

INDEX _____ 321

INTERACTIVE CD-ROM CONTENTS

- Quick Quizzes®
- Illustrated Glossary
- Flash Cards
- Virtual Meters
- Troubleshooting Simulations
- Media Clips
- ATPeResources.com

INTRODUCTION

Electrical Systems for Facilities Maintenance Personnel is a comprehensive resource for those responsible for maintaining, testing, and troubleshooting electrical systems in commercial buildings. Maintenance personnel are expected to understand the operating principles of electrical systems and develop the necessary skills for effective troubleshooting. This book is designed to be a practical resource for personnel working on electrical systems in a facility that have some background in electrical theory.

Electrical Systems for Facilities Maintenance Personnel begins with a detailed overview of power transmission and distribution systems, electrical safety, electrical theory, and test instrument use. Distribution system, lighting system, HVAC system, and motor control circuit principles, testing, and troubleshooting concepts required by facilities maintenance personnel are presented through a hands-on approach. Objectives at the beginning of each chapter provide learning goals for the topics presented. Case studies present and help reinforce system troubleshooting and problem solving skills through real-world scenarios. Review questions at the end of each chapter test for comprehension of information presented.

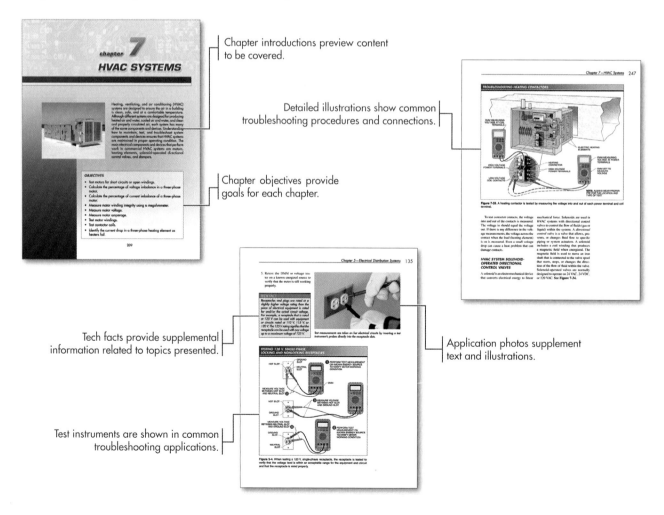

INTERACTIVE CD-ROM FEATURES

The interactive CD-ROM included with the book is a study aid that includes the following:

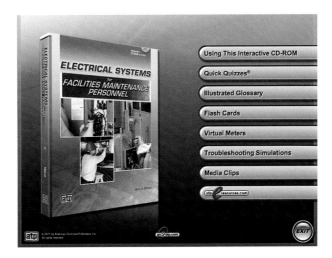

- Quick Quizzes® with 10 questions per chapter to reinforce fundamental concepts
- an Illustrated Glossary of industry terms with links to illustrations, video clips, and animated graphics
- Flash Cards for review of common industry terms and definitions
- Troubleshooting Simulations that present common electrical system problems and their solutions
- Media Clips that depict electrical concepts through video clips and animated graphics
- access to ATPeResources.com, which provides a comprehensive array of instructional resources

To obtain information on related training products, visit the ATP web site at www.go2atp.com

The Publisher

chapter 1
POWER TRANSMISSION AND DISTRIBUTION SYSTEMS

Once produced, electricity must be safely transmitted and distributed for use in residential, commercial, and industrial applications. Power distribution systems include power production equipment, transformers, switchgear, and branch circuits. At the point of use, electrical energy is converted into light, heat, rotary motion, linear motion, and sound for entertainment, comfort, and business and production operations.

OBJECTIVES

- Explain the difference between a wye and delta distribution system.
- Explain the purpose of transformers in a distribution system and their kVA rating.
- Explain the difference between single-phase and three-phase electric services and describe their uses.
- Define switchgear, feeder panels, busway systems, motor control centers (MCCs), and panelboards in a building power distribution system.
- Identify the types of single-phase and three-phase receptacles used in a building.
- Explain the differences between pictorial diagrams, single-line diagrams, line (ladder) diagrams, wiring diagrams, schematic diagrams, assembly drawings, floor plans, plot plans, utility plans, and elevation drawings.

POWER PRODUCTION

Generators produce the vast majority of electrical power used in the United States. Utilities must produce enough power for every load connected to the utility grid. For example, a single-family home requires that thousands of watts be available at any one time. A 240 V/200 A residential service panel delivers 13,800 W when loaded to 30% of its total rating. A commercial building may require tens of thousands of watts.

A *watt (W)* is the unit of electrical power equal to the power produced by a current of 1 A across a potential difference of 1 V. A watt is the product of the voltage and current in a circuit. The amount of power required for a commercial building is based on the power required for the individual loads located in the building. Some loads require a higher starting power than operating power. **See Figure 1-1.**

COMMON ELECTRICAL DEVICE POWER RATINGS*		
Load	Operating Power	Starting (Surge) Power
Coffee maker	750	750
Dish washer (electric heat dry)	1400	1600
Clock or radio	5–15	5–15
Microwave oven	650	1000
Portable space heater	1000–1500	1000–1500
3 A computer with monitor	360	430
Office light fixture	80	96
Refrigerator	560	1500
Pendant light fixture	40	50
Copier	720	1400

* in W

Figure 1-1. The amount of power required for a building is based on the power required for the individual loads in the building.

TECH FACT
Many sources of energy are available that are used to power generators that produce electricity. Some of these sources of energy include nonrenewable sources such as coal, nuclear power, natural gas, and oil, while others include renewable sources such as wind, solar power, and water.

Generators

A *generator* is a device that converts mechanical energy into electrical energy by means of electromagnetic induction. *Electromagnetic induction* is the production of electricity by the interaction of a conductor cutting through a magnetic field. Alternating current (AC) generators convert mechanical energy into AC voltage and current. AC generators are the most common generators used to produce electrical power.

AC generators consist of field windings, an armature, slip rings, and brushes. **See Figure 1-2.** A *field winding* is a group of wires used to produce the magnetic field in the stator of a generator. The magnetic field can be produced by electromagnets or permanent magnets. Most generators use electromagnets. An *electromagnet* is a device consisting of a core and coil that generates magnetism only when an electric current is present in the coil.

An armature is the movable coil of wire in a generator that rotates through the magnetic field. As the armature rotates, each half of a loop cuts across the magnetic lines of force, inducing the same strength voltage in each side of the armature. The voltage in half of the loop enables current flow in one direction, and the voltage in the other half enables current flow in the opposite direction. However, since the two halves of the loop are connected, the voltages add together. The result is that the total voltage of a full rotation of the armature is twice the voltage of each loop or coil half.

The ends of the armature coil are connected to slip rings. A *slip ring* is a metallic ring connected to the end of an armature loop and is used to connect the induced voltage to a brush. A *brush* is the sliding contact that rides against the slip rings and is used to connect the armature to the external circuit (load). The slip rings do not reverse the polarity of the output voltage produced by the generator.

Single-Phase AC Generators. Each complete rotation of the armature in a single-phase AC generator produces one complete alternating current cycle. In the 0° position, before the armature begins to rotate, there is no voltage and no current in the external (load) circuit because the armature is not cutting across any magnetic lines of force. As the armature rotates from the 0° position to the 90° position, each half of the armature cuts across the magnetic lines of force producing current in the armature and external circuit. The current increases from zero to its maximum value in the positive direction and is represented by the first quarter (0° to 90°) of the sine wave. **See Figure 1-3.**

As the armature rotates from the 90° position to the 180° position, current continues to be produced. The current output decreases from its maximum value to zero. This changing value of current is represented by the second quarter (91° to 180° of rotation) of the sine wave.

As the armature continues to rotate to the 270° position, each half of the coil cuts across the magnetic lines of force in the opposite direction, which changes the direction of the current produced. During this time, the current produced increases from zero to its maximum negative value and is represented by the third quarter (181° to 270° of rotation) of the sine wave.

As the armature continues to rotate to the 360° position (same as the 0° position), the current produced decreases to zero, which completes one 360° cycle of the sine wave. Single-phase generators are used for low wattage applications such as portable generators on construction sites and back-up power generators.

The voltage produced by a sine wave can be measured and stated as a peak or rms value. The *peak value* is the maximum instantaneous value of the sine wave. The *root-mean-square voltage (Vrms)*, also known as the effective voltage, of a sine wave is the AC voltage value that produces the same amount of heat in a pure resistive circuit as DC voltage of the same value. The rms value is equal to the peak value times 0.707 and is the value stated on most electrical devices. For example, a light bulb rated at 120 V (rms) is designed for 170 V peak. Peak values are used when rating surge suppressors and troubleshooting power quality problems.

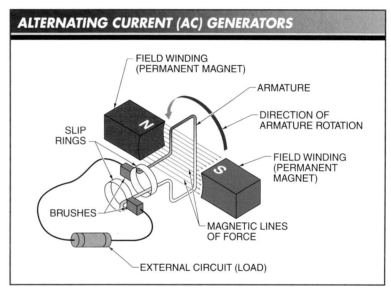

Figure 1-2. AC generators consist of field windings, an armature, slip rings, and brushes.

Wind turbines are generators that convert the mechanical energy of their rotating blades into electrical energy.

ELECTRICAL SYSTEMS for FACILITIES MAINTENANCE PERSONNEL

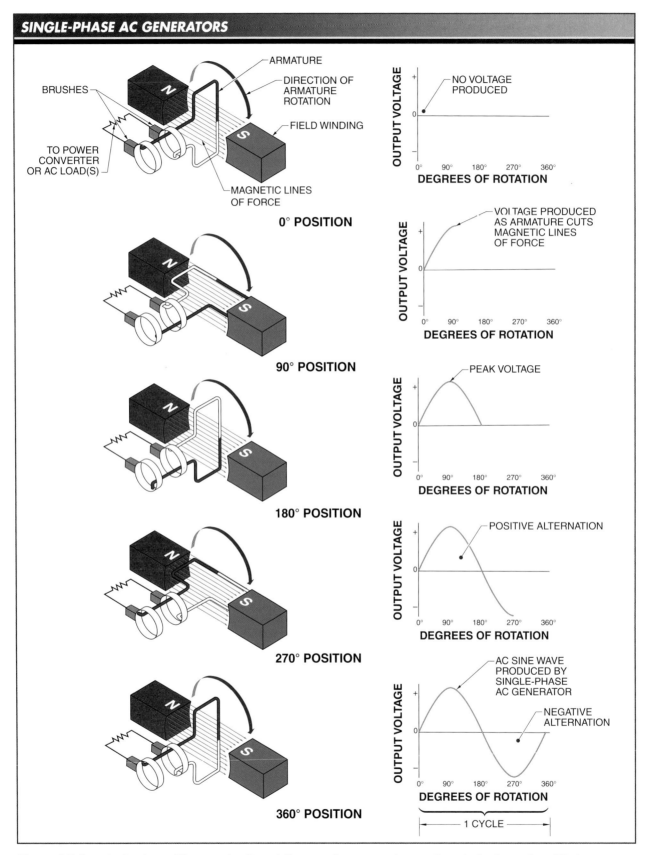

Figure 1-3. In a single-phase AC generator, the rotating armature generates a voltage in the form of an AC sine wave.

Three-Phase AC Generators. Three-phase AC generators produce power more efficiently than single-phase generators and are used by utility companies. A *utility* is an organization that installs, operates, or maintains the electrical, communication, gas, water, or related services in a community. A three-phase generator produces three individual phases (phases A, B, and C). Phases A, B, and C can be wye or delta connected to the distribution system. A *wye connection* is a connection that has one end of each coil connected together and the other end of each coil left open for external connections. A *delta connection* is a connection that has each coil connected end to end to form a closed loop. **See Figure 1-4.**

The three power lines (phases A, B, and C) are brought out of the generator and connected to transformers at the beginning of the utility power distribution system. The manner in which the leads are connected determines the electrical characteristics of the generator output.

UTILITY POWER TRANSMISSION AND DISTRIBUTION SYSTEMS

Electrical power used in residential, commercial, and industrial buildings is normally generated by a utility at a central point and transmitted and distributed to where it is required through the utility power transmission and distribution system. A utility power transmission and distribution system controls, protects, transforms, and regulates electrical power so it can be safely delivered to the user. The utility power transmission and distribution system begins at the point of power production and normally ends at a building metered service entrance point, which is where the building distribution system begins. A utility power transmission and distribution system consists of transmission substations (step-up transformers), transmission lines, distribution substations (step-down transformers), and distribution lines. **See Figure 1-5.**

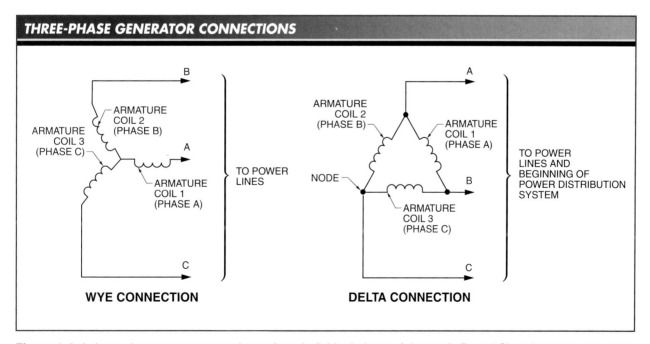

Figure 1-4. A three-phase generator produces three individual phases (phases A, B, and C) and can be connected to the distribution system using wye or delta connections.

ELECTRICAL SYSTEMS for FACILITIES MAINTENANCE PERSONNEL

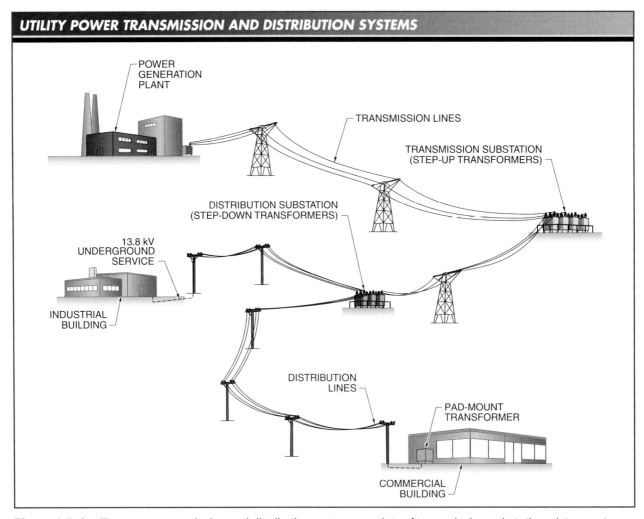

Figure 1-5. A utility power transmission and distribution system consists of transmission substations (step-up transformers), transmission lines, distribution substations (step-down transformers), and distribution lines.

Transmission Substations

A *transmission substation* is an outdoor facility located along a utility system that is used to change voltage levels, provide a central place for system switching, monitoring, protection, and redistribute power. Transmission substations normally operate at high voltage (HV), 69 kV to 345 kV, and extra-high voltage (EHV), voltage over 345 kV. Transmission substations are also used to make changes in the size and number of lines sent out from the station.

Transformers. A *transformer* is an electrical device that uses electromagnetism to change voltage from one level to another or to isolate one circuit from another. Transformers are used in electrical distribution systems to increase or decrease the voltage safely and efficiently. Although a transformer can be used to increase or decrease voltage, transformers cannot be used to increase or decrease the amount of power available. Except for some minor power loss caused primarily by heat loss, the amount of power entering a transformer is the same amount of power leaving the transformer. Transformers allow utilities to distribute large amounts of power at a reasonable cost. **See Figure 1-6.** Transformers are rated in kVA, which specifies their power output capability.

TRANSFORMERS

Figure 1-6. Transformers are used in electrical distribution systems to safely and efficiently increase or decrease voltage to allow utilities to distribute large amounts of power at a reasonable cost.

The main advantage of increasing voltage and reducing current is that power may be transmitted through small gauge conductors, which reduces the cost of power lines. For this reason, the generated voltages are stepped up to high levels for distribution across large distances and then stepped down to meet user requirements. Although both the voltage and current can be stepped up or down, the terms "step-up" and "step-down," when used with transformers, always apply to voltage.

Transmission Lines

A *transmission line* is an aerial conductor that carries large amounts of electrical power at high voltages over long distances. Transmission lines must be spaced far enough apart in order to be safe. The transmission voltage level varies depending on the required transmission distance and amount of power carried. The longer the distance or higher the transmitted power, the higher the transmitted voltage. **See Figure 1-7.**

Figure 1-7. Transmission lines safely carry large amounts of electrical power at high voltages over long distances.

Transmission line voltages can vary from a few kilovolts to hundreds of kilovolts. Transmission-line voltage is stepped up to allow large amounts of power to be transmitted using smaller conductors. Since conductor sizes are based on the amount of current they can safely carry without overheating, low current levels can be carried over small size conductors. For a given power level, the amount of current varies inversely with the amount of voltage. **See Figure 1-8.**

POWER, VOLTAGE, AND CURRENT RELATIONSHIP		
Power*	Voltage†	Current‡
10,000	20,000	0.5
10,000	10,000	1
10,000	5000	2
10,000	2500	4
10,000	1250	8
10,000	480	20.83
10,000	240	41.66
10,000	120	83.33

* in W
† in V
‡ in A

Figure 1-8. For a given power level, the amount of current varies inversely with the amount of voltage.

In addition, increasing the transmitted voltage lowers the power losses between the utility generator and final delivery point. Doubling the transmitted voltage can reduce the power loss by up to 75%. Because transmitting power at high voltages reduces the required size and weight of the conductors, the poles and towers that support the conductors can be smaller and spaced farther apart. Therefore, higher transmitted voltages allow for reduced conductor size, allow more power to be transmitted, and result in lower construction and material cost.

Utility generators that output three-phase power have high-voltage distribution lines arranged in groups of three. In addition to the power lines, a neutral/ground conductor is also routed with the power lines. The neutral/ground conductor is routed on top of power lines and used as a grounding wire to help dissipate lightning strikes. The neutral/ground conductor is grounded at every power pole and at the transmission and distribution substations. The voltage on the power lines is stepped up and down many times before it reaches the end use.

Distribution Substations

A *distribution substation* is an outdoor facility located close to the point of electrical service use and is used to change voltage levels, provide a central place for system switching, monitoring, and protection, and redistribute power. Distribution substations take high-transmitted voltages and reduce the voltage for further distribution. Distribution substations normally operate at lower voltages than transmission substations. Distribution substation output voltages are normally between 12 kV and 13.8 kV.

Distribution substations provide a location along the distribution system near the end user to easily test the system, adjust voltage output, add new lines, disconnect lines, and redirect power during distribution system problems such as power outages caused by lightning strikes. **See Figure 1-9.** Distribution substations take the incoming power and, after changing the voltage level, produce multiple outputs with different voltages on each line.

Distribution Lines

Distribution lines are used to carry electrical power from a distribution substation to the building service entrance. Distribution lines connect parts of the system together and are often run in multiple lines so that electrical power can be switched to meet changing power requirements and switched between different utilities. The term "grid" is used to describe the network of interconnected transmission and distribution lines.

DISTRIBUTION SUBSTATIONS

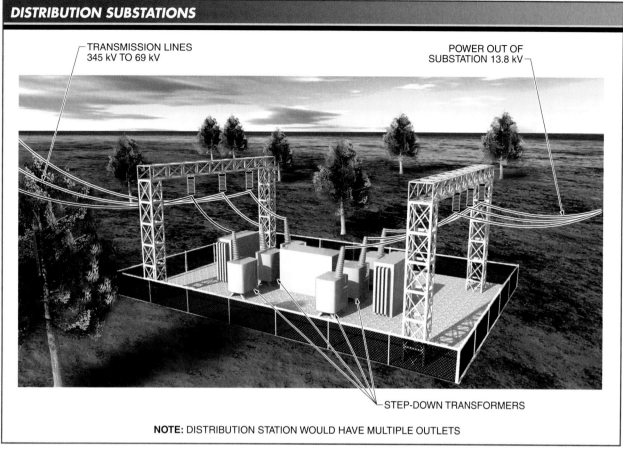

Figure 1-9. Distribution substations provide a convenient place along the distribution system for maintenance, checks, and line adjustments.

Electrical Services

An electrical service may be overhead or lateral. An *overhead service* is an electrical service in which service-entrance conductors are run from the utility pole through the air and to the building. A *lateral service* is an electrical service in which service-entrance conductors are run underground from the utility service to the building. **See Figure 1-10.**

A 120/240 V, single-phase, three-wire service is used to supply power to end users that require 120 V and 240 V single-phase power. This level of service provides 120 V single-phase circuits, 240 V single-phase circuits, and 120/240 V single-phase circuits. Because the neutral wire is grounded, it is not fused or switched at any point. **See Figure 1-11.** A 120/240 V, single-phase, three-wire service is commonly used for interior wiring for lighting and small-appliance use. This service is used to supply most residential buildings and for small commercial applications, although a large power panel or additional panels can be used.

TECH FACT

The U.S. Occupational Safety and Health Administration (OSHA) requires that rooms or spaces that contain electrical supply lines or equipment, such as distribution lines from substations, be enclosed within fences, screens, partitions, or walls to deter unqualified persons from entering.

ELECTRICAL SYSTEMS for FACILITIES MAINTENANCE PERSONNEL

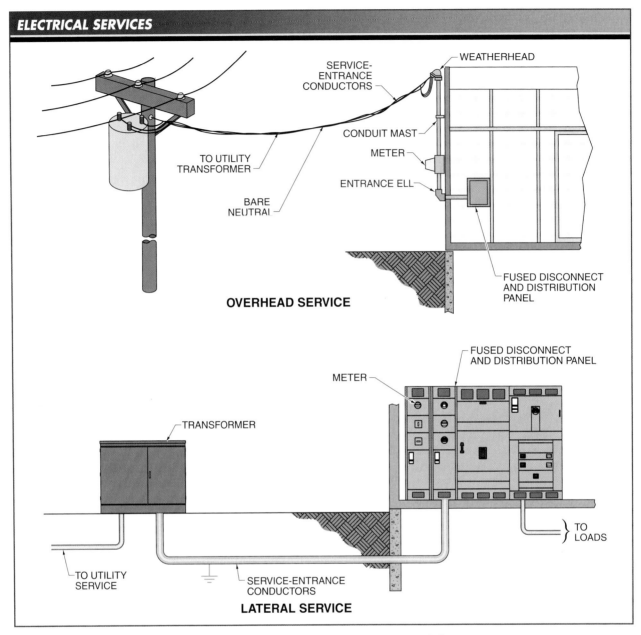

Figure 1-10. Overhead or lateral services may be used to supply power to buildings.

A 120/208 V, three-phase, four-wire, wye-connected service is used to supply commercial customers that require large amounts of 120 V single-phase power, 208 V single-phase power, and low-voltage three-phase power. This service level includes three ungrounded (hot) lines and one grounded (neutral) line. Each hot line has 120 V to ground when connected to the neutral line. **See Figure 1-12.**

A 120/208 V, three-phase, four-wire service is used to provide large amounts of low-voltage, 120 V single-phase power. The 120 V circuits are balanced to distribute the power among the three hot lines equally. This is accomplished by alternately connecting the 120 V circuits to the power panel so each phase (A to N, B to N, and C to N) is divided among the loads.

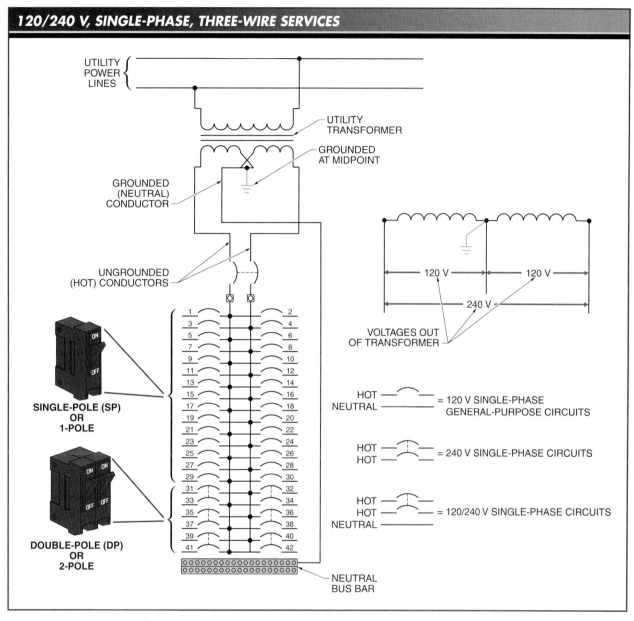

Figure 1-11. A 120/240 V, single-phase, three-wire service is used to supply power to customers that require 120 V and 240 V single-phase power.

Likewise, 208 V single-phase loads, such as 208 V lamps and heating appliances, should also be balanced between phases (A to B, B to C, and C to A). Three-phase loads, such as heating elements and three-phase motors, can be connected to phases A, B, and C. The loads connected to a transformer should be connected so that the transformer is as electrically balanced as possible. Electrical balance occurs when loads on a transformer are placed so that each coil of the transformer carries the same amount of current. **See Figure 1-13.**

TECH FACT
Regardless of the color-code system used to identify different phase conductors (phases A, B, and C), it is also recommended to mark each conductor with bands of electrical tape or wire markers.

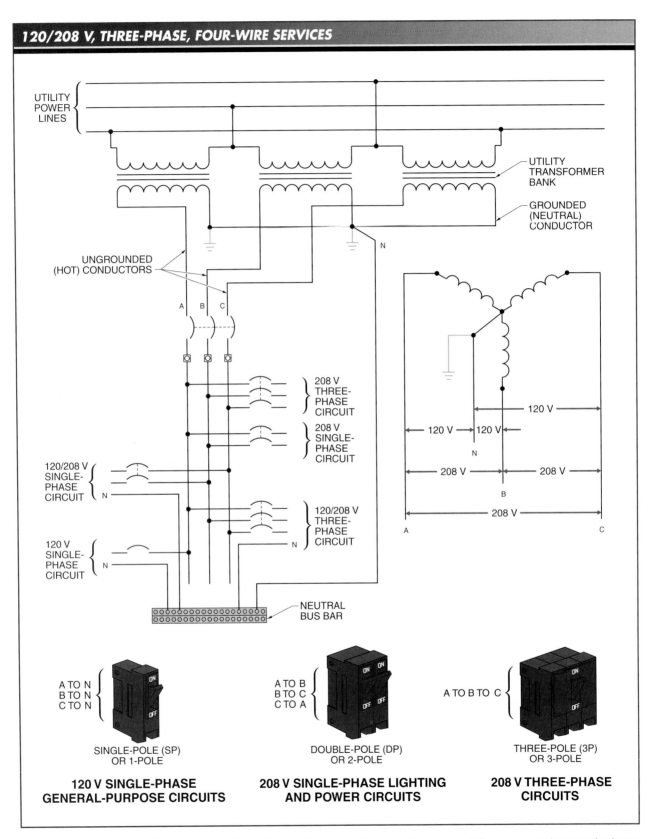

Figure 1-12. A 120/208 V, three-phase, four-wire service is used to supply commercial customers that require large amounts of 120 V single-phase power, 208 V single-phase power, and low-voltage three-phase power.

LOAD BALANCING

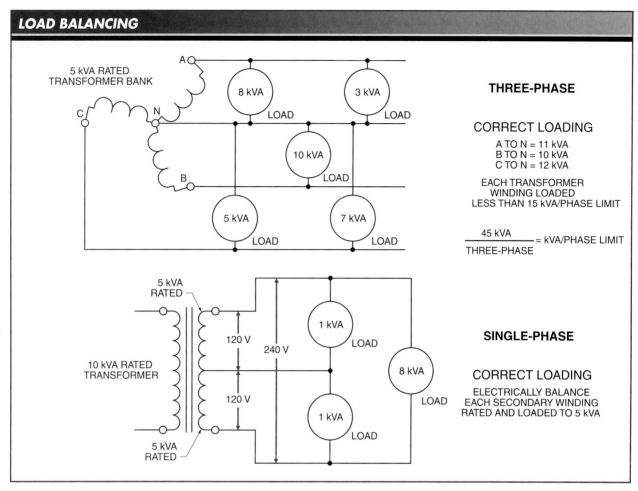

Figure 1-13. The loads connected to a transformer should be connected so that the transformer is as electrically balanced as possible.

A 120/240 V, three-phase, four-wire, delta-connected service is used to supply commercial and industrial customers that require large amounts of three-phase power with some 120 V and 240 V single-phase power. This service supplies single-phase power delivered by one of the three transformers and three-phase power delivered by using all three transformers. Single-phase power is provided by center tapping one of the transformers. **See Figure 1-14.**

A 277/480 V, three-phase, four-wire, wye-connected service is the same as the 120/208 V, three-phase, four-wire service except the voltage levels are higher. This service includes three ungrounded (hot) lines and one grounded (neutral) line. Each hot line has 277 V to ground when connected to the neutral or 480 V when connected between any two hot lines (A to B, B to C, or C to A). **See Figure 1-15.**

This service provides 277 V or 480 V single-phase power but not 120 V single-phase power. For this reason, a 277/480 V, three-phase, four-wire service is not used to supply 120 V single-phase general-lighting and appliance circuits. However, this service can be used to supply 277 V and 480 V single-phase lighting circuits. Such high-voltage lighting circuits are used in commercial fluorescent and high-intensity discharge (HID) lighting circuits, which are a major part of many commercial applications.

14 ELECTRICAL SYSTEMS for FACILITIES MAINTENANCE PERSONNEL

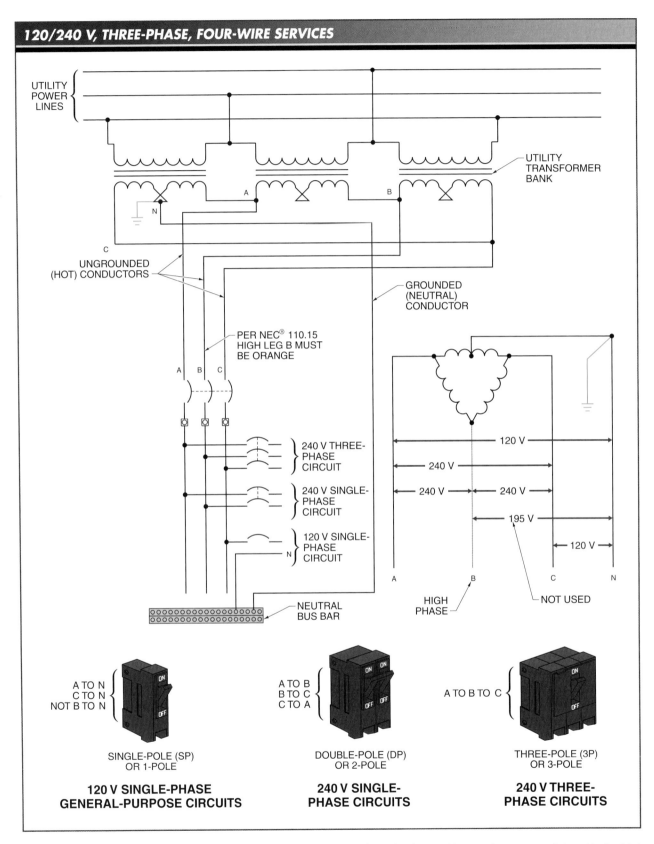

Figure 1-14. A 120/240 V, three-phase, four-wire, delta-connected service is used to supply commercial and industrial customers that require large amounts of three-phase power and some 120 V and 240 V single-phase power.

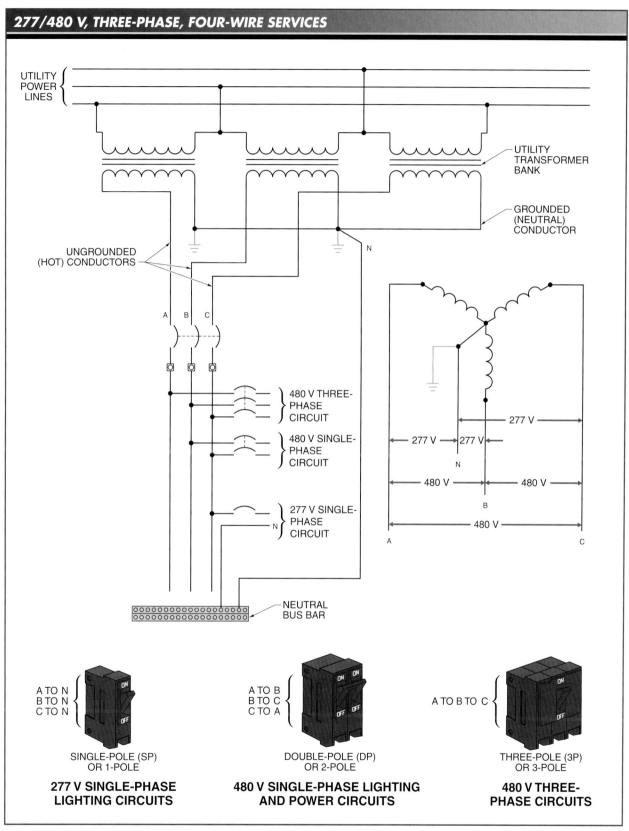

Figure 1-15. A 277/480 V, three-phase, four-wire, wye-connected service is used for buildings that require high voltage for high-power loads.

BUILDING POWER DISTRIBUTION SYSTEMS

The utility power transmission and distribution system ends at the building service entrance. The building power distribution system begins where the utility system ends. The purpose of the utility is to deliver quality power to the building and meter the power for billing. Quality power should be as free as possible from power losses, voltage fluctuations, transient voltages (high-voltage spikes), and other problems that could cause damage to equipment, equipment downtime, or personal injury.

Building Power Distribution Systems Media Clip

Building Power Requirements

The amount of power a building requires depends on the building size, electrical use within the building, and any possible changes or additions within or to the building. Electricity is used in commercial applications to provide energy for office equipment, lighting, heating, cooling, large-scale cooking, and pumping applications.

Building Voltage-Level Requirements

Commercial buildings require electrical power to operate electrical loads. The amount of power required determines the appropriate voltage level. Since power is determined by the amount of voltage and current, higher voltages are best for reducing the amount of required current. For example, a typical 120 V, 250 W high-pressure sodium lamp draws an operating current of 2.70 A, a 240 V, 250 W lamp draws 1.36 A, and a 480 V, 250 W lamp draws 0.65 A. Therefore, the higher the rated lamp voltage, the less operating current required. Since most lamps in a commercial building are normally connected to a lighting circuit, the higher voltage and lower current reduces the required wire and conduit size.

The size of the building distribution system and available power depends on the physical size of the building, intended use of the building (office, manufacturing, warehouse, etc.), and any future requirements. Buildings with multiple tenants, such as commercial retail, office, or apartment buildings, require many separately metered panels customized to meet individual requirements. Commercial and industrial buildings have customized systems to meet specific requirements. **See Figure 1-16.**

Switchgear/Switchboards

Electrical power is delivered to commercial and industrial buildings through a transmission and distribution system. S*witchgear* is a high power electrical device that switches or interrupts devices or circuits in a building distribution system. A *switchboard* is a panel or an assembly of panels containing electrical switches, meters, busses, and other overcurrent protection devices (OCPDs). Once power is delivered to a building, switchgears and switchboards further distribute the power to where it is required within the building.

TECH FACT
The terms "switchgear" and "switchboard" are sometimes used interchangeably. Traditionally, the term switchboard was applied to a single unit that housed fuses or circuit breakers, high-voltage power switches, and monitoring/metering devices. The term switchgear was applied to separate units that also included high-voltage switches, overcurrent protection, and monitoring/metering devices that may or may not be housed as one unit. Also, the term switchgear was normally applied to most high-voltage electrical switches that did not house overload protection devices in the same enclosure. Today it is more common to simply refer to both as switchgear.

BUILDING POWER DISTRIBUTION SYSTEMS

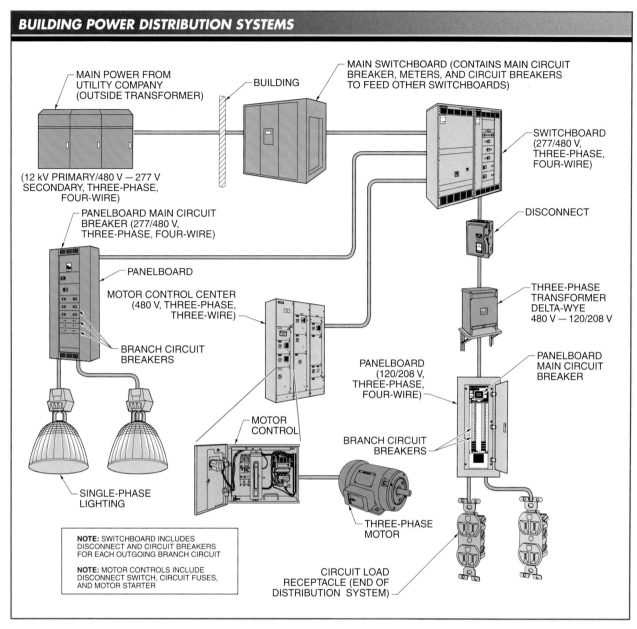

Figure 1-16. The design of a building's power distribution system depends on the physical size of the building, intended use, and any future needs.

In most buildings, the switchgear is the last point on the power distribution system for the utility company and the beginning of the building distribution system. A switchgear is rated by the manufacturer for a maximum voltage and current output. For example, a switchgear may have a 600 V rating and a bus rating of up to 5000 A. The 600 V rating is the maximum voltage rating and is normally used with a lower voltage to the building, such as 480 V. In addition to dividing the incoming power, switchgear normally contain all equipment required to control, monitor, protect, and record power usage. Switchgear is normally designed as service-entrance and distribution switchgear.

ELECTRICAL SYSTEMS for FACILITIES MAINTENANCE PERSONNEL

Service-entrance switchgear have space and mounting provisions for metering equipment as required by the local power company, fuses or circuit breakers, and disconnecting means for the service conductors. **See Figure 1-17.** Provisions for grounding the service neutral conductor when a ground is required are also provided.

Distribution switchgear contain protective devices, such as circuit breakers or fuses, and feeder circuit conductors required to distribute the power throughout a building. Distribution switchgear may include metering and monitoring devices depending on the size and function of the equipment.

Transformers

Transformers are used to step up, step down, or isolate voltages. Step-up transformers increase the voltage out of the transformer secondary. Step-down transformers decrease the voltage out of the transformer secondary. An isolation transformer is used to separate the input voltage from the output voltage electrically without increasing or decreasing the voltage.

Isolation transformers are used to help correct power distribution problems such as transient voltages, noise, and harmonics. In a building power distribution system, transformers are used to step down the incoming high voltage to a low voltage that operates components such as lamps, motors, appliances, electric heating elements, security systems, and computers.

Transformers are depicted on single-line diagrams to illustrate the voltage levels at different points along the building power distribution system. A *single-line diagram* is an electrical drawing that uses a single line and basic symbols to illustrate the current path, voltage values, circuit disconnects, fuses or circuit breakers, transformers, and panelboards for a power distribution system. **See Figure 1-18.**

Single-line diagrams are used when troubleshooting distribution system problems such as loss of power, low voltage, blown fuses, tripped circuit breakers, and power-quality problems. They are also used to determine power shut-off points, future expansion capacity, and where emergency back-up generators or secondary power systems are connected into the system.

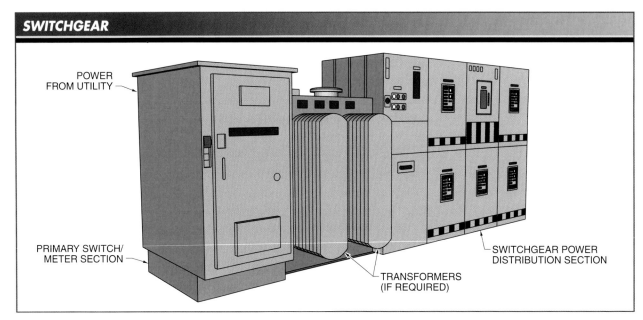

Figure 1-17. Switchgear is a high power electrical device that switches or interrupts devices or circuits in a building distribution system.

SINGLE-LINE DIAGRAMS

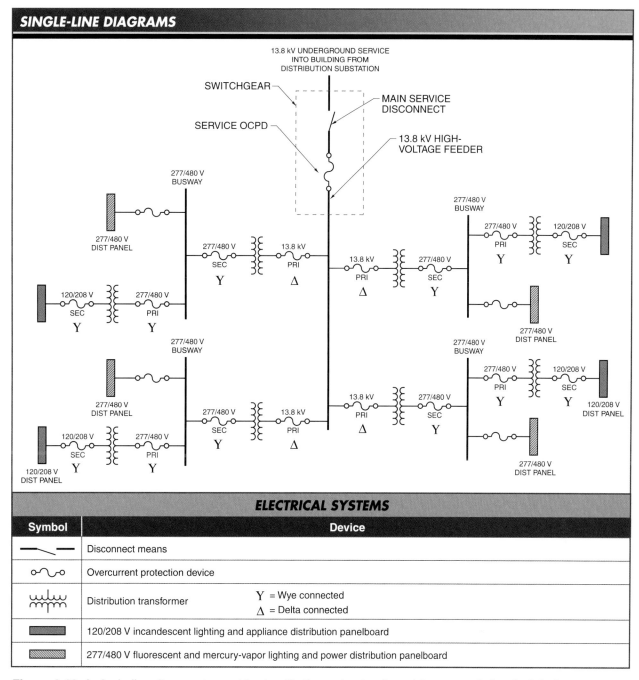

Figure 1-18. A single-line diagram is used to simplify the understanding of the types of electrical devices used in a building power distribution system.

Single-line diagrams use the most basic symbols because the intent of the drawing is to illustrate as clearly as possible the flow of current throughout the building distribution system and where each component or device connects into the system. Single-line diagrams are also used when designing large commercial and industrial installations to show the flow path of electrical power throughout a building.

Transformer banks can be arranged in either a wye or delta connection. The four possible transformer connections are wye primary to wye secondary, delta primary to delta secondary, wye primary to delta secondary, and delta primary to wye secondary.

Wye-connected transformers provide three different types of voltages: single-phase low voltage, single-phase high voltage, and three-phase voltage. Single-phase low voltage is available between any one power line (phases A, B, or C) and the grounded neutral. Single-phase high voltage is available between any two power lines (phases A, B, or C), and three-phase voltage is available from all three power lines (phases A, B, and C). For example, a wye-connected secondary can provide three different types of service: 208 V three-phase service, 208 V single-phase service, and 120 V single-phase service. A wye-connected transformer bank is commonly used in schools, commercial stores, and offices. **See Figure 1-19.**

The advantage of a wye-connected secondary is that the single-phase power draw may be divided equally over the three transformers. Each transformer carries one-third of the single-phase and three-phase power if the loads are divided equally. A disadvantage of a wye-to-wye connection is that interference with telephone circuits may result.

On the primary side of the transformer bank, the grounded neutral wire is connected to the common points of all three high-voltage primary coil windings. The high-voltage side is marked "H1" and "H2" on each transformer to indicate the high-voltage side of the transformer. The low-voltage side is marked "X1" and "X2." The voltage from the neutral to any phase of the three power lines is 2400 V. The voltage across the three power lines is 4152 V (1.73 × 2400 V = 4152 V).

As on the primary side, the grounded neutral is connected to the common points of all three low-voltage secondary coil windings. This allows for a 120 V output on each of the secondary coils. The voltage across the three secondary power lines is 208 V (1.73 × 120 V = 208 V).

TECH FACT
Low-voltage conditions can cause computers and other pieces of equipment to reset or turn off. This condition may be present when indicators for low voltage are displayed on HVAC drives or when voltage measurements determine a problem. If there is a low-voltage condition in the power distribution system, the taps on transformers are tested to determine if they can be used to raise (or lower for high-voltage problems) the voltage in the building distribution system.

Power Protection and Monitoring

Power protection devices are essential to prevent personal injury and property damage due to electrical shock and fire. However, power protection devices that activate during a ground fault can also be used to find where problems exist along a system. These problems indicate locations to monitor and systems to modify to help prevent future problems.

Monitoring a system provides information about how the system is operating and any faults that occur. When a fault occurs, protection devices must respond. In addition, signaling devices notify operators of a problem. Signaling devices may be local, such as alarms, or remote, such as wireless or hard-wired devices that can transmit information back to central monitoring stations.

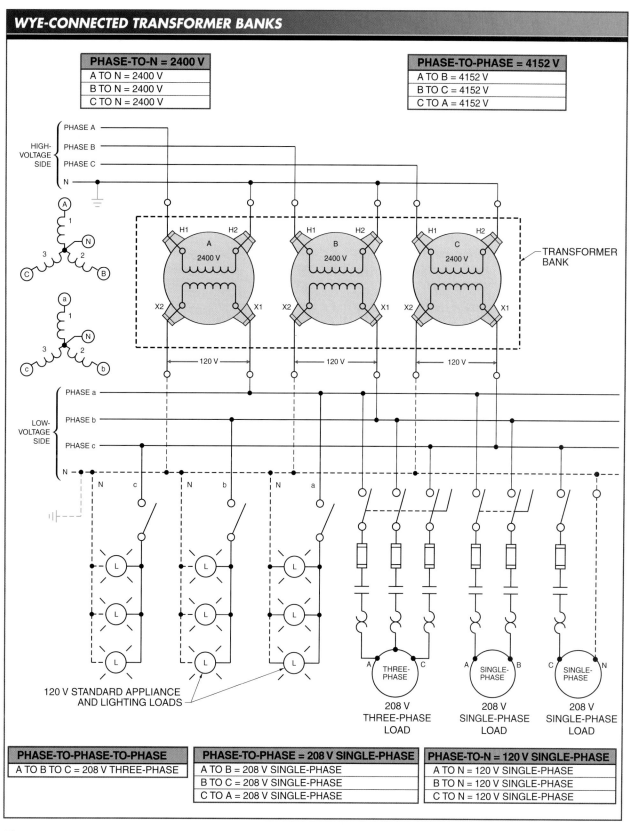

Figure 1-19. A wye-connected transformer bank is used to produce single-phase and three-phase voltages and is used mostly where large amounts of single-phase power are required.

Monitoring devices that record electrical operating conditions, such as voltage, current, and power, are essential for maintaining and improving a distribution system as well as handling legal issues that arise from customer complaints. System monitoring also allows power to be switched to different areas of a system that require more or less power than other areas in certain situations. **See Figure 1-20.**

Automatic Transfer Switches

An *automatic transfer switch* is an electrical device that transfers the load of a building from public utility circuits to the output of a standby generator during a power failure. Automatic transfer switches, also known as bypass switches, are critical devices that are part of a building emergency and standby power system. Automatic transfer switches can be stand-alone units or can be integral devices in switchgear, switchboards, or panelboards. The location of the transfer switch depends on the loads that must be powered after a loss of the main power. The transfer switch and backup power source must be rated to power the loads that remain connected to the system after the transfer. **See Figure 1-21.**

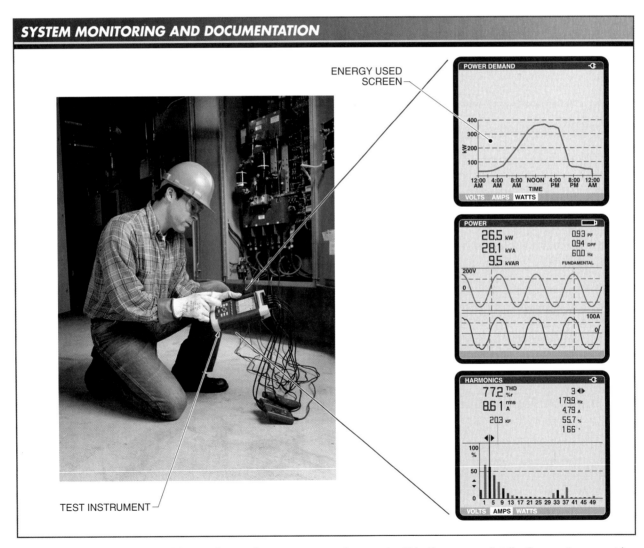

Figure 1-20. Electrical quantities such as voltage, power, and current within the power distribution system must be measured, monitored, and documented.

Transfer switches can be automatic or manual. Automatic transfer switches can be either open or closed transition. An *open transition switch,* also known as a break-before-make switch, is an automatic transfer switch that opens one power source before closing another power source. A *closed transition switch,* also known as a make-before-break switch, is an automatic transfer switch that closes one power source before opening another power source. Some transfer switches can allow both power sources to remain connected in parallel permitting either or both sources to be connected to supply power.

Feeder Panels

A *feeder panel* is a junction box used to house electrical circuit breakers and main connections. Electrical systems are interconnected using conductors. Interconnecting electrical devices and loads to a power source requires that conductors be spliced. Splicing is accomplished when connecting individual sections between the power sources and points of use or when circuits are divided. Splices are not permitted inside devices such as conduit, panelboards, branch-circuit panels, and motor control centers. When splices and branching are required along the system, enclosures can be added to house the splices. Such enclosures can be small or large depending on the conductor size and number of conductors in the enclosure.

Conduit and Busway Systems

A building electrical distribution system must transport electrical power from the source of power to the loads. In large buildings, this may consist of distribution over large areas with many different electrical requirements throughout the building. Electrical power can be routed throughout a building using different methods, such as wire in conduit or busways. A *feeder* is a conductor between the service equipment and the last branch-circuit OCPD. The most common method of interconnecting electrical devices is through the use of conduit or busways. **See Figure 1-22.**

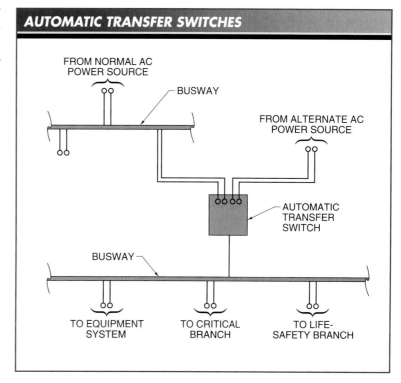

Figure 1-21. Automatic transfer switches monitor building infeed power and transfer power between normal and alternate power sources.

Conduit is a tube or pipe used to support and protect electrical conductors. Conduit is either metal or plastic and may be rigid or flexible. Most conduit used for electrical wiring is metal. The two main types of metal conduit used in commercial installations are rigid metal conduit (RMC) and electrical metallic tubing (EMT). RMC uses threaded fittings while EMT uses compression fittings. In commercial applications, RMC is normally used for underground installations. Specific applications for RMC or EMT are determined by local codes and accepted standards.

CONDUIT AND BUSWAY SYSTEMS

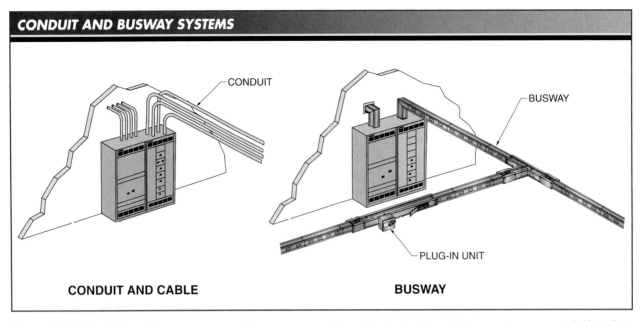

Figure 1-22. Conduit and busways are used to carry power throughout a building. A busway is composed of prefabricated fittings, tees, elbows, and crosses to simplify the connection and reconnection of distribution systems.

A *busway* is a grounded metal enclosure that contains factory-mounted bare or insulated conductors. Prefabricated fittings, tees, elbows, and crosses simplify the installation of the distribution system. By connecting different components of the busway together, electrical power can be made available at many locations throughout a building. For example, busways enable manufacturing plants to move machinery without major changes in the electrical distribution system.

Motor Control Centers

A *motor control center (MCC)* is a central location for troubleshooting and servicing motor control circuits. Electric motors are the most common loads requiring simple and complex control. An MCC combines individual control units into standard individual modules. Power for an MCC is normally supplied from a panelboard or switchgear. An MCC is different from a switchboard or switchgear containing motor panels in that the MCC is a modular structure designed specifically for plug-in control units and motor control. **See Figure 1-23.**

Simplifying and consolidating motor control circuits is required because electric motors are used in most commercial and industrial applications including HVAC, pumping, and production operations. To do this, an MCC combines the incoming power, control circuitry, required overload and overcurrent protection, and any transformation of power into one control center.

An MCC receives incoming power and delivers it to the control circuit and motors. The incoming power is connected to the MCC horizontal bus, which is rated for the maximum amount of current available to all units connected to the MCC. Horizontal bus sections have ratings such as 600 A, 800 A, 1200 A, 1600 A, 2000 A, and 2500 A. Connected to the horizontal bus are vertical bus sections that deliver power to the individual modules connected to that section.

MOTOR CONTROL CENTERS (MCCs)

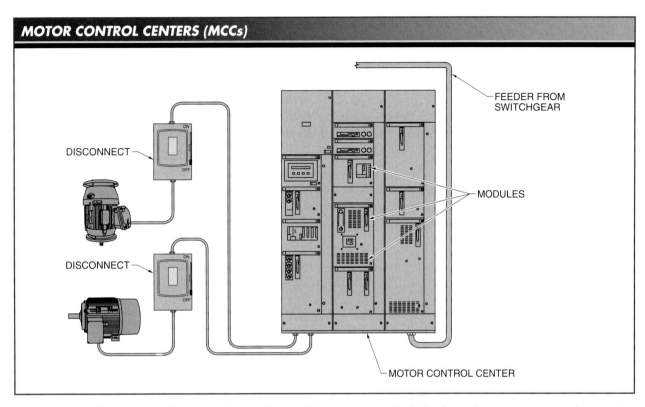

Figure 1-23. MCCs are used to centralize and consolidate motor control circuits and components, such as motor starters, motor drives, circuit breakers, and control transformers.

A vertical bus has a lower rating than the horizontal bus. Vertical bus sections have ratings such as 300 A and 600 A. In addition to current ratings, MCCs have a maximum voltage rating, such as 2.5 kV or 5 kV. Some specifications also include a horsepower rating, such as 200 HP/150 kW. Both horsepower and kilowatt ratings are commonly used because National Electrical Manufacturers Association (NEMA) devices are rated in horsepower and International Electrotechnical Commission (IEC) devices are rated in kilowatts.

An MCC provides space for the control and load wiring in addition to providing space for required control devices. The control inputs into the MCC include control devices such as pushbuttons, liquid-level and limit switches, and other devices that provide a signal. The MCC outputs are the conductors connected to the motors. Other control devices that are located in the MCC include relays, control transformers, motor starters, overload devices, fuses or circuit breakers, timers, counters, and any other required control devices.

An advantage of an MCC is that it provides one location for installing and troubleshooting motor control circuits. This is useful in applications that require multiple individual control circuits that are related to other control circuits, such as an assembly line with one machine that feeds an adjacent machine. Another advantage is that individual plug-in units can be easily removed, replaced, added to, and interlocked at one central location. MCC manufacturers produce preassembled units to control standard motor functions, such as start/stop, reversing, reduced-voltage starting, and speed control. The preassembled units enable the end user to only have to connect the control devices and the motors to the MCC. **See Figure 1-24.**

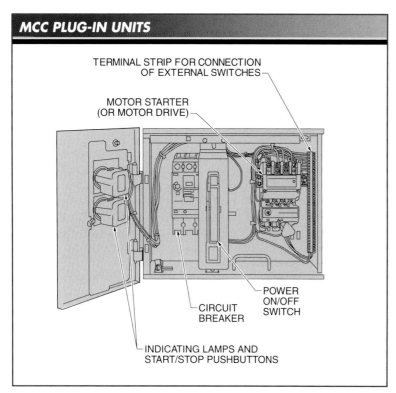

Figure 1-24. MCCs include plug-in units (modules) that are designed to simplify the servicing and troubleshooting of motor control circuits.

TECH FACT

Interchanging any two of the three power lines (A/L1, B/L2, or C/L3) reverses the rotation of a three-phase motor. A phase sequence tester can be used to ensure that the power lines are correctly marked (phases A, B, and C) and the motor rotates in the correct direction.

UE Systems, Inc.
Troubleshooting and servicing are easily performed at motor control centers (MCCs) by qualified persons.

Panelboards and Branch Circuits

A *panelboard* is a wall-mounted distribution cabinet containing a group of OCPDs and short-circuit protection devices for lighting, appliance, or power distribution branch circuits. The wall-mount feature distinguishes the panelboard from a switchboard, which is normally freestanding.

In residential applications, panelboards are supplied with power directly from the utility after the power is metered. In commercial and industrial applications, panelboards are normally supplied from switchboards that further divide the power distribution system into smaller sections.

Panelboards are the part of the distribution system that provides the final centrally located protection for the power run to the load and its control circuitry. Power out of a panelboard can be delivered through branch circuits by conductors routed through conduit or nonmetallic cable as permitted by local codes. A *branch circuit* is the part of an electrical wiring system between the final set of circuit breakers or fuses and the fixtures and receptacles they protect. Panelboards are also referred to as load centers in residential dwellings and small commercial buildings. **See Figure 1-25.**

Fuses and Circuit Breakers. In a properly operating circuit, current is confined to the conductive paths provided by conductors and other components when a load is turned on. Every load draws a normal amount of current when switched on. This normal amount of current is the current level for which the load, conductors, switches, and system are designed to carry safely. Under normal operating conditions, the current in a circuit must be equal to or less than the normal current level. However, sometimes an electrical circuit may have a higher-than-normal current flow (overcurrent).

Chapter 1—Power Transmission and Distribution Systems

PANELBOARDS AND BRANCH CIRCUITS

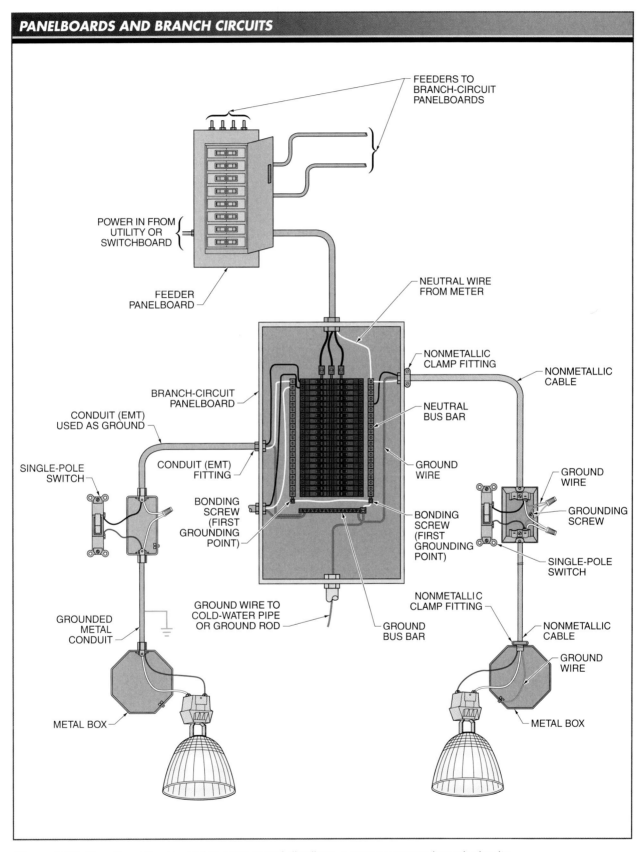

Figure 1-25. Panelboards are used to protect and distribute power to system branch circuits.

Overcurrent Protection Media Clip

Circuit Breaker Operation Media Clip

An *overcurrent* is electrical current in excess of the equipment limit, total amperage load of the circuit, or conductor or equipment rating. An overcurrent may be a short circuit current or an overload current. A *short circuit* is a condition that occurs when two ungrounded conductors (hot wires), or an ungrounded and a grounded conductor of a single-phase circuit, come in contact with each other. A short circuit causes current to rise hundreds of times higher than normal at a fast rate. Short circuits may cause a fire, shock, or explosion and may damage equipment. All circuits must be protected from short circuits.

> **GREEN TECH FACT**
>
> *If fuses blow or circuit breakers trip when power is applied to a circuit, a clamp-on ammeter can be used to test the current draw on the circuit. If the circuit is normally loaded more than 75% of its current rating, the circuit conductors are producing excess heat. Increasing the conductors by one size reduces the amount of heat produced by the conductors and may eliminate nuisance tripping caused by brief overloads, such as when a motor starts.*

An *overload* is a small-magnitude overcurrent that, over a period of time, leads to an overcurrent, which may trip the fuse or circuit breaker. Overloads are caused by defective equipment, overloaded equipment, or excessive loads on one circuit. Overloaded equipment draws a higher-than-normal current based on the degree to which the equipment is overloaded. The more overloaded the equipment, the higher the current draw. As with short circuits, overloads must also be removed from the system.

An *overcurrent protective device (OCPD)* is a fuse or circuit breaker (CB) that disconnects or discontinues current flow when the amount of current exceeds the design load. An OCPD must be used to provide protection from short circuits and overloads. Fuses and circuit breakers are OCPDs designed to automatically stop the flow of current in a circuit that has a short circuit or is overloaded.

A *fuse* is an OCPD with a fusible link that melts and opens a circuit on an overcurrent condition. A *circuit breaker* is an OCPD with a mechanical mechanism that can manually or automatically open the circuit when an overload condition or short circuit occurs.

Fuses and circuit breakers are installed at various points in a power distribution system and in individual pieces of equipment. Fuses and circuit breakers include current and voltage ratings. The current rating is the maximum amount of current the OCPD can carry without opening or tripping. OCPD current ratings are determined by the size and type of conductors, control devices used, and loads connected to the circuit. The voltage rating is the maximum amount of voltage that can be applied to an OCPD.

Every ungrounded (hot) power line must be protected from short circuits and overloads. A fuse or circuit breaker is installed in every ungrounded power line. One OCPD is required for low-voltage single-phase circuits (120 V or less) and all DC circuits. The neutral line in AC circuits or the negative line in DC circuits does not include an OCPD. Two OCPDs are required for high-voltage single-phase circuits (208 V or 240 V). Both ungrounded power lines include an OCPD. Three OCPDs are required for all three-phase circuits regardless of the voltage level. Each of the three ungrounded power lines includes an OCPD. **See Figure 1-26.**

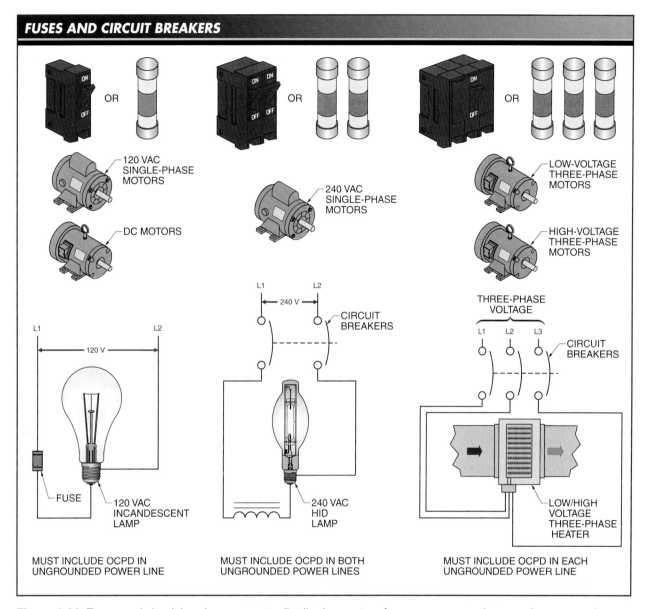

Figure 1-26. Fuses and circuit breakers protect a distribution system from overcurrents that can damage equipment and injure personnel.

Conductors. Electrical circuits and components are connected using conductors. A *conductor* is a material that has little resistance and permits electrons to move through it easily. Conductors are available as individual wire or in groups, such as cable and cord. **See Figure 1-27.** A *wire* is any individual conductor. A *cable* is two or more conductors grouped together within a common protective cover and used to connect individual components. A *cord* is a group of two or more conductors in one cover used to deliver power to a load by means of a plug. Most individual conductors are enclosed in an insulated cover to protect the conductor, increase safety, and meet code requirements. Some individual conductors, usually ground wires, may be bare.

Conductors
Media Clip

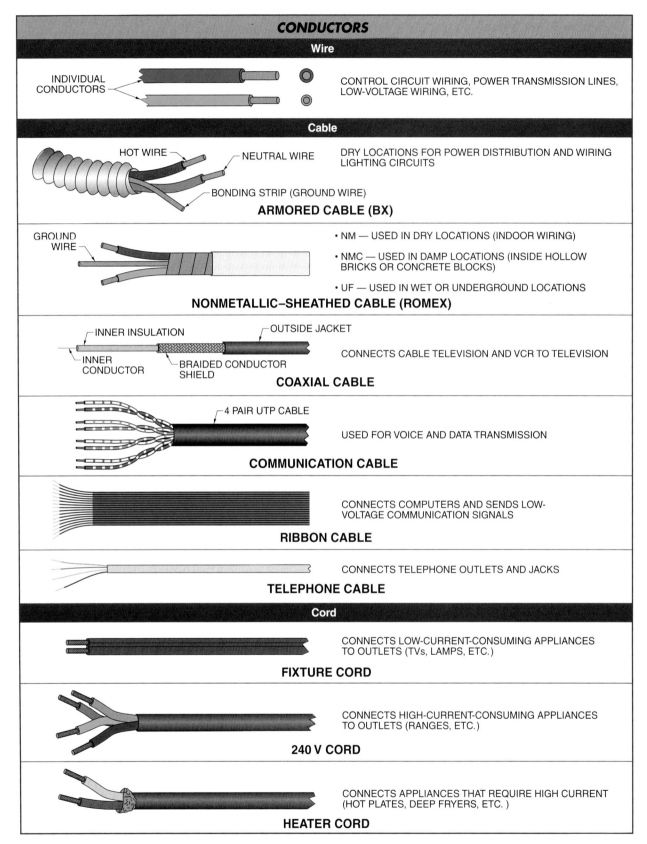

Figure 1-27. Conductors carry electrical current to different equipment and loads within the distribution system.

Conductor material can be copper, aluminum, or copper-clad aluminum. Because of cost, copper (Cu) and aluminum (Al) are the most commonly used materials for conductors. Copper is the most common because it has a lower resistance than aluminum for any given wire size. For this reason, aluminum conductors must be sized one or more sizes larger than copper conductors. *Copper-clad aluminum* is a conductor that has copper bonded to an aluminum core. The total amount of copper used is less than 10% of the conductor. Copper is used to counter the disadvantages of aluminum.

Conductors are sized by using a number, such as No. 12 AWG or No. 14 AWG. The conductor size number is based on the American Wire Gauge (AWG) numbering system. **See Figure 1-28.** The lower the AWG number, the larger the diameter of the conductor and the higher the current-carrying capacity. For example, a No. 12 copper conductor is larger in diameter than a No. 14 copper conductor and can carry 5 A more current than a No. 14 wire. The AWG size used for a circuit depends on the maximum current that the conductor must carry and the conductor material.

Receptacles

A *receptacle,* also known as an outlet, is an electrical contact device for connecting electrical equipment to a power source. The most common 120 V receptacles are duplex receptacles. A *duplex receptacle* is an electrical contact device containing two receptacles. The three basic types of receptacles are standard, isolated-ground, and ground-fault circuit interrupter (GFCI) receptacles. **See Figure 1-29.**

Standard receptacles are the most common receptacles used for most wiring applications. Standard receptacles include a long (neutral) slot, a short (hot) slot, and a U-shaped ground hole. Wires are attached to the receptacle at screw terminals or push-in fittings. A connecting tab between the two hot and two neutral screw terminals provides an electrical connection between the terminals. This electrical connection allows for both terminals to be powered when one wire is connected to either terminal. Receptacles are marked with ratings for maximum current and voltage. Standard receptacles are marked 15 A, 125 V or 20 A, 125 V.

With a standard receptacle, the receptacle ground is connected to the common grounding system when the receptacle is installed in a metal outlet box. The common grounding system normally includes all metal parts of the system such as conductors, boxes, conduit, water pipes, and non-current-carrying metal parts of most electrical equipment. The receptacle ground is included as part of the larger grounding system when electrical equipment is plugged into the receptacle. The common grounding system may act as a large antenna and conduct electrical noise, which may cause interference in computer, medical, security, and communication equipment.

An isolated-ground receptacle is used to minimize problems in sensitive applications or in areas of high electrical noise. An *isolated-ground receptacle* is an electrical connection device in which the grounding terminal is isolated from the device yoke or strap. An isolated-ground receptacle minimizes electrical noise by providing a separate grounding path for each receptacle and is identified by an orange triangle on the face of the receptacle and/or an orange-colored receptacle. A separate ground conductor is run with the circuit conductors to an isolated-ground receptacle.

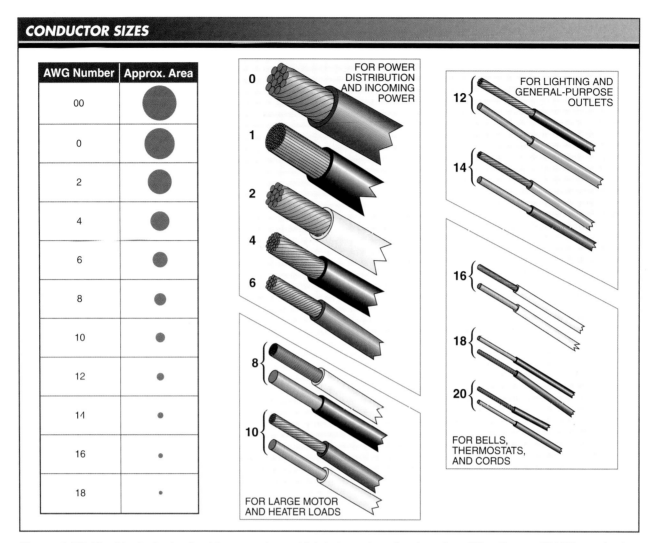

Figure 1-28. Electrical wire is sized by a number, which is based on the American Wire Gauge (AWG) numbering system.

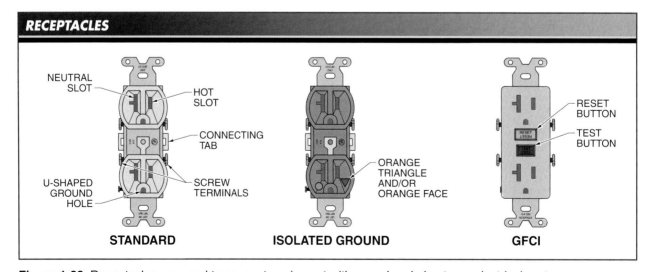

Figure 1-29. Receptacles are used to connect equipment with a cord and plug to an electrical system.

A *ground-fault circuit interrupter (GFCI)* is a fast-acting electrical device that automatically deenergizes a circuit by opening the circuit in response to the grounded current exceeding a predetermined value. A *ground fault* is any amount of current above a certain level that may deliver a dangerous shock. GFCIs provide greater protection than standard and isolated-ground receptacles.

Three-Wire, 120 V and 240 V Single-Phase Receptacles. Standard amperage ratings are available for 120 V single-phase receptacles. Amperage ratings range from 15 A to 50 A, with 15 A and 20 A as the most common. Each receptacle is designed so that only the correctly rated plug can be plugged into the receptacle. Receptacles and plugs rated for 120 V have three slots/prongs, the hot (ungrounded), neutral, and ground. These 120 V plugs may be nonlocking or locking. Locking plugs must be turned after they are inserted into the receptacle to lock the plug in place.

Standard amperage ratings are also available for 240 V single-phase receptacles. Amperage ratings range from 15 A to 50 A. As with 120 V receptacles and plugs, 240 V single-phase receptacles are designed so that only the correctly rated plug can be plugged into a receptacle. The advantage of a 240 V receptacle over a 120 V receptacle is that a 240 V receptacle can deliver twice the power at the same amperage rating. **See Figure 1-30.**

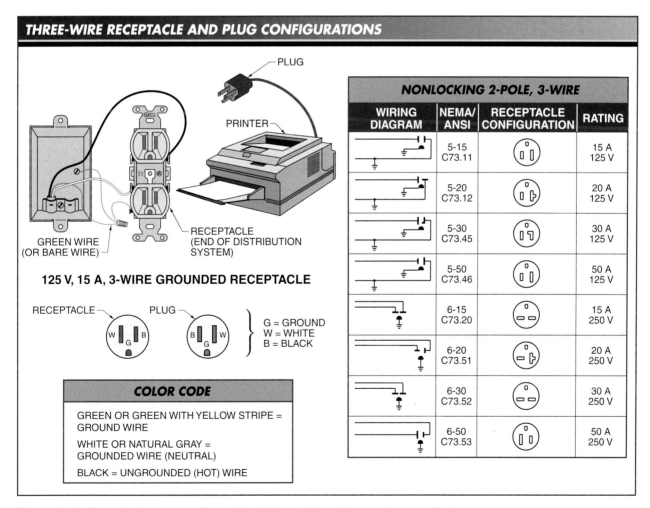

Figure 1-30. Receptacles are configured to indicate their voltage and current limits.

Four-Wire, 120/240 V Single-Phase Receptacles. Some electrical loads require both a 120 V and 240 V power source. Dual-voltage allows for a change in power output and/or multiple electrical loads inside the appliance.

For example, an electric range produces different heat outputs on the heating element by changing the voltage applied to the heating element. The heating element produces higher heat when connected to 240 V than when connected to 120 V. Dual-voltage heating elements have their temperature controlled by an infinite switch (sometimes referred to as a surface burner switch). Theoretically, they can be set at an infinite number of settings. By turning the switch between low and high settings, the current to the electric burner can be increased or decreased to adjust the temperature of the heating element.

Likewise, a crane uses 120 V for the truck motors and 240 V for the hoist motors. A 120/240 V receptacle and plug has four slots, two hot (ungrounded), one neutral, and a ground. **See Figure 1-31.**

Three-Phase Receptacles. Commercial and industrial portable equipment, such as lathes, drill presses, and welding equipment, can have high power ratings and are designed to operate on three-phase power. Three-phase components and loads can be connected to power by hard wiring to a disconnect switch or control center, or they can be connected to a receptacle when furnished with a plug. Hard wiring is more common and standard for large power-consuming devices. Plug and receptacle connections are used with small portable devices and equipment used in laboratories, schools, and other commercial locations.

As with single-phase receptacles, three-phase receptacles are designed to only accept plugs that are rated for the correct voltage and current. Three-phase receptacles and plugs have different configurations of slots and prongs depending on the type of three-phase power supplied. For example, a three-phase receptacle may have three hot lines and one neutral or ground line. **See Figure 1-32.**

Building Distribution System Drawings

Building distribution system drawings consist of electrical prints, drawings, and diagrams used to convey facts, such as directions, measurements, and information on the location and operation of devices, components, and circuits. Electrical prints, drawings, and diagrams use standard symbols to simplify information presented. A *symbol* is a graphic representation of a device, component, or object on a print. Proper identification of the symbol for each electrical device or component in a building diagram or print is required when installing, troubleshooting, servicing, or modifying electrical circuits and systems. **See Appendix.**

Most office printers or copiers are common commercial loads connected to 120 V single-phase circuits.

Chapter 1 — Power Transmission and Distribution Systems 35

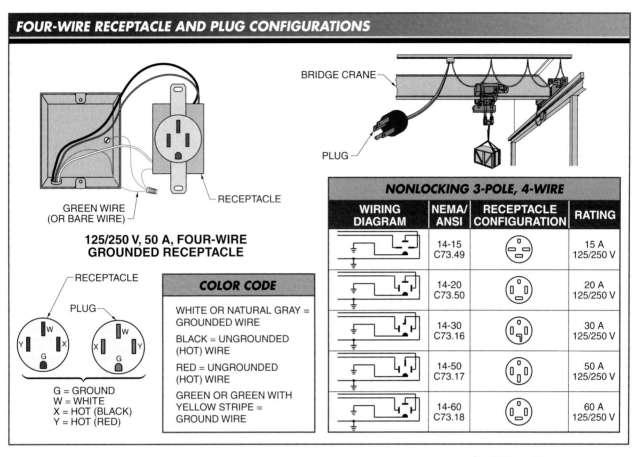

Figure 1-31. Dual-voltage receptacles allow for greater voltage, current, and power flexibility with the loads connected to them.

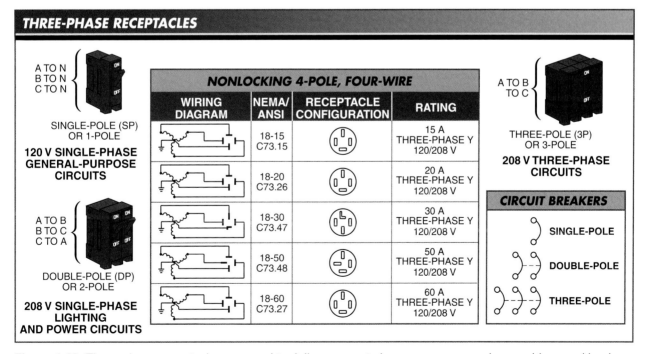

Figure 1-32. Three-phase receptacles are used to deliver power to large power-consuming machines and loads.

Many types of electrical prints, drawings, and diagrams are used by individuals performing electrical work. The major types of building distribution system drawings are pictorial diagrams, single-line diagrams, line (ladder) diagrams, wiring diagrams, schematic diagrams, assembly drawings, floor plans, plot (site) plans, utility plans, and elevation drawings. **See Figure 1-33.**

A *pictorial diagram* is an electrical wiring sketch showing actual positions of system devices and wiring. Pictorial diagrams are used to provide a detailed visual image.

A *single-line diagram* is an electrical drawing that uses a single line and basic symbols to illustrate the current path, voltage values, circuit disconnects, OCPDs, transformers, and panelboards for a power distribution system. Single-line diagrams are used when troubleshooting distribution system problems.

A *line diagram,* also known as a ladder diagram, is a drawing that has a series of single lines (rungs) that indicate the logic of the control circuit and how the control devices are interconnected. Line diagrams are the standard diagram used to show most commercial and industrial control circuits.

A *wiring diagram* is a drawing that indicates the connections of all devices and components in an electrical system. Wiring diagrams are included with most devices and components to indicate how manufacturers' specific equipment are connected into a circuit or system.

A *schematic diagram* is a drawing that shows electrical system circuitry with symbols that depict electrical devices and lines representing the conductors. Schematic diagrams are used to trace a circuit's operation without regard to the actual size, shape, or location of the devices.

An *assembly drawing* is a drawing that shows an entire machine or device with all detailed parts. Assembly drawings must show as much information as possible to allow for proper assembly of a machine or device but must not be overwhelming in their detail.

A *floor plan* is a plan view of a structure that shows the arrangement of walls and partitions as they appear in an imaginary section taken horizontally and approximately 5'-0" above floor level. Floor plans are used to identify and install the required devices at the proper location.

A *plot plan* is a drawing that shows property lines of a building lot, elevation information, compass directions, lengths of property lines, and locations of structures to be built on the lot. Plot plans may also show utility services, landscaping, and other information required to perform the required work.

A *utility plan* is a drawing that indicates the location and intended path of utility lines such as electrical, water, sewage, gas, and communication cables. Utility plans are used along with other plans to ensure proper location and construction of a project.

An *elevation drawing* is an orthographic view of a vertical surface without allowance for perspective. Elevation drawings are used to define the architectural style, structural material, and features of electrical equipment.

Refer to Chapter 1 Quick Quiz® on CD-ROM.

TECH FACT
Printreading skills are necessary for individuals performing work on electrical systems. Printreading skills involve accurately reading and interpreting the available prints, drawings, and diagrams for a project, as well as recognizing standard symbols and abbreviations. Individuals must learn how to visualize an entire system and the relationship of all of its components from a set of prints so that work can be accomplished in a safe and efficient manner and the system can operate as designed.

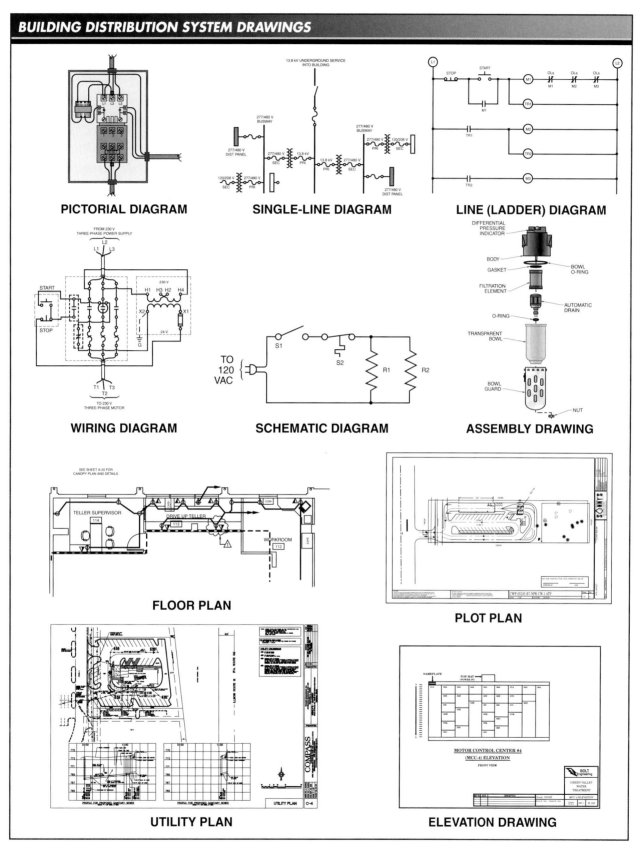

Figure 1-33. Many types of electrical prints, drawings, and diagrams are used by individuals performing electrical work.

Case Study—Building Power Distribution Preventive Maintenance

A building owner hired an outside thermal imaging contractor to help with a building energy savings program. The contractor was charged with looking at building heat losses and other potential problem areas that a thermal scan would detect. As part of the service call, the contractor also scanned part of the building electrical distribution equipment looking for any overheating problems that may be a fire hazard or contribute to wasted energy.

In general, the building distribution system components showed no signs of abnormal heating. However, one disconnect switch showed excessive heat being produced from the box.

Because the disconnect switch showed overheating on the thermal imager, it was looked at closer. The consultant asked the building maintenance staff to open the disconnect switch door. Wearing proper PPE, the maintenance staff opened the disconnect switch so additional thermal images could be taken.

The thermal image taken of the open disconnect shows an overheating problem on the phase B (line 2) fuse output lug and a potential problem on phase A (line 1) fuse input lug. Fuses and lug connectors should not show signs of overheating unless they have abnormally high resistance caused by a loose connection, corrosion, or undersized conductors or fuse.

Based on the thermal imaging contractors findings, the building maintenance staff scheduled a service call for the disconnect switch as part of their preventive maintenance program. As part of the service call, the maintenance staff measured the voltage drop across each phase (phases A, B, and C) of the lugs and fuses. The higher the measured voltage drop, the greater the power (heat) produced across the device.

The voltage measurements show a higher than normal voltage drop (approximately 2 V) across the phase B fuse output lug. The part of the electrical system feeding the disconnect switch was turned off and locked out. Voltage measurements were taken to ensure power was off at the disconnect switch and all lugs were checked and tightened. Power was restored and the new voltage measurements taken indicated that there was no longer a voltage drop (heat) problem.

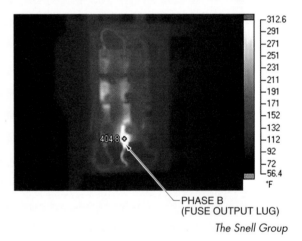

PHASE B (FUSE OUTPUT LUG)

The Snell Group

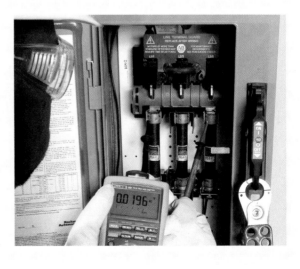

Definitions

- An *assembly drawing* is a drawing that shows an entire machine or device with all detailed parts.
- An *automatic transfer switch* is an electrical device that transfers the load of a building from public utility circuits to the output of a standby generator during a power failure.
- A *branch circuit* is the part of an electrical wiring system between the final set of circuit breakers or fuses and the fixtures and receptacles.
- A *brush* is the sliding contact that rides against the slip rings and is used to connect the armature to the external circuit (load).
- A *busway* is a grounded metal enclosure that contains factory-mounted bare or insulated conductors.
- A *cable* is two or more conductors grouped together within a common protective cover and used to connect individual components.
- A *circuit breaker* is an OCPD with a mechanical mechanism that can manually or automatically open the circuit when an overload condition or short circuit occurs.
- A *closed transition switch* is an automatic transfer switch that closes one power source before opening another power source.
- A *conductor* is a material that has little resistance and permits electrons to move through it easily.
- *Conduit* is a tube or pipe used to support and protect electrical conductors.
- *Copper-clad aluminum* is a conductor that has copper bonded to an aluminum core.
- A *cord* is a group of two or more conductors in one cover used to deliver power to a load by means of a plug.
- A *delta connection* is a connection that has each coil connected end to end to form a closed loop.
- A *distribution substation* is an outdoor facility located close to the point of electrical service use and is used to change voltage levels, provide a central place for system switching, monitoring, and protection, and redistribute power.
- A *duplex receptacle* is an electrical contact device containing two receptacles.
- An *electromagnet* is a device consisting of a core and coil that generates magnetism only when an electric current is present in the coil.
- *Electromagnetic induction* is the production of electricity by the interaction of a conductor cutting through a magnetic field.
- An *elevation drawing* is an orthographic view of a vertical surface without allowance for perspective.
- A *feeder* is a conductor between the service equipment and the last branch-circuit OCPD.
- A *feeder panel* is a junction box used to house electrical circuit breakers and main connections.
- A *field winding* is a group of wires used to produce the magnetic field in the stator of a generator.
- A *floor plan* is a plan view of a structure that shows the arrangement of walls and partitions as they appear in an imaginary section taken horizontally and approximately 5′-0″ above floor level.
- A *fuse* is an OCPD with a fusible link that melts and opens a circuit on an overcurrent condition.
- A *generator* is a device that converts mechanical energy into electrical energy by means of electromagnetic induction.
- A *ground fault* is any amount of current above a certain level that may deliver a dangerous shock.
- A *ground-fault circuit interrupter (GFCI)* is a fast-acting electrical device that automatically deenergizes a circuit by opening the circuit in response to the grounded current exceeding a predetermined value.

- An *isolated-ground receptacle* is an electrical connection device in which the grounding terminal is isolated from the device yoke or strap.
- A *lateral service* is an electrical service in which service-entrance conductors are run underground from the utility service to the building.
- A *line diagram* is a drawing that has as a series of single lines (rungs) that indicate the logic of the control circuit and how the control devices are interconnected.
- A *motor control center (MCC)* is a central location for troubleshooting and servicing motor control circuits.
- An *open transition switch* is an automatic transfer switch that opens one power source before closing another power source.
- An *overcurrent* is electrical current in excess of the equipment limit, total amperage load of the circuit, or conductor or equipment rating.
- An *overcurrent protective device (OCPD)* is a fuse or circuit breaker (CB) that disconnects or discontinues current flow when the amount of current exceeds the design load.
- An *overhead service* is an electrical service in which service-entrance conductors are run from the utility pole through the air and to the building.
- An *overload* is a small-magnitude overcurrent that, over a period of time, leads to an overcurrent, which may trip the OCPD.
- A *panelboard* is a wall-mounted distribution cabinet containing a group of OCPDs and short-circuit protection devices for lighting, appliance, or power distribution branch circuits.
- The *peak value* is the maximum instantaneous value of the sine wave.
- A *pictorial diagram* is an electrical wiring sketch showing actual positions of system devices and wiring.
- A *plot plan* is a drawing that shows property lines of a building lot, elevation information, compass directions, lengths of property lines, and locations of structures to be built on the lot.
- A *receptacle* is an electrical contact device for connecting electrical equipment to a power source.
- The *root-mean-square voltage (Vrms)* of a sine wave is the AC voltage value that produces the same amount of heat in a pure resistive circuit as DC voltage of the same value.
- A *schematic diagram* is a drawing that shows electrical system circuitry with symbols that depict electrical devices and lines representing the conductors.
- A *short circuit* is a condition that occurs when two ungrounded conductors (hot wires), or an ungrounded and a grounded conductor of a single-phase circuit, come in contact with each other.
- A *single-line diagram* is an electrical drawing that uses a single line and basic symbols to illustrate the current path, voltage values, circuit disconnects, OCPDs, transformers, and panelboards for a power distribution system.
- A *slip ring* is a metallic ring connected to the end of an armature loop and is used to connect the induced voltage to a brush.
- A *switchboard* is a panel or an assembly of panels containing electrical switches, meters, busses, and other overcurrent protection devices (OCPDs).
- A *switchgear* is a high power electrical device that switches or interrupts devices or circuits in a building distribution system.
- A *symbol* is a graphic representation of a device, component, or object on a print.
- A *transformer* is an electrical device that uses electromagnetism to change voltage from one level to another or to isolate one voltage from another.

- A *transmission line* is an aerial conductor that carries large amounts of electrical power at high voltages over long distances.
- A *transmission substation* is an outdoor facility located along a utility system that is used to change voltage levels, provide a central place for system switching, monitoring, and protection, and redistribute power.
- A *utility* is an organization that installs, operates, or maintains the electrical, communication, gas, water, or related services in a community.
- A *utility plan* is a drawing that indicates the location and intended path of utility lines such as electrical, water, sewage, gas, and communication cables.
- A *watt (W)* is the unit of measure equal to the power produced by a current of 1 A across a potential difference of 1 V.
- A *wire* is any individual conductor.
- A *wiring diagram* is a drawing that indicates the connections of all devices and components in an electrical system.
- A *wye connection* is a connection that has one end of each coil connected together and the other end of each coil left open for external connections.

Review Questions

1. How does an armature generate current?
2. What are the connection types used by three-phase AC generators, and how do they differ?
3. What are the common operating voltage levels of transmission substations, and what are the voltages associated with those levels?
4. What is a transformer and what is its purpose in an electrical distribution system?
5. Why are voltages stepped up or stepped down in an electrical distribution system?
6. What are the different varieties of electrical services?
7. What is the advantage and disadvantage of wye-connected transformers in building power distribution systems?
8. What is an automatic transfer switch, and why is it used in a building power distribution system?
9. What is a feeder panel?
10. Why are motor control centers (MCCs) used to simplify and consolidate motor control circuits?
11. How are panelboards used in power distribution systems?
12. What are fuses and circuit breakers designed to do?
13. Why are copper and aluminum used as conductor material?
14. What are the three basic types of receptacles and how do they differ?
15. What are four of the major types of building distribution system drawings?

chapter 2
PRACTICING ELECTRICAL SAFETY

The two major areas that must be addressed to ensure the safest possible working conditions are ensuring personal safety and protecting property and equipment. Personal safety is achieved by using the proper safety devices and procedures. Protecting property and equipment includes maintaining the property and equipment in as safe a condition as possible. This can also reduce accidents and ensure personal safety.

OBJECTIVES

- Identify common safety labels.
- Describe the procedure for treating electrical shock.
- Define arc blast and arc flash.
- Identify the proper personal protective equipment (PPE) required for working near electrical circuits.
- Describe common lockout/tagout safety procedures.
- Describe the use of electrical safety precautions for property and equipment.

PERSONAL SAFETY

Proper safety equipment and procedures are designed to keep individuals safe in any possible hazardous condition. Safety equipment includes safety labels and personal protective equipment (PPE). Safety procedures include treating electrical shock and applying basic first aid.

Safety Labels

A *safety label* is a label that indicates areas or tasks that can pose a hazard to personnel and/or equipment. Safety labels help to ensure a safe work environment and are designed to provide written warnings of potential hazards. Safety labels are required in areas and on equipment that can be hazardous. Safety labels should be placed in a location that helps ensure maximum understanding of the potential danger. The safety label should be located near the potential hazard but the placement should also allow enough time and distance to heed the warning.

Safety labels should be displayed to ensure maximum understanding of the potential danger.

Each safety label encountered on equipment should be read and understood before using the equipment. Each safety label is designed to give a specific or general warning. Some warning labels are designed to warn of personal dangers, and others warn of equipment and environmental dangers.

Safety labels appear several ways on equipment and in equipment manuals. Safety labels use signal words to communicate the severity of a potential problem. The three most common signal words are danger, warning, and caution. **See Figure 2-1.**

Danger Signal Word. *Danger* is a signal word used to indicate an imminently hazardous situation that, if not avoided, will result in death or serious injury. The information indicated by the danger signal word indicates the most extreme type of potential situation and must be followed. The danger symbol is an exclamation mark enclosed in a triangle followed by the word "danger" written boldly in a red box.

Warning Signal Word. *Warning* is a signal word used to indicate a potentially hazardous situation that, if not avoided, could result in death or serious injury. The information indicated by the warning signal word indicates a potentially hazardous situation and must be followed. The warning symbol is an exclamation mark enclosed in a triangle followed by the word "warning" written boldly in an orange box.

Caution Signal Word. *Caution* is a signal word used to indicate a potentially hazardous situation that, if not avoided, may result in minor or moderate injury. The information indicated by the caution signal word indicates a potential situation that may pose a threat to individuals and/or equipment. A caution signal word also warns of problems due to unsafe work practices. The caution symbol is an exclamation mark enclosed in a triangle followed by the word "caution" written boldly in a yellow box.

DANGER, WARNING, AND CAUTION SIGNAL WORDS

Safety Label	Box Color	Symbol	Significance
⚠ DANGER **HAZARDOUS VOLTAGE** • Ground equipment using screw provided. Electric panel must be properly grounded before applying power. • Do not use metallic conduits as a ground conductor. Failure to observe these precautions will cause shock or burn, resulting in severe personal injury or death!	RED	⚠	**DANGER** – INDICATES AN IMMINENTLY HAZARDOUS SITUATION WHICH, IF NOT AVOIDED, WILL RESULT IN DEATH OR SERIOUS INJURY
⚠ WARNING **MEASUREMENT HAZARD** When taking measurements inside the electric panel, make sure that only the test lead tips touch internal metal parts. Keep hands behind the protective finger guard provided on the test leads.	ORANGE	⚠	**WARNING** – INDICATES A POTENTIALLY HAZARDOUS SITUATION WHICH, IF NOT AVOIDED, COULD RESULT IN DEATH OR SERIOUS INJURY
⚠ CAUTION **MOTOR OVERHEATING** This controller does not provide direct thermal protection for the motor. Use of a thermal sensor in the motor may be required for protection at all speeds and loading conditions. Consult motor manufacturer for thermal capability of motor when operated over desired speed range. Failure to observe this precaution can result in equipment damage.	YELLOW	⚠	**CAUTION** – INDICATES A POTENTIALLY HAZARDOUS SITUATION WHICH, IF NOT AVOIDED, MAY RESULT IN MINOR OR MODERATE INJURY, OR DAMAGE TO EQUIPMENT; MAY ALSO BE USED TO ALERT AGAINST UNSAFE WORK PRACTICES

Figure 2-1. Danger, warning, and caution signal words are used to communicate the severity of a potential problem.

Other symbols or signal words may also appear with the danger, warning, and caution signal words. ANSI Z535.4, *Product Safety Signs and Labels,* provides additional information concerning safety labels. Common electrical system signal words include electrical warnings and explosion warnings. **See Figure 2-2.**

Electrical Warning. An electrical warning indicates a high-voltage location and condition that could result in death or serious personal injury from an electrical shock if proper precautions are not taken. The electrical warning safety label is usually placed where there is a potential for contact with live electrical wires, terminals, or parts. The electrical warning symbol is a lightning bolt enclosed in a triangle. The electrical warning safety label may be shown with no words or may be preceded by the word "warning" written boldly.

Explosion Warning. An explosion warning indicates a location and condition where exploding parts may cause death or serious personal injury if proper precautions and procedures are not followed. The explosion warning symbol is an explosion enclosed in a triangle. The explosion warning safety label may be shown with no words or may be preceded by the word "warning" written boldly.

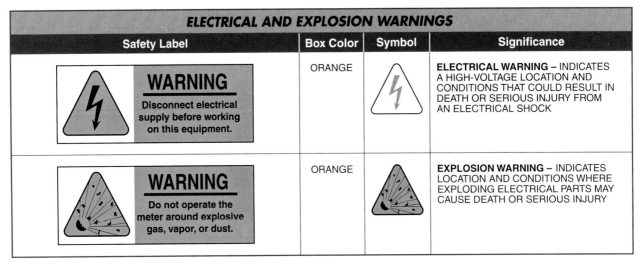

Figure 2-2. Electrical warning and explosion warning labels are used on electrical circuits and equipment to indicate a location and condition that could result in death or serious personal injury.

Electrical Shock

According to the National Safety Council, over 1000 individuals are killed by electrical shock in the United States each year. An electrical shock results any time a body part comes in contact with an electrical circuit.

Electrical shock varies from a mild shock to a fatal shock. The severity of an electrical shock depends on the amount of electric current (measured in milliamperes) that flows through the body, the length of time the body is exposed to the current, the path the current takes through the body, the physical size and condition of the body, and the amount of body area exposed to the electric current. Much less current is required in order to cause fatality when current flows through the heart and lungs because electrical shock can cause cardiac arrest.

The amount of current that passes through a circuit depends on the voltage and resistance of the circuit. During an electrical shock, the body becomes part of the circuit. The amount of current flow through a body depends on the resistance to the flow of current. The resistance a body offers to the flow of current varies. Sweaty hands have less resistance than dry hands. A wet floor has less resistance than a dry floor. The body's resistance can be increased by wearing electrical gloves, dry shoes or boots, and all required PPE. Wearing the required PPE also helps to prevent, or reduce the severity of, an electrical shock. The lower the resistance, the greater the current flow. The greater the current flow, the more severe the electrical shock. **See Figure 2-3.**

When handling an individual injured from an electrical shock accident, the following procedure should be applied:

1. Think and react fast because time is important. However, ensure you do not become part of the circuit.
2. Break the circuit to free the individual immediately and safely. Never touch any part of the individual's body when in contact with the circuit. When the circuit cannot be turned off, use any nonconducting device to free the individual. Resist the urge to touch the individual when power is on.
3. Send for help and determine if the injured individual is breathing after

the individual is free from the circuit. If there is no breathing or pulse, start cardiopulmonary resuscitation (CPR) if trained to do so. *Note:* Always seek immediate medical attention for an individual injured by electrical shock.

4. Check for burns and cuts if the individual is breathing and has a pulse. *Note:* Burns are caused by contact with a live circuit and are found at the points where the electricity entered and exited the body. The higher the voltage and/or the longer the exposure to an electrical shock, the greater the internal tissue burning.

5. Treat the entrance and exit burns as thermal burns and seek immediate medical help. *Note:* Medical help is required because most of the damage will be internal burns between the entrance and exit points and will not be visible.

When electricity enters the body, it travels through areas of least resistance. Nerves and blood vessels have lower resistance than muscle and bone. The longer the body is exposed to an electrical shock, the greater the damage caused by the heat produced by the current flow. Current flow from hand to hand results in current flow across the heart and lungs.

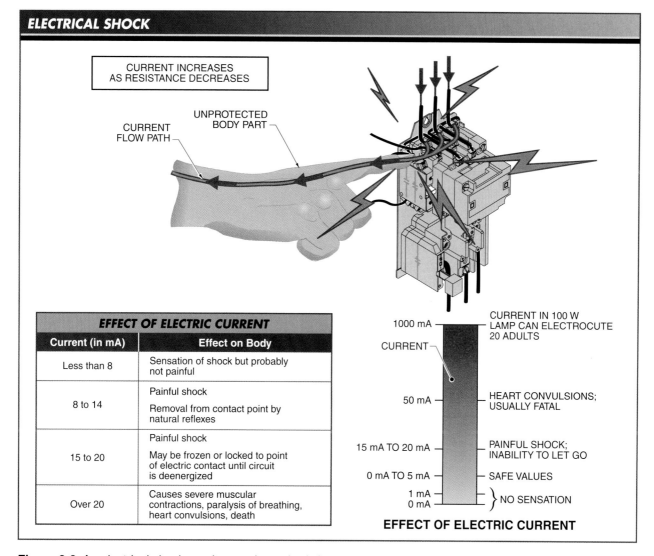

Figure 2-3. An electrical shock results any time a body becomes part of an electrical circuit.

48 ELECTRICAL SYSTEMS for FACILITIES MAINTENANCE PERSONNEL

TECH FACT

Per NFPA 70E, individuals working on or adjacent to a piece of equipment that could produce an arc flash must wear long-sleeve flame-resistant (FR) clothing, properly rated electrical gloves with leather protectors, an electrically rated hard hat, safety glasses or goggles, and an arc-resistant face shield.

Grounding Systems Media Clip

Extra care should be taken around high-voltage DC. High-voltage DC tends to cause continuous muscular contractions that can prolong the individual's attachment to the circuit. Standard low-frequency AC (60 Hz power) tends to travel through the body. High-frequency AC tends to run along the skin.

Grounding to Prevent Electrical Shock. Electrical circuits are grounded to safeguard personnel and equipment against the hazards of electrical shock. Proper grounding of electrical tools, machines, equipment, and distribution systems is required to prevent hazardous conditions. *Grounding* is the connection of all exposed non-current-carrying metal parts to the earth. Grounding provides a direct path for unwanted (fault) current to the earth without causing harm to personnel or equipment. Grounding is accomplished by connecting the ground circuit to a metal underground water pipe, a metal frame of a building, a concrete-encased electrode, or a ground ring. **See Figure 2-4.**

Non-current-carrying metal parts that are connected to ground include all metal boxes, raceways, enclosures, electric tools and appliances, and any equipment that may have an electrical fault that would allow unwanted current to flow into an individual. *Fault current* is any current that travels a path other than the normal operating path for which a system is designed.

Fault current exists because of insulation failure or because a current-carrying conductor makes contact with a non-current-carrying part of a system that is not grounded. In a properly grounded system, the unwanted current flows through the low-resistance ground system, which causes a high current flow and trips fuses or circuit breakers. Once the fuse or circuit breaker is tripped, the circuit is opened and no additional current flows.

Arc Flash and Arc Blast

An *electric arc* is a discharge of electric current across an air gap. An electric arc is caused by excessive voltage ionizing the air gap between two conductors or by two conductors that accidentally contact each other and then separate. When an electric arc occurs, there is the possibility of an arc flash or an arc blast. An *arc flash* is an extremely high-temperature discharge produced by an electrical fault in the air. Arc flash temperatures reach 35,000°F. An *arc blast* is an explosion that occurs when the air surrounding electrical equipment becomes ionized and conductive.

Arc flashes and arc blasts are always a possibility when working with electrical equipment. Only qualified personnel are allowed to work on energized circuits of 50 V or higher. Arc blasts are possible in systems of lesser voltage, but are not likely to be as destructive as in a high-voltage system. The threat of an arc blast is greatest from electrical systems of 480 V and higher. To prevent an arc flash or an arc blast, an electrical system must be deenergized, locked out, and tagged out before performing work.

Personal Protective Equipment

Working around electrical circuits and equipment can be dangerous because electricity can cause electrical shock, electrocution, fires, and explosions. According to the National Safety Council, more than 100,000 individuals are killed in electrical fires each year. The dangers of working around electricity are reduced when proper safety procedures are followed and PPE is being worn and is in good condition. Whenever possible, electrical maintenance and repair should only be performed on deenergized equipment and circuits.

Personal protective equipment (PPE) is clothing and/or equipment worn by an individual to reduce the possibility of injury in the workplace. PPE is required whenever work is done on or near energized exposed electrical circuits. PPE includes rubber insulating gloves, flame-resistant (FR) clothing, eye protection, head protection, and ear protection. **See Figure 2-5.**

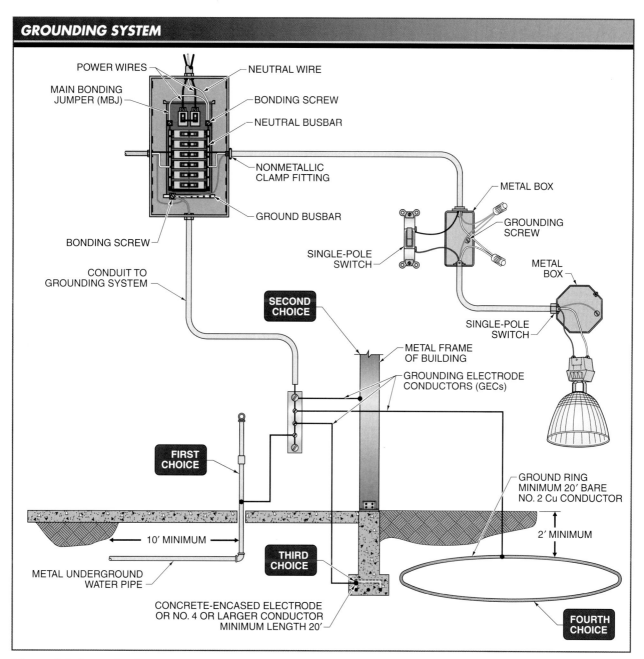

Figure 2-4. Grounding is accomplished by connecting the ground circuit to a metal underground water pipe, the metal frame of a building, a concrete-encased electrode, or a ground ring.

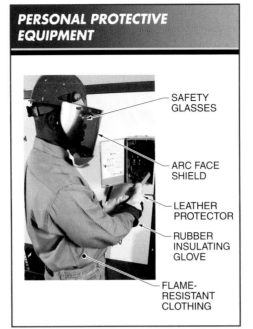

Figure 2-5. Personal protective equipment (PPE) includes rubber insulating gloves, flame-resistant (FR) clothing, and eye protection.

OSHA
Media Clip

NFPA 70E. For maximum safety, PPE must be worn and all safety requirements followed when working on or around electrical equipment or taking electrical measurements. The *National Fire Protection Association (NFPA)* is a national organization that provides guidance in assessing the hazards of the products of combustion. The NFPA publishes the National Electrical Code® (NEC®) and NFPA 70E.

The purpose of the NEC® is the practical safeguarding of individuals and property from the hazards that arise from the use of electricity. The NEC® is primarily intended for those who design, install, and inspect electrical systems such as in new construction or remodeling. The goal of the NEC® is to have electrical systems designed and installed safely.

NFPA 70E is intended to cover electrical safety concerns related to safe work practices, equipment maintenance, and taking electrical measurements. The goal of NFPA 70E is to ensure that no one is hurt while working with or around electrical equipment. Per NFPA 70E, "Only qualified persons shall perform testing on or near live parts operating at 50 V or more."

All PPE and tools are selected to be appropriate for the operating voltage of the equipment or circuits being worked on. Equipment, devices, tools, and test instruments must be suited for the work being performed. In most cases, voltage-rated insulating gloves and tools are required. Voltage rated insulating gloves and tools are rated by maximum line-to-line voltage and tested accordingly. Protective gloves must be inspected or tested as required for maximum safety before each task.

The *Occupational Safety and Health Administration (OSHA)* is a federal agency that requires all employers to provide a safe working environment for their employees. The NFPA 70E attempts to meet OSHA's requirements to provide a safe work environment for individuals. Following NFPA 70E regulations helps employers meet OSHA mandates that employees must follow to avoid electrical hazards such as electrical shock, electrocution, arc flashes, and arc blasts.

Rubber Insulating Gloves. Safety requirements for the usage of rubber insulating gloves and leather protectors must be followed at all times. The primary purpose of rubber insulating gloves and leather protectors is to insulate the hands and the lower arms from possible contact with live conductors. Rubber insulating gloves offer a high resistance to current flow to help prevent an electrical shock, and the leather protectors protect the rubber gloves and add additional insulation. Rubber insulating gloves are rated and labeled for the maximum voltage allowed.

Rubber insulating gloves must be field-tested by visual inspection and air-tested prior to each use. Rubber insulating gloves must also be tested by an approved laboratory every six months. Rubber insulating gloves have color-coded labels with applicable voltage ratings. **See Figure 2-6.**

Gloves must be air-tested when there is cause to suspect any damage. The entire surface must be inspected by rolling the cuff tightly toward the palm in such a manner that air is trapped inside the glove or by using a mechanical inflation device. When using a mechanical inflation device, care must be taken to avoid overinflation. The glove is examined for punctures and other defects. Puncture detection may be enhanced by listening for escaping air. Some brands of rubber insulating gloves are available with two colored layers. When one colored layer becomes visible, the color serves as notification to the wearer that the gloves must be replaced.

Visual inspection of rubber insulating gloves is performed by stretching a small area, particularly the fingertips, and checking for defects such as punctures or pin holes, embedded or foreign material, deep scratches or cracks, cuts or snags, and/or deterioration. When the gloves have passed visual inspection, an air test is performed using the following procedure:
1. Grasp the cuff of the glove between thumb and forefinger.
2. Wave the glove to fill it with air.
3. Trap the air by squeezing the cuff with one hand while using the other hand to squeeze the palm, fingers, and thumb to look for defects.
4. Hold the glove to the face or ear to detect escaping air. Tag gloves that fail the air test "unsafe" and return them to a supervisor or properly dispose of them.

Proper care of leather protector gloves is essential for wearer safety. Leather protector gloves are checked for cuts, tears, holes, abrasions, defective or worn stitching, oil contamination, or any other condition that might prevent the leather glove from adequately protecting the rubber insulating gloves. Any substance that can physically damage rubber gloves must be removed before testing. Rubber insulating gloves and leather protector gloves found to be defective should not be discarded or destroyed in the field but should be tagged "unsafe" and returned to a supervisor.

RUBBER INSULATING GLOVE RATINGS*		
Class	Maximum Use	Label Color
00	500 V	Beige
0	1 kV (1000 V)	Red
1	7.5 kV (7500 V)	White
2	17 kV (17,000 V)	Yellow
3	26.5 kV (26,500 V)	Green
4	36 kV (36,000 V)	Orange

* Refer to ASTM D 120-09, *Standard Specification for Rubber Insulating Gloves.*

Figure 2-6. Rubber insulating gloves have color-coded labels with applicable voltage ratings.

Flame-Resistant Clothing. Sparks from an electrical circuit can cause a fire. Approved FR clothing must be worn in conjunction with rubber insulating gloves for protection from electrical arcs when performing certain operations on or near energized equipment or circuits. FR clothing must be kept as clean and sanitary as possible and must be inspected prior to each use. Defective FR clothing must be tagged "unsafe" and returned to a supervisor.

To protect against arc flashes and arc blasts, all FR clothing must meet NFPA 70E and OSHA standards. FR clothing is divided into categories based on the minimum arc rating of the PPE, which is measured in cal/cm^2. The category and cal/cm^2 rating describe the level of protection provided. The higher the category, the greater the protection. The cal/cm^2 rating is the unit given to indicate the amount of incident energy the PPE can safely withstand at a given distance from an arc flash. **See Figure 2-7.**

FLAME-RESISTANT PROTECTIVE EQUIPMENT REQUIREMENTS					
Flame-Retardant Clothing Type	NFPA 70E Category Number (1 = Least Hazardous)				
	1	2	2*	3	4
Required minimum arc rating of PPE (cal/cm^2)	4	8	8	25	40
Flash suit jacket					X
Flash suit pants					X
Head protection (hard hat)	X	X	X	X	X
Flame-retardant hard hat liner				X	X
Safety glasses w/side shields or goggles	X	X	X	X	X
Arc face shield w/wrap-around face, forehead, ear, and neck protection		X	X	X	X
Face protection (2-layer hood)			X	X	X
Hearing protection (ear canal inserts)			X	X	X
Rubber gloves w/leather protectors	X	X	X	X	X
Leather shoes w/rubber soles	X	X	X	X	X

* is a higher energy environment than Category 2

Figure 2-7. The cal/cm^2 rating of FR clothing indicates the amount of incident energy the clothing can safely withstand at a given distance from an arc flash.

When working with energized electrical circuits, the proper PPE must always be worn for protection against an arc blast or arc flash. Clothing made of synthetic materials such as nylon, polyester, or rayon, alone or combined with cotton, must never be worn because synthetic materials burn and melt to the skin.

The minimum PPE requirements for electrical work are an untreated, natural-material, long-sleeve shirt, long pants, safety glasses with side shields, and rubber insulating shoes or boots. Additional PPE includes FR coveralls, FR long-sleeve shirt and pants, a hard hat with an FR liner, hearing protection, and double-layer flash suit jacket and pants. Flash suits are similar to firefighter turnout gear and must be worn when working near a Category 4 hazard/risk area. The minimum PPE requirement for working on deenergized circuits is NFPA Category 1. For work that is to be performed on energized circuits, PPE that is appropriate for the risk area must be worn.

Eye Protection. Eye protection must be worn to prevent eye and face injuries caused by flying particles, contact arcing, and radiant energy. Eye protection must comply with OSHA 29 CFR 1910.133, *Eye and Face Protection*. Eye protection standards are specified in ANSI Z87.1, *Occupational and Educational Eye and Face Protection*. Eye protection includes safety glasses, face shields, and arc blast hoods. **See Figure 2-8.**

Safety glasses are an eye protection device with special impact-resistant glass or plastic lenses, reinforced frames, and side shields. Plastic frames are designed to keep the lenses secured in the frame when an impact occurs in order to minimize the shock hazard when working with electrical equipment. Side shields provide additional protection from flying objects. Tinted-lens safety glasses protect against low-voltage arc hazards.

A *face shield* is any eye and face protection device that covers the entire face with a plastic shield and is used for protection from flying objects. Tinted face shields protect against low-voltage arc hazards. An *arc blast hood* is an eye and face protection device that consists of a flame-resistant hood and face shield.

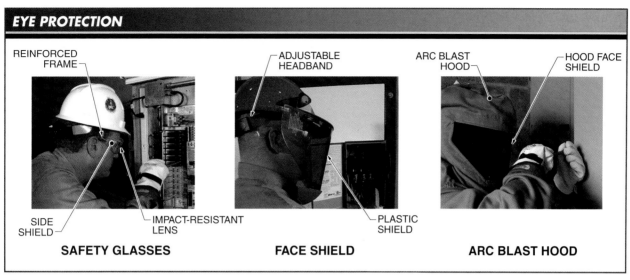

Figure 2-8. Eye protection must be worn to prevent eye and face injuries caused by flying particles, contact arcing, and radiant energy.

Safety glasses, face shields, and arc blast hoods must be properly maintained to provide protection and clear visibility. Lens cleaners that clean without risk of lens damage are available. Pitted, scratched, and crazed lenses (lenses with microscopic cracks caused by exposure to aggressive solvents, chemicals, or heat) reduce vision and may cause lenses to fail on impact.

Insulating Matting. *Insulating matting* is a floor covering that provides personnel with protection from electrical shock when working on live electrical circuits. Dielectric black-fluted rubber insulating matting is specifically designed for use in front of open cabinets or high-voltage equipment. It is used to protect personnel when voltages are over 50 V and is designated as Type I natural rubber and Type II elastomeric compound matting. **See Figure 2-9.**

Lockout/Tagout

Lockout is the process of removing the source of electrical power and installing a lock that prevents the power from being turned on. To ensure the safety of personnel working with equipment, all electrical power must be removed and the equipment must be locked out and tagged out.

Tagout is the process of placing a danger tag on the source of electrical power to indicate that the equipment may not be operated until the danger tag is removed by the individual who installed the tag. Per OSHA standards, equipment must be locked out and tagged out before any installation or preventive maintenance is performed. **See Figure 2-10.**

A danger tag has the same importance and purpose as a lock and is used alone only when a lock does not fit the disconnect device. A danger tag must be attached at the disconnect device with a tag tie or equivalent attaching device and must have space for the individual's name, craft, and other company-required information. A danger tag must withstand the elements and expected atmosphere for the maximum period of time that exposure is expected. Lockouts/tagouts should also be applied to fluid pressure systems or any other source of danger that does not need to be operational at the time of service.

Lockout Procedures Media Clip

TECH FACT
OSHA requires that employees exposed to workplace hazards be given in-house safety training at regular intervals. In-house safety training must be documented by the company, and the documentation must be signed by the employees who received the training.

RUBBER INSULATING MATTING RATINGS

Safety Standard	Material Thickness		Material Width*	Test Voltage	Maximum Working Voltage
	Inches	Millimeters			
BS921†	0.236	6	36	11,000	450
BS921†	0.236	6	48	11,000	450
BS921†	0.354	9	36	15,000	650
BS921†	0.354	9	48	15,000	650
VDE0680‡	0.118	3	39	10,000	1000
ASTM D178§	0.236	6	24	25,000	17,000
ASTM D178§	0.236	6	30	25,000	17,000
ASTM D178§	0.236	6	36	25,000	17,000
ASTM D178§	0.236	6	48	25,000	17,000

* in in.
† BSI–British Standards Institute
‡ VDE–Verband Deutscher Elektrotechniker Testing and Certification Institute
§ ASTM International

Figure 2-9. Rubber insulating matting provides personnel with protection from electrical shock when working on live electrical circuits.

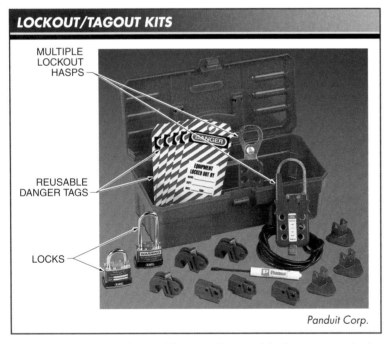

Figure 2-10. Lockout/tagout kits contain reusable danger tags, locks, multiple lockout hasps, and information on lockout/tagout procedures.

Lockout/tagout is used in the following situations:
- when power is not required to be on
- when removing or bypassing machine guards or other safety devices
- when there is a possibility of being injured or caught in moving machinery
- when jammed equipment is being cleared
- when there is a danger of being injured if equipment power is turned on

Lockout and tagouts do not, by themselves, remove power from a machine or its circuitry. OSHA provides a standard procedure for equipment lockout/tagout. Lockout is performed and tags are attached only after the equipment is turned off and has been tested. Typical company lockout/tagout procedures are as follows:

1. Notify all affected individuals that a lockout/tagout is required. Notification must include the reason for the lockout/tagout and the expected duration.
2. Shut down operating equipment using the normal equipment shutdown procedure.
3. Operate the energy-isolating devices so that the equipment is isolated from all energy sources. Stored energy in springs, elevated machine members, and capacitors must be dissipated or restrained by blocking, discharging, or other appropriate methods.
4. Lockout/tagout the energy-isolating devices with assigned locks and danger tags.

5. After ensuring that no personnel are exposed to potential moving equipment, attempt to operate the normal controls to verify that the equipment is inoperable and that all energy sources have been isolated.
6. Inspect and test the equipment with appropriate test instruments to verify that all energy sources are disconnected. *Note:* Multiphase electrical power requires that each phase be tested.

A lockout/tagout must not be removed by any individual other than the authorized personnel who installed the lockout/tagout unless there is an emergency. In an emergency, only supervisory personnel may remove a lockout/tagout and only upon notification of the authorized lockout/tagout personnel. A list of company rules and procedures should be given to authorized personnel and any individual who may be affected by a lockout/tagout.

When more than one individual is required to perform a task on a piece of equipment, each individual must place a lockout/tagout on the energy-isolating device. A multiple lockout/tagout device must be used because the energy-isolating device typically cannot accept more than one lockout/tagout at a time. A hasp is a multiple lockout/tagout device. Lockout devices resist chemicals, cracking, abrasion, and temperature changes. **See Figure 2-11.**

Basic First Aid

Accidents that cause injury to personnel can occur at any time in any place. Electrical shock accidents can involve both the electrical damage caused when current flows through the body and the damage from secondary responses to the electrical shock (falling, jumping back, etc.). Immediate medical treatment is required for the injured individual, regardless of the extent of the injury. Often, first aid given immediately at the scene of an accident can improve the individual's chances of survival and recovery. *First aid* is help for an injured individual immediately after an injury and before professional medical help arrives. If someone is injured, steps taken to keep that individual as safe as possible until professional help arrives include the following:

- Remain calm.
- Call 911 or the workplace emergency number immediately.
- Only move an injured individual if a fire or explosives are involved. Otherwise, do not move the individual. Moving an injured individual may make the injury worse.
- Assess the injured individual carefully and perform basic first aid procedures if trained to do so.
- Maintain first aid procedures until professional medical help arrives.
- Report all injuries to a supervisor.

A conscious injured individual must provide consent before care can be administered. To obtain consent, the individual must be asked if help can be provided. Once the individual gives consent, the appropriate care can be provided. Care should not be provided if the individual does not give consent. In such a case, 911 or the workplace emergency number should be called. Consent is implied in cases where an injured individual is unconscious, confused, or seriously injured. Implied consent allows care to be provided because it is assumed the injured individual would agree to care if possible.

Various states have enacted good Samaritan laws to give legal protection to individuals who provide emergency care to injured individuals. These laws are designed to encourage individuals to help others in emergencies and vary from state to state. A legal professional or the local library should be consulted for information regarding the good Samaritan laws in a particular state.

56 ELECTRICAL SYSTEMS for FACILITIES MAINTENANCE PERSONNEL

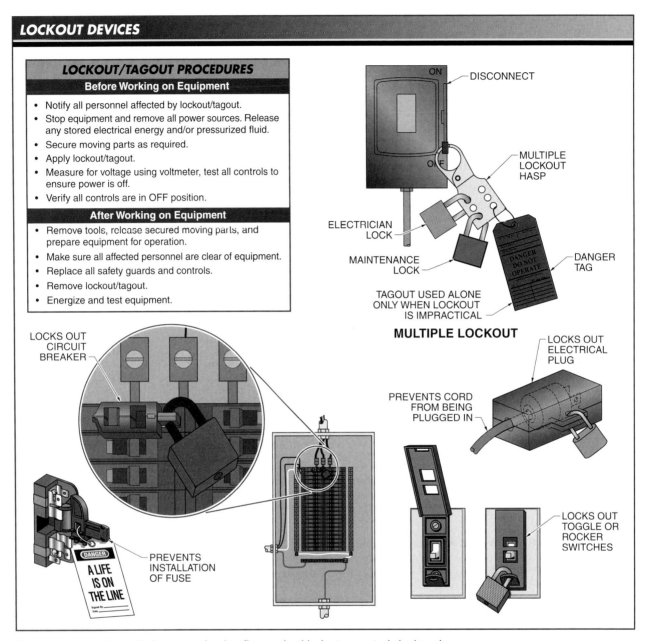

Figure 2-11. Lockout devices are sized to fit standard industry control-device sizes.

An individual should not attempt to provide care that they have not been trained to give. The American Red Cross may be consulted for training programs covering basic first aid and cardiopulmonary resuscitation (CPR). Basic first aid procedures can be used to help treat shock and burns.

Shock. Shock usually accompanies severe injury. Shock can threaten the life of an injured individual if not treated quickly. Shock occurs when the body's vital functions are threatened by lack of blood or when the major organs and tissues do not receive enough oxygen. Symptoms of shock include cold, clammy, or pale skin, chills, confusion, nausea or vomiting, shallow breathing, unusual thirst, cardiac arrest, seizures, and unconsciousness. When shock is suspected or if any of these signs or symptoms occur, 911 or the local emergency number should be called

immediately. The following measures should be taken to treat shock:
- Get professional medical help as soon as possible.
- Look first before touching the individual to ensure they are not still in contact with live power.
- Check for signs of breathing. If absent, begin CPR immediately if trained to do so.
- Lay the individual down with his or her legs elevated if there are no signs of broken bones or spinal injury.
- Cover the individual to prevent chills or loss of body heat.
- Control any obvious signs of bleeding. An individual who is unconscious or bleeding from the mouth should lie on one side so breathing is easier.

Burns. Burns can be caused by heat, chemicals, or electricity. For burns caused by heat or chemicals, the burn should be immediately flushed with cool water for a minimum of 30 min. Ice should not be applied because it may cause further damage to the burned area. This treatment should be maintained until the pain or burning stops. After the burn is flushed with cool water, it should be covered with a clean cotton cloth. If a clean cotton cloth is not available, the burn should not be covered.

Clothing that is stuck to a burn should not be removed. The burn should not be scrubbed, and soap, ointments, greases, or powders should not be applied. The breaking of blisters that appear should be avoided. Also, the burn victim should not be offered anything to drink or eat. The victim should be covered with a blanket to maintain a normal body temperature. All burn victims should be treated for shock, and professional medical help should be summoned as soon as possible.

A burn caused by electricity requires first ensuring that the victim is no longer in contact with the power source. Once the victim is clear of the power source, the victim should be checked for any airway obstructions and checked for breathing and circulation. CPR should be administered if necessary. Once the victim is stable, the burn should be flushed with cool water for a minimum of 30 min. The victim should not be moved. The burn should not be scrubbed, and soap, ointments, greases, or powders should not be applied. After the burn is flushed with cool water for 30 min, the burn should be covered with a clean cotton cloth. If a clean cotton cloth is not available, the burn should not be covered.

The victim should be treated for shock, and a normal body temperature should be maintained. Professional medical help should be summoned as soon as possible because electrical shock can cause internal burns which may not be apparent for some time because they may be deep inside the body.

PROPERTY AND EQUIPMENT SAFETY

Electrical energy can cause fires, explosions, and damage to property and equipment. Property and equipment must be properly protected and maintained to ensure safe and efficient operation. Just as grounding is required to protect individuals from electrical shock, property and equipment grounding is also required to prevent fires, explosions, and equipment damage. In addition to grounding, electrical equipment and circuits must be housed in enclosures that provide protection from outside environmental elements such as dust, rain, and snow. Double-insulated electric safety tools should be used whenever possible to minimize the risk of electrical shock.

To ensure that electrical circuits are not overloaded to the point of causing a fire hazard, electrical circuits must be protected using fuses, circuit breakers, and overload protection devices. This precaution removes power before damage occurs. When a fire occurs, the correct fire extinguisher must be used to eliminate the fire quickly and safely.

Fire Extinguishers

The five classes of fires are A, B, C, D, and K. Class A fires include burning wood, paper, textiles, and other ordinary combustible materials containing carbon. Class B fires include burning oil, gas, grease, paint, and other liquids that convert to a gas when heated. Class C fires include burning electrical devices, motors, and transformers. Class D is a specialized class of fires that includes burning metals such as zirconium, titanium, magnesium, sodium, and potassium. Class K fires include grease fires in commercial cooking equipment. **See Figure 2-12.** The correct fire extinguisher must be used on the appropriate fire. Fire extinguishers are used by aiming the extinguisher at the base of the fire and sweeping from side to side.

TECH FACT
Lightning produces a short burst of current ranging from 10 kA to 200 kA. Proper lightning protection systems ensure that this current does not cause damage to building equipment.

Grounding for Property and Equipment Safety

All non-current-carrying metal parts should be grounded to prevent electrical shock, damage to property and equipment, and electrical fires. Property and equipment grounding includes protecting all equipment and the building from electrical damage. Building grounding includes protecting the building and any outside structures from lightning strikes by providing the lightning with a path to earth ground. Equipment grounding reduces the chance of an electrical shock from a fault current within the electrical system and also helps prevent electrostatic discharge from static electricity and static buildup in equipment. Static electricity can also cause fires and explosions when allowed to build up.

Equipment grounding of electronic devices such as computers, communication and entertainment equipment, and medical equipment is used to prevent damage from electrostatic discharge. Equipment grounding is also used to prevent and/or reduce electromagnetic interference from compromising the electronic control/process signals. **See Figure 2-13.**

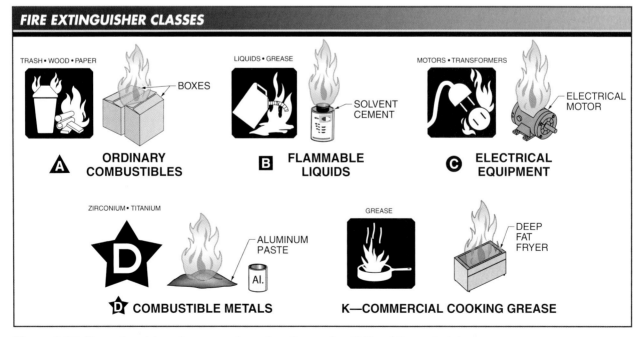

Figure 2-12. Fire extinguisher classes are based on the combustibility of the material.

Chapter 2—Practicing Electrical Safety 59

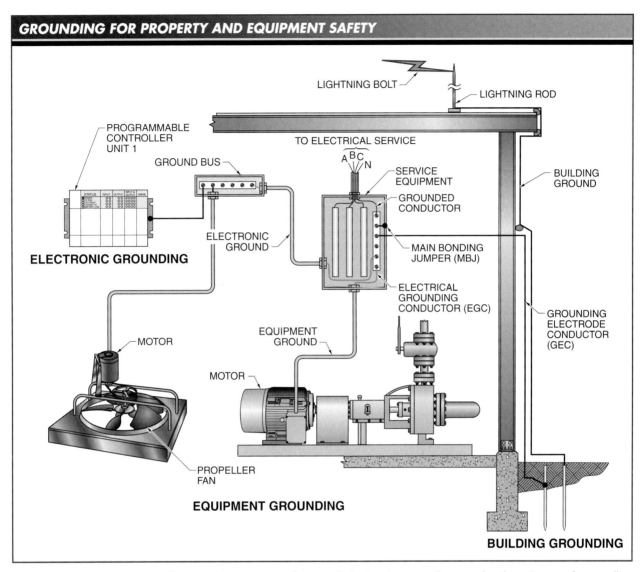

Figure 2-13. Building grounding, equipment grounding, and electronic grounding are the three types of grounding systems used to create a safe working environment for individuals.

Double-Insulated (1000 V) Electric Safety Tools

Double-insulated electric safety tools have a recommended maximum use of 1000 V and are designed to meet NFPA 70E and OSHA requirements for protecting individuals working with electrical equipment. The handles of double-insulated electric safety tools are covered with a double-bonded, flame- and impact-resistant insulation material. Certain tools, such as pliers, also have finger guards to prevent accidental metal contact. Wrenches, hex keys, nutdrivers, screwdrivers, pliers, wire strippers, and socket wrenches are available as double-insulated electric safety tools.

Nonconductive Ladders. All ladders used when installing any commercial electrical wiring must be made from wood or have nonconductive (fiberglass) side rails because there is a possibility of contact with exposed energized parts. **See Figure 2-14.** The purpose of the nonconductive side rails is to minimize the possibility of electrical shock to personnel using the ladder.

Receiving a mild electric shock while standing on a ladder can cause personnel using the ladder to fall.

The advantages of wood ladders include relatively low cost, durability, no conductivity, and good temperature-insulating qualities. Because they are nonconductive, wood ladders are safe to use when working around power lines and service-entrance conductors. Wood ladder temperature-insulating qualities also prevent the transmission of excessive heat or cold to a worker.

The disadvantages of wood ladders include deterioration with age, shrinkage causing loose rungs when the wood becomes dry and warm, and the need to maintain wood integrity by regularly coating it with clear shellac or linseed oil. Additionally, wood ladders must never be painted because paint covers any visible defects.

Wood ladders must be stored on their side away from excessive dampness, dryness, and heat to reduce the possibility of warping. Wood ladders must be stored by hanging horizontally on hooks spaced a minimum distance of 4′ to 6′ apart. Minimum and maximum storage hook spacing prevents sagging and offers easy access to ladders for inspection.

TECH FACT
Portable ladders must be able to support at least four times the maximum intended load. Ladders also must not be tied or fastened together to create longer sections unless they are specifically designed for such use.

Figure 2-14. Ladders used for performing work on or near electrical equipment must be made from wood or have fiberglass side rails.

Advantages of fiberglass ladders include no conductivity when dry, durability, and no surface finish requirements. Disadvantages of fiberglass ladders include cracking and failure when overloaded and cracks and chips when severely impacted. Safety guidelines for safe portable ladder usage include the following:
- Use wood ladders or ladders with fiberglass side rails when installing or working on electrical equipment.
- Prior to use, inspect the ladder for damage such as cracks, broken or missing rungs, and rotted wood.
- Verify that the rubber feet on the bottom of the ladder are secure to prevent slippage.
- Place the ladder on a firm foundation.
- Verify that the pitch does not exceed 1′ of horizontal distance for each 4′ of vertical rise.
- Use a rope to secure the ladder at the top of work area as required.
- Barricade the work area as required to prevent incidental contact with the ladder while work is in progress.
- Place extension ladders used to gain roof access so they protrude at least 3′ above the point of support.
- Do not allow more than one individual on a ladder at one time unless the ladder is rated for such use.
- Do not go beyond the highest step allowed for a specific ladder.
- Maintain ladders in good working condition.

Fuses, Circuit Breakers, and Overload Protection

All electrical circuits include multiple points at which fuses or circuit breakers are located. For example, the main power panel in a residential building includes a main fuse or circuit breaker and branch circuit fuses or circuit breakers. The branch circuit fuses or circuit breakers protect individual branch circuits from an overcurrent condition. They are sized based on the wire size used in the branch circuit. The main fuse or circuit breaker protects the entire building electrical system. It is sized based on the total rating of the power panel and the size of the utility feed into the residence.

In commercial and industrial electrical systems, electrical circuits and equipment are protected at multiple points. **See Figure 2-15.** For example, in the duct resistance heating element system, the elements are protected from an overload or short circuit by fuses that are sized up to 25% higher than the actual current draw of the elements. This prevents the fuses from tripping every time the heating elements are turned on. However, the size of the fuses cannot be rated any higher than the rating of the wires used to connect the heating elements to the power circuit.

The local heating element disconnect includes a switch that when opened (placed in the OFF position) removes all power to the heating elements and provides a point for lockout/tagout. The disconnect may also include fuses or circuit breakers that protect the heating elements and act as a backup to the panelboard circuit breakers.

Likewise, the supply-air blower motor is protected from an overload by the supply-air motor starter located in the motor control center. The supply-air motor starter includes overload protection (thermal overload heaters, electronic overloads, or combination of both), that is sized slightly higher than the current rating listed on the blower motor nameplate. When the thermal overload heaters and/or electronic overloads detect an overload, they open the normally closed overload contacts on the motor starter. This shuts the motor off.

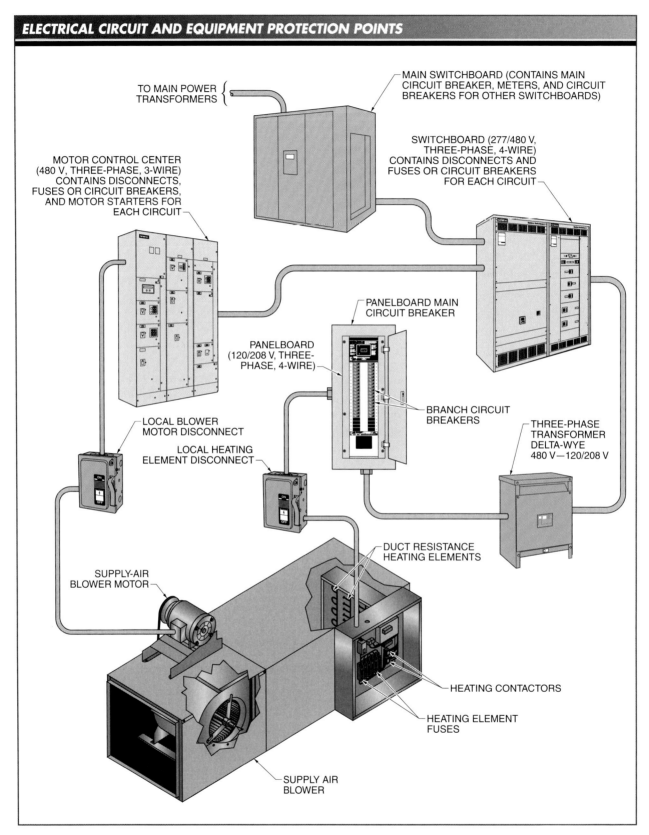

Figure 2-15. In commercial and industrial electrical systems, all electrical circuits and equipment are protected at multiple points.

The motor control center also contains disconnects, fuses or circuit breakers, and motor starters to protect each load connected to it and provide a centralized location for turning on and off all loads. The switchboard feeding the motor control center also contains disconnects and fuses or circuit breakers for each individual load connected to it. In addition, the main switchboard contains a main circuit breaker, meters, and circuit breakers for other switchboards.

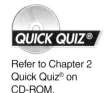

Refer to Chapter 2 Quick Quiz® on CD-ROM.

Case Study

Testing Fuses

A service call requires testing fuses in a disconnect switch servicing single-phase, 240 VAC equipment. Before taking any measurements, the maintenance staff must understand which PPE is required for the service call and all safety procedures that must be followed.

The minimum required PPE for the service call includes the following:

- protective long-sleeve shirt and pants or overalls rated at 8 cal/cm^2
- class 00 (500 V) rubber insulating gloves with leather protectors
- safety goggles or safety glasses with side shields
- wrap-around arc faceshield with a minimum rating of 8 cal/cm^2
- hard hat
- ear protection
- rubber insulating shoes or boots

Knowledge required for the service call includes the following:

- an understanding of all safety labels and signs located in the area and on the disconnect switch that houses the fuses to be tested
- an understanding of lockout/tagout usage, the procedures to be followed when locking out and tagging out the disconnect switch, and how to determine if there is a lockout/tagout already installed by someone else
- an understanding of the safety requirements when working in the area and on electrical equipment, such as when two individuals are required to be present during the service call
- an understanding of the tools and test instruments required for the service call and how to use them properly
- an understanding of any required emergency and first aid procedures that may be needed

Case Study (continued)

Measuring Circuit Current

As part of a facility preventive maintenance program, motor operating current draw must be measured and recorded. Measuring current draw over time can help indicate potential problems that may be developing. Motors draw more current as their loads increase, when insulation breaks down, when alignment changes, as bearings fail, and if there are mechanical problems such as excessively tight belts.

While wearing the proper PPE, the maintenance staff uses a clamp-on meter to measure the operating current of a 2 HP, 240 VAC motor. The motor nameplate current rating is 6.8 A. The individual uses the MIN MAX meter recording function to record the motor current draw through one complete operating cycle of the machine. The recorded current values are as follows:

	Minimum Current*	Maximum Current*
Motor placed in service	5.42	6.34
After 6 months	5.63	6.46
After 12 months	5.68	6.52
After 18 months	5.71	6.59
After 24 months	5.82	6.61
After 30 months (today's readings)	6.1	6.92

*in A

The maintenance staff records the new readings on the service log and notes that additional testing and inspection are required because the recorded values show an increase in motor current draw higher than expected.

Definitions

- An **_arc blast_** is an explosion that occurs when the air surrounding electrical equipment becomes ionized and conductive.

- An **_arc blast hood_** is an eye and face protection device that consists of a flame-resistant hood and face shield.

- An **_arc flash_** is an extremely high-temperature discharge produced by an electrical fault in the air.

- **_Caution_** is a signal word used to indicate a potentially hazardous situation that, if not avoided, may result in minor or moderate injury.

- **_Danger_** is a signal word used to indicate an imminently hazardous situation that, if not avoided, will result in death or serious injury.

- An **_electric arc_** is a discharge of electric current across an air gap.

- A **_face shield_** is any eye and face protection device that covers the entire face with a plastic shield and is used for protection from flying objects.

- **_Fault current_** is any current that travels a path other than the normal operating path for which a system is designed.

- **_First aid_** is help for an injured individual immediately after an injury and before professional medical help arrives.

- **_Grounding_** is the connection of all exposed non-current-carrying metal parts to the earth.

- **_Insulating matting_** is a floor covering that provides personnel with protection from electrical shock when working on live electrical circuits.

- **_Lockout_** is the process of removing the source of electrical power and installing a lock that prevents the power from being turned on.

- The **_National Fire Protection Association (NFPA)_** is a national organization that provides guidance in assessing the hazards of the products of combustion.

- The **_Occupational Safety and Health Administration (OSHA)_** is a federal agency that requires all employers to provide a safe working environment for their employees.

- **_Personal protective equipment (PPE)_** is clothing and/or equipment worn by an individual to reduce the possibility of injury in the workplace.

- **_Safety glasses_** are an eye protection device with special impact-resistant glass or plastic lenses, reinforced frames, and side shields.

- A **_safety label_** is a label that indicates areas or tasks that can pose a hazard to personnel and/or equipment.

- **_Tagout_** is the process of placing a danger tag on the source of electrical power to indicate that the equipment may not be operated until the danger tag is removed by the individual who installed the tag.

- **_Warning_** is a signal word used to indicate a potentially hazardous situation that, if not avoided, could result in death or serious injury.

Review Questions

1. What hazards are indicated when electrical warning and explosion warning signs are present?
2. What are the factors that affect the severity of electrical shock?
3. List the precautions that can be taken to prevent electrical shock.
4. What is fault current, why does it exist, and how is it controlled?
5. What causes the risk of an arc blast or arc flash, and what precautions must be taken to prevent an arc blast or arc flash?
6. What are the personal safety considerations that need to be taken before working on electrical circuits?
7. What is the purpose of using rubber insulating gloves and leather protectors?
8. What are the situations in which lockout/tagout is used?
9. What are the steps for a typical lockout/tagout procedure?
10. What are specific precautions that can be taken to prevent property and equipment damage due to electrical energy?

chapter 3
ELECTRICAL QUANTITIES AND CIRCUITS

Installation, maintenance, and troubleshooting of electrical circuits and equipment require an understanding of electrical quantities such as voltage, current, and resistance and their relationships. When two electrical quantities are known, Ohm's law and the power formula can be used to determine unknown electrical quantities. Electrical circuits can be series, parallel, or series-parallel circuits. Understanding how electrical circuits are affected by varying electrical quantities is required to install, maintain, and troubleshoot a circuit.

OBJECTIVES

- Define the two types of voltage used in electrical circuits.
- Identify the different types of voltage problems.
- Differentiate between resistive and inductive circuits and give examples of equipment in each.
- Describe the difference between true, reactive, and apparent power and give examples of where each applies.
- Calculate voltage, current, and resistance in series and parallel circuits.
- Discuss how magnetism applies to electrical motor operation.
- Identify electrical inputs and outputs in an electrical circuit.

VOLTAGE

Voltage Media Clip

Voltage (E) is the amount of electrical pressure in a circuit and is measured in volts (V). Voltage is normally produced by generators (electromagnetism), batteries (chemicals), or photocells (light). Voltage is either direct current (DC) or alternating current (AC). Common voltage problems that can affect commercial circuits include excessive voltage (overvoltage), too low of a voltage (undervoltage), and stray voltage (transient voltages).

DC Voltage

DC voltage is voltage that flows in one direction only. DC voltage sources have a positive and negative terminal. The positive and negative terminals establish polarity in a circuit. *Polarity* is the positive (+) or negative (−) state of an object. Each point in a DC circuit has polarity. The polarity can be used to determine if a circuit is DC or AC. If the circuit is DC, the polarity of the test leads determines the positive and negative circuit point. Understanding the polarity is required when installing components because most circuits designed for DC must be connected with the correct polarity in order to operate properly. **See Figure 3-1.**

AC Voltage

AC voltage is voltage that reverses its direction of flow at regular intervals. AC voltage is either single phase (1ϕ) or three phase (3ϕ). Single-phase AC voltage contains only one alternating voltage waveform. In an AC sine wave, the wave reaches its peak positive value at 90°, returns to 0 V at 180°, increases to its peak negative value at 270°, and returns to 0 V at 360°. Three-phase AC voltage is a combination of three alternating voltage waveforms, each displaced 120° (one-third of a cycle) apart. **See Figure 3-2.**

AC voltage can be stated as peak, peak-to-peak, average, or root-mean-square voltage. *Peak voltage (V_{max}, V_p)* of a sine wave is the maximum value of either the positive or negative alternation. *Peak-to-peak voltage (V_{p-p})* of a sine wave is the voltage measured from the maximum positive alternation to the maximum negative alternation. *Average voltage (V_{avg})* is the mathematical mean of instantaneous voltage values in the sine wave and is equal to 0.637 of the peak value. *Root-mean-square voltage (V_{rms})* is the AC voltage value that produces the same amount of heat in a pure resistive circuit as DC voltage of the same value. **See Figure 3-3.**

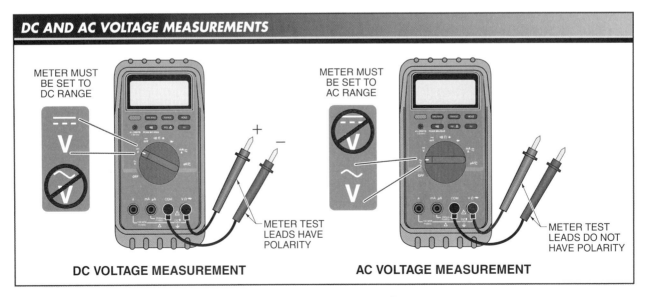

Figure 3-1. Electrical and electronic circuits require DC voltage and AC voltage measurements.

SINGLE-PHASE AND THREE-PHASE AC VOLTAGE

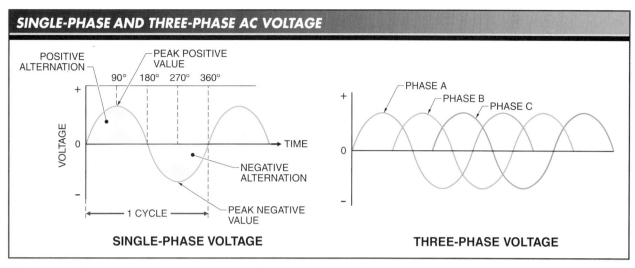

Figure 3-2. AC voltage is either single phase (1ϕ) or three phase (3ϕ).

AC VOLTAGE VALUES

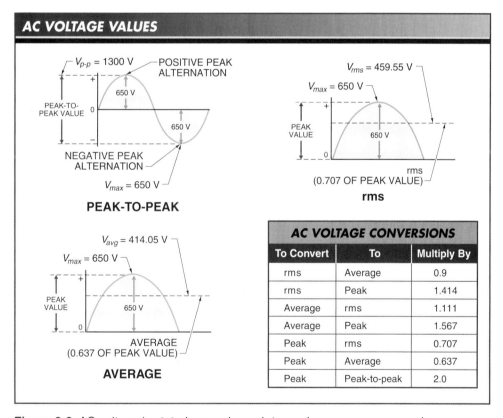

Figure 3-3. AC voltage is stated as peak, peak-to-peak, average, or rms values.

Each voltage value is used to describe the operation of a circuit and can be used when troubleshooting a circuit. For example, the peak voltage measurement is the most necessary measurement when testing for high transient voltages that can be damaging to computers and other electronic components connected to the electrical system.

Both DC and AC voltages are commonly used and have their advantages and disadvantages. For example, advantages of AC voltage are that it is easily transmitted and distributed over long distances and can be used directly by AC components such as motors and incandescent lamps with little or no circuit changes required. Disadvantages of AC voltage are that it cannot be used in portable components such as flashlights and portable tools and cannot be used with most electronic components such as computers and printers.

Although computers, printers, and other electronic components are connected to an AC supply (receptacle), the AC is changed (rectified) into DC and is used to operate the electronic components. A *rectifier* is a device that converts AC voltage to DC voltage by allowing the voltage and current to flow in one direction only. A greater amount of electrical components are now designed for operation on DC than in the past, but are powered by rectified AC. These components include building fire alarm systems, security systems, and light emitting diode (LED) lamps that are replacing incandescent and compact fluorescent lamps (CFLs) to save energy.

Voltage Problems

All electrical and electronic components are rated for operation at a specific voltage. The rated voltage is actually a voltage range. In the past, when taking voltage measurements at the connection point of components, the acceptable range was ±10% of the rated voltage. However, with many modern components derated to save energy and operating cost, the acceptable voltage range is +5% to –10%. This range is used because an overvoltage condition is generally more damaging than an undervoltage condition. For best load and equipment operation, the manufacturer-listed voltage range should be used when a range is predetermined, and the +5% to –10% range should be used for branch circuit, receptacle, and power panel measurements.

Undervoltage and Overvoltage Problems. *Undervoltage* is a voltage decrease more than 10% below the normal rated line voltage for a period of longer than 1 min. Undervoltages are often caused by overloaded transformers, undersized conductors, excessively long conductors, too many loads being placed on a circuit, and peak power usage periods.

Overvoltage is a voltage increase more than 10% above the normal rated line voltage for a period of longer than 1 min. Overvoltages are often caused when loads are near the beginning of a power distribution system or if taps on a transformer are set improperly. A *tap* is a connection point provided along the transformer coil. Taps are commonly provided at 2.5% increments along one end of the transformer coil. Undervoltage (low voltage) is more common than overvoltage (high voltage).

Even though the applied voltage to a load or on a circuit is within an acceptable range, the operation of the load is affected by a change in voltage. For example, resistance heating elements are rated for a predetermined voltage. The voltage

TECH FACT

Changing building exit sign lamps from incandescent lamps to LED lamps lowers energy use by approximately 90%. In addition, LED lamps last approximately 25 years.

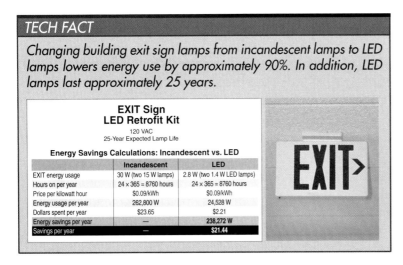

EXIT Sign LED Retrofit Kit
120 VAC
25-Year Expected Lamp Life

Energy Savings Calculations: Incandescent vs. LED

	Incandescent	LED
EXIT energy usage	30 W (two 15 W lamps)	2.8 W (two 1.4 W LED lamps)
Hours on per year	24 × 365 = 8760 hours	24 × 365 = 8760 hours
Price per kilowatt hour	$0.09/kWh	$0.09/kWh
Energy usage per year	262,800 W	24,528 W
Dollars spent per year	$23.65	$2.21
Energy savings per year	—	238,272 W
Savings per year	—	$21.44

applied to the heating elements can vary because the heating elements are basically resistors used to dissipate heat.

The applied voltage can range from 1% to 105% of the heating element rated voltage for both DC- and AC- rated heating elements. The lower the applied voltage, the less heat output produced. For example, a heating element rated for 230 V and connected to an applied voltage of 220 V delivers 91% of its rated power output. The same heating element connected to an applied voltage of 115 V delivers 25% of its rated power output. **See Figure 3-4.**

The voltage requirements of lamps vary depending on the type of lamp. With certain lamp types, the output power and amount of light produced is reduced when the voltage to the lamp is reduced. For other lamp types, such as fluorescent and HID lamps, the lamp does not operate when the voltage drops below a minimum value. Lamp life is reduced when the voltage is higher than the lamp rated value. The acceptable voltage range of the manufacturer as listed in the lamp data, lamp box, or on the manufacturer web site should be checked to ensure optimum lamp life. **See Figure 3-5.**

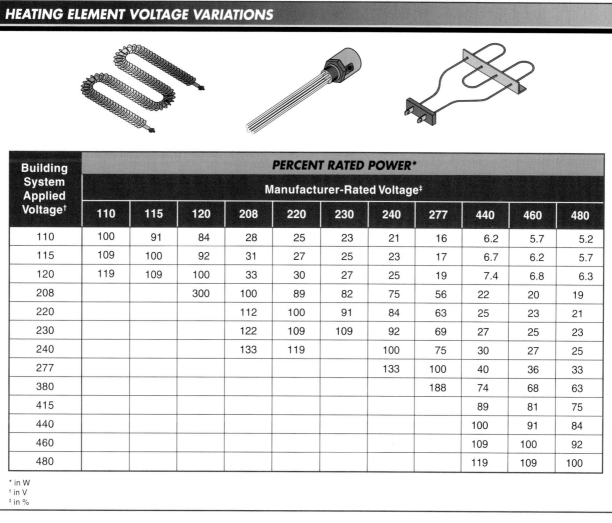

HEATING ELEMENT VOLTAGE VARIATIONS

Building System Applied Voltage†	PERCENT RATED POWER* Manufacturer-Rated Voltage‡										
	110	115	120	208	220	230	240	277	440	460	480
110	100	91	84	28	25	23	21	16	6.2	5.7	5.2
115	109	100	92	31	27	25	23	17	6.7	6.2	5.7
120	119	109	100	33	30	27	25	19	7.4	6.8	6.3
208			300	100	89	82	75	56	22	20	19
220				112	100	91	84	63	25	23	21
230				122	109	109	92	69	27	25	23
240				133	119		100	75	30	27	25
277							133	100	40	36	33
380								188	74	68	63
415									89	81	75
440									100	91	84
460									109	100	92
480									119	109	100

* in W
† in V
‡ in %

Figure 3-4. Heating element output varies when a heating element is supplied with a voltage level that is lower or higher than the manufacturer-rated voltage level.

MANUFACTURER LAMP RATINGS

VOLTAGE RANGES*

A (5%)			B (10%)		
Maximum	Rated	Minimum	Maximum	Rated	Minimum
126	120	114	132	120	108
218	208	198	229	208	187
252	240	228	264	240	216
291	277	263	305	277	249

* in V

METAL-HALIDE LAMP CHARACTERISTICS

Lamp Wattage	Voltage Rating	Fuse*	Starting Current*	Operational Current*	Line Input† (Includes Ballast)	% Allowable Variation Operating Voltage Range	% Allowable Variation Lamp Wattage Output
50	120	3	0.60	0.65	72	A	±8
	277	3	0.25	0.30			
100	120	8	1.15	1.15	129	A	±12
	208	5	0.66	0.66			
	240	3	0.58	0.58			
	277	3	0.50	0.50			
175	120	5	1.30	1.80	210	B	±10
	208	3	0.75	1.05			
	240	3	0.65	0.90			
	277	3	0.55	0.80			
	480	3	0.35	0.45			
250	120	8	2.10	2.50	285	B	±9
	208	5	1.40	1.45			
	240	5	1.10	1.25			
	277	3	1.00	1.10			
	480	3	0.60	0.60			
1000	120	20	8.0	9.0	1080	B	±10
	208	15	4.6	5.2			
	240	10	4.0	4.5			
	277	10	3.5	3.9			
	480	10	2.0	2.3			

* in A
† in W

Figure 3-5. Lamp output varies when a lamp is supplied with a voltage level that varies from the manufacturer rating.

As with any load, motors have an acceptable voltage operating range. For most motors, the acceptable range is ±10% of the motor nameplate rating. Any voltage value above or below the nameplate rating affects motor operating torque (output power), current draw, temperature, and speed. **See Figure 3-6.**

Transient Voltage Problems. Electrical loads can tolerate small increases and decreases in voltage as long as they are within the acceptable limits of operation. Some loads, such as heating elements, solenoids, magnetic motor starter coils, and most motors, can tolerate a higher voltage as long as it is not higher than the specified maximum voltage level and only last for a short time. However, almost no electronic component or load, such as a heating element, solenoid, coil, or motor, can withstand a high transient voltage.

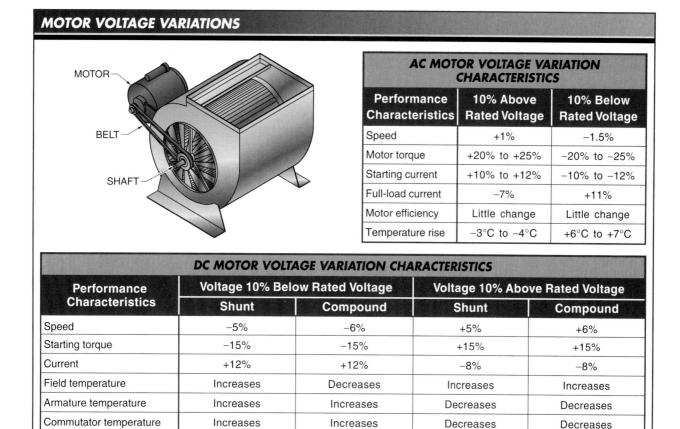

Figure 3-6. Motor operating torque (output power), current draw, temperature, and speed vary when a motor is supplied with a voltage level that varies from the manufacturer rating.

Research by the Institute of Electrical and Electronics Engineers (IEEE), service technicians, and equipment manufacturers indicates that solid-state circuits and components do not tolerate momentary voltage surges exceeding twice their normal operating voltage. Voltage surges are produced in electrical systems from outside sources, such as lightning strikes and utility switching, and within a building system by turning off loads that include coils, such as solenoids, motor starter coils, lighting/heating contactor coils, and motors. The voltage surge produced is referred to as a transient voltage.

A *transient voltage*, also known as a voltage spike, is a short, temporary, undesirable voltage in an electrical circuit. Transient voltages range from a few volts to several thousand volts and last from a few microseconds up to a few milliseconds. Transient voltages are the primary cause of electronic circuit damage. An *oscillatory transient voltage* is a transient voltage that includes both positive and negative polarity values. Oscillatory transient voltages are commonly caused by turning off high inductive loads, such as coils, and by switching large utility power factor correction capacitors. An *impulse transient voltage* is a short-duration transient voltage commonly caused by a lightning strike, which results in a short, unwanted voltage placed on the power distribution system. Oscillatory and impulse transient voltages both cause damage to equipment, but impulse transients cause the most damage and are responsible for most of the damaged electronic equipment after a lightning strike. **See Figure 3-7.**

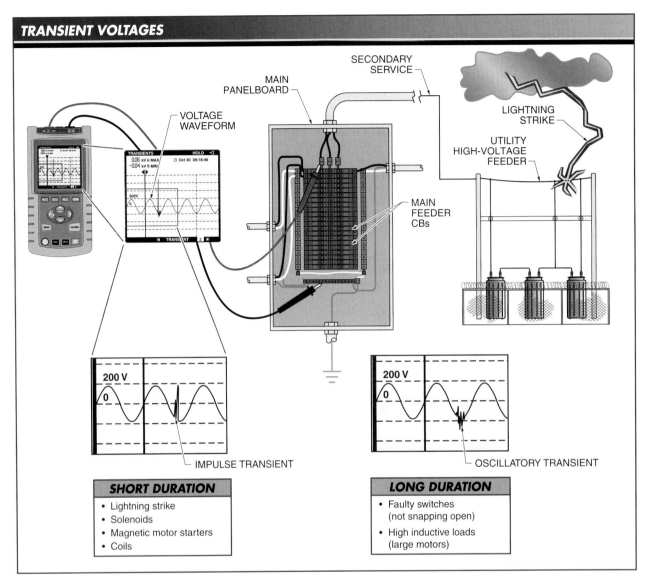

Figure 3-7. Transient voltages cause damage to electrical and electronic equipment.

Current
Media Clip

TECH FACT

Transient voltages caused from utility capacitor switching outside a building to correct poor power factor on utility lines travel long distances. They are the cause of overvoltage trips on motor drives because they overcharge the drive capacitors. Transient voltages caused by equipment switching within a building usually do not cause overvoltage trips on motor drives, but do cause short data/signal loss and damage to solid-state electronic devices within the drive.

CURRENT

Current (I) is the amount of electrons flowing through an electrical circuit. Current is measured in amperes (A). Current flows through a circuit when a source of electrical power is connected to a component such as a lamp or motor that uses electricity. An *ampere* is the amount of electrons passing a predetermined point in a circuit in 1 sec. Current, as with voltage, can be DC or AC. DC flows in any circuit connected to a power supply producing a DC voltage such as a battery, solar cell, or DC generator. AC flows in any circuit connected to a power supply producing an AC voltage such as an AC generator.

Direct Current

DC measurements are taken to check any DC load or circuit. Ammeters can be used to test battery-powered equipment to determine the state of their DC charging system and batteries. Measuring current is the best way to determine if a DC power source or battery is overloaded, since different voltage sources can produce the same amount of voltage, but not the same amount of current output. For example, standard AAA, AA, C, and D size batteries produce 1.5 VDC, but each size is capable of delivering different amounts of current without becoming overloaded.

Alternating Current

Power distribution systems within a building are always AC systems. Alternating current is delivered to equipment and loads through branch circuits, receptacles (outlets), disconnects, control centers, and hard wiring. Taking AC measurements on the power distribution system indicates the working load on each circuit, panel, or transformer. The higher the current, the greater the loading. Loads that operate directly from AC voltage, such as AC motors and lamps, can be checked for proper operation by measuring the amount of current drawn by the load and comparing it to the load rated current.

Current Problems

Every load, such as a lamp, motor, or heating element, that converts electrical energy into some other form of energy, such as light, rotating motion, or heat, uses current. The more electrical energy required, the higher the current use. Although voltage remains relatively constant in an electrical circuit and at the load, current varies. For example, the voltage applied to a heating element can remain the same if the element is operating normally, is producing its rated heat output, or is open and not producing heat. The voltage can also remain the same if the resistance of the element has increased due to corrosion or a bad connection and is producing only a fraction of its rated heat output. Measuring voltage at a heating element only indicates that voltage is present. Measuring current drawn by a heating element indicates the range in which the heating element is operating. **See Figure 3-8.** Common current problems in commercial circuits include overcurrents, overloads, and short circuits.

Direct Current Media Clip

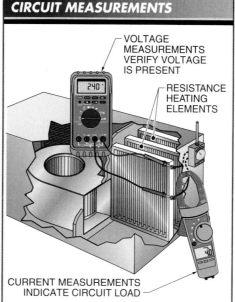

Alternating Current Media Clip

Figure 3-8. Voltage and current measurements should be taken when testing equipment and circuits for the clearest indication of circuit operation.

Overcurrent and Overload Problems. When a motor is overloaded, the motor tries to rotate its shaft. The harder it is to rotate the shaft, the greater the current drawn by the motor and the hotter the motor becomes. If the shaft cannot rotate (locked rotor condition), the current drawn and heat produced by the motor is the highest. Motor current measurements should be taken at different operating times when testing an AC or DC motor.

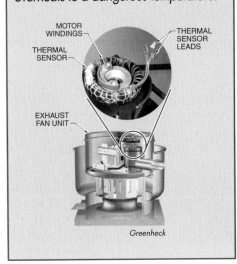

TECH FACT

Codes and standards are set to help prevent an overheated motor from causing a fire. For example, UL Standard 507, Electric Fans, states that any motor used as a fan motor in a building such as a bathroom exhaust fan, in-line wall or ceiling insert fan, or other air moving fan that is likely to operate unattended must include a thermal cut-off device (typically a sensor that is wound into the motor windings) that automatically removes power from the motor before the motor overheats to a dangerous temperature.

Greenheck

Short Circuits. A *short circuit* is an overcurrent that leaves the normal current-carrying path by going around the load and back to the power source or ground. Short circuits occur when loose connections allow conductors to touch or conductor insulation is nicked or damaged.

A short circuit causes circuit current to rise hundreds of times higher than normal at a fast rate. Short circuits are a major cause of fires in buildings and can cause shocks, explosions, and equipment damage. Circuits must be protected against short circuits. Fuses and circuit breakers are designed to remove power when a short circuit occurs. However, regardless of how quickly power is removed, there is still a danger of arcs being produced by short circuits. Common practices applied to help avoid short circuits include the following:

- applying national, state, and local electrical codes
- using the correct size fuse/circuit breaker
- looking for loose connections and damaged insulation
- checking for other possible faults and damage after a short circuit condition is corrected

RESISTANCE

Resistance (R) is the opposition to the flow of electrons in a circuit and is measured in ohms (Ω). Resistance limits the flow of current in an electrical circuit. The higher the resistance, the lower the current flow. The lower the resistance, the higher the current flow. Resistive circuits are a result of components that produce resistance.

Materials designed to insulate, such as rubber or plastic, have high resistance. Materials designed to conduct, such as copper and aluminum conductors and silver switch contacts, have low resistance. Loads that are designed to convert electrical energy into heat, light, sound, and motion have resistance values of a few ohms, thousands of ohms ($k\Omega$), or millions of ohms ($M\Omega$). For example, a 1200 W heating element has a resistance of approximately 12 Ω, and a 10 W heating element has a resistance of approximately 1440 Ω.

Resistive Circuits

A *resistive circuit* is a circuit that contains components that produce only resistance, such as heating elements and incandescent lamps. Electrical circuits include some resistance because conductors, switch contacts, connections, splices, and loads have resistance. Rubber mats and rubber insulating gloves have as high of a resistance as possible to reduce the chance of current flow through the body when contact is made with an energized circuit. Conversely, electrical conductors (wires), splices, and switch contacts should have as low of a resistance as possible, so they offer as little opposition to current flow as possible.

INDUCTIVE REACTANCE

Inductance (L) is the property of a circuit that causes it to oppose a change in current due to energy stored in a magnetic field. Any time current flows through a conductor, a magnetic field is produced around the conductor. A conductor formed into a coil produces a stronger magnetic field than a straight conductor. DC flow produces a constant magnetic field around a coil. AC flow produces an alternating magnetic field around a coil.

Inductive reactance (X_L) is the opposition to the flow of alternating current in a circuit due to inductance. As with resistance, inductive reactance is expressed in ohms. Without inductive reactance, electric motors, transformers, and coils would not be practical due to their excessive current draw. This is because a wire coil has little resistance due to the actual conductor resistance. For this reason, if this resistance is the only resistance between the power lines when power is applied to the coil, the current draw would be so high, the circuit fuses/circuit breakers would open/trip. However, because inductive reactance is higher than the actual conductor resistance in a coil, current is limited and the magnetic field produced by transformer coils, motor coils, and solenoid coils can be used to accomplish work.

CAPACITIVE REACTANCE

Capacitive reactance (X_C) is the opposition to current flow by a capacitor. A *capacitor* is an electric device that stores electrical energy by means of an electrostatic field. *Capacitance (C)* is the ability of a component or circuit to store energy in the form of an electric charge. The unit of capacitance is the farad (F). Capacitors offer opposition to current flow in an electrical circuit similar to the resistance produced by heating elements and inductance produced by coils. As with inductive reactance, capacitive reactance is expressed in ohms.

OHM'S LAW

Ohm's law is the relationship between voltage (*E*), current (*I*), and resistance (*R*) in a circuit. Ohm's law states that current in a circuit is proportional to the voltage and inversely proportional to the resistance. If the resistance in a circuit remains constant, a change in current is directly proportional to a change in voltage. If voltage in a circuit remains constant, current in a circuit decreases with an increase in resistance, and current in a circuit increases with a decrease in resistance. **See Figure 3-9.**

Ohm's Law Media Clip

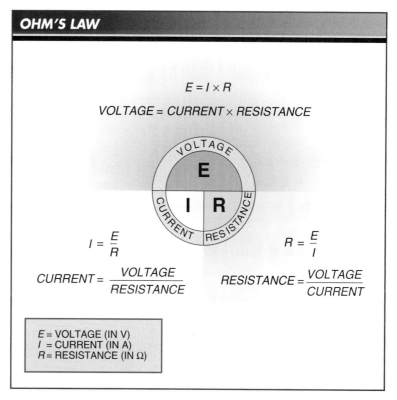

Figure 3-9. Ohm's law is the relationship between voltage (*E*), current (*I*), and resistance (*R*) in a circuit.

Understanding the application of Ohm's law is required because when any two values of voltage, current, or resistance are known, the third value can be determined. Ohm's law is used when troubleshooting components such as electric heating elements. The resistance of the heating element can be measured, and the expected current draw can be calculated for a predetermined voltage. Ohm's law explains why heating elements draw less current as the heating element resistance increases with age and corrosion and why heating elements draw more current as the heating element resistance decreases when moisture enters the insulated heating unit. Ohm's law is used to help explain the effects of impedance on an electrical circuit.

In any circuit that includes inductance or capacitance, the opposition to the flow of current is reactance (X). In any circuit that contains resistance and reactance, the combined opposition to the flow of current is impedance (Z). Impedance is stated in ohms.

In DC circuits and AC circuits that include only resistance, the word resistance is used when describing a circuit or load resistance. In AC circuits that include motors, transformers, coils, and solenoids, the word impedance is used when describing a circuit or load resistance. For most AC electrical circuits, it is more accurate to refer to the circuit resistance to the flow of current as impedance and not resistance. **See Figure 3-10.** Ohm's law can be used to determine circuit characteristics in circuits that contain impedance, however, Z is substituted for R in the formula. Z represents the total resistive force (resistance and reactance) opposing current flow.

TECH FACT
In an electric resistance heating element application, when the resistance of the element is held constant, an increase in voltage increases the amount of current and heat produced, and a decrease in voltage decreases the amount of current and heat produced.

The amount of heat produced by electric heating elements can be controlled by varying the voltage applied to the heating elements.

Power
Media Clip

Ohm's Law and Impedance
Ohm's law is limited to circuits in which electrical resistance is the only significant opposition to the flow of current, such as in circuits that include heating elements. The circuit limitation includes DC circuits and any AC circuits that do not contain a significant amount of capacitance or inductance.

POWER
Electrical energy is converted into other forms of energy, such as light, heat, sound, linear motion, or rotary motion, any time current flows in a circuit. *Power (P)* is the rate of doing work or using energy. Power can be expressed as true power, reactive power, or apparent power. For example, electrical heating elements and lamps include a power rating in watts (true power), and transformers include a power rating in volt amps (apparent power). **See Figure 3-11.**

Chapter 3—Electrical Quantities and Circuits

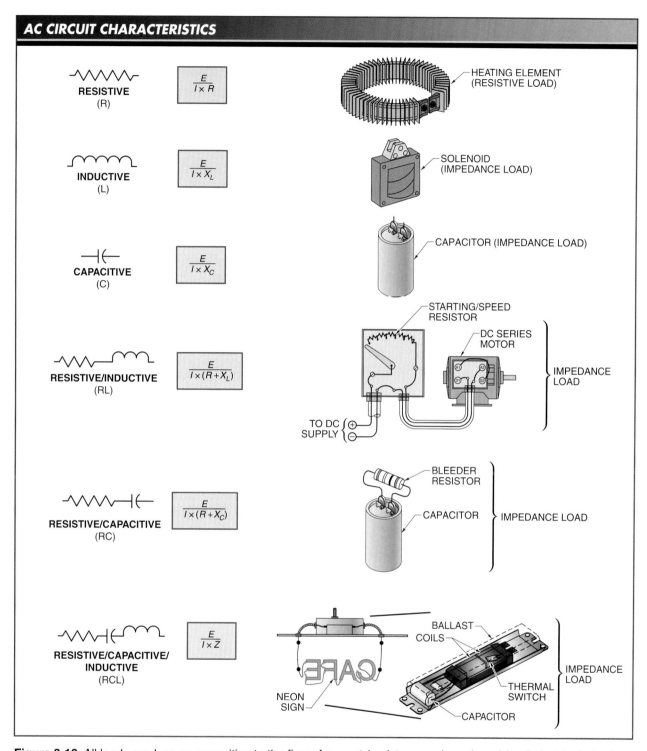

Figure 3-10. All loads produce an opposition to the flow of current (resistance or impedance) in all electrical circuits.

True Power

True power (P_T) is the actual power used in an electrical circuit to produce work, such as heat, light, sound, and motion. True power represents the pure resistive components and parts of a circuit. True power is measured in watts (W), kilowatts (kW), or megawatts (MW). Electrical loads such as heating elements and lamps list their power ratings in watts.

POWER

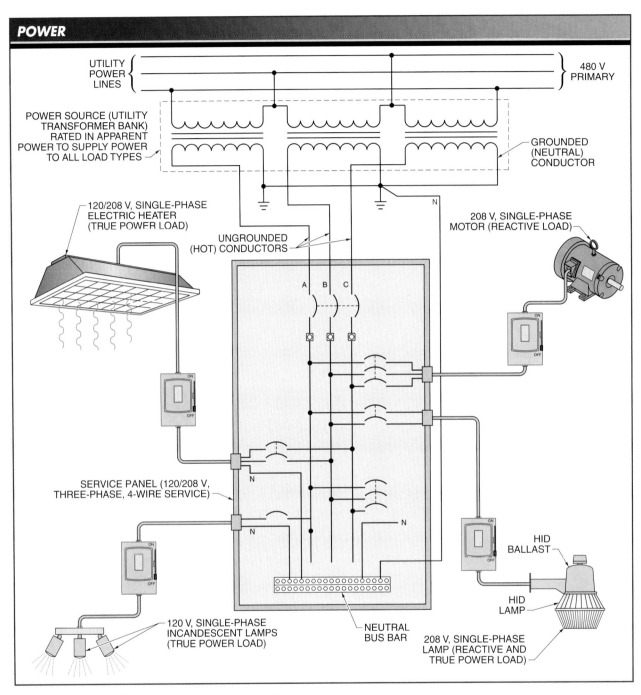

Figure 3-11. Power is stated as true power, reactive power, or apparent power.

Reactive Power

Reactive power (VAR) is power supplied to a reactive load, such as a transformer coil, motor winding, or solenoid coil. Reactive power represents the inductive and capacitive components of a circuit. Although some electrical circuits include only true power, most electrical circuits include true power and reactive power. These types of electrical circuits include loads such as fluorescent lamps, HID lamps, solenoids, transformers, computers, printers, and motors.

Apparent Power

Apparent power (P_A) is the product of the voltage and current in a circuit calculated without considering the phase shift that can be present between the voltage and current in a circuit. Apparent power is expressed in volt amps (VA), kilovolt amps (kVA), or megavolt amps (MVA).

Phase Shift

Phase shift is the state when voltage and current in a circuit do not reach their maximum amplitude and zero level simultaneously. There is little to no phase shift in AC circuits that contain resistive components. A phase shift exists in AC circuits that contain devices that include inductance and/or capacitance. **See Figure 3-12.**

Phase shift is present in electrical circuits that include inductance and/or capacitance and causes the circuit power factor to be less than perfect (less than 100%). *Power factor (PF)* is the ratio of true power used in an AC circuit to apparent power delivered to the circuit. Power factor is expressed as a percentage. True power equals apparent power when the power factor is 100% or 1. When the power factor is less than 100% or 1, the circuit is less efficient and has a higher operating cost because only part of the current is performing work due to the phase shift between voltage and current.

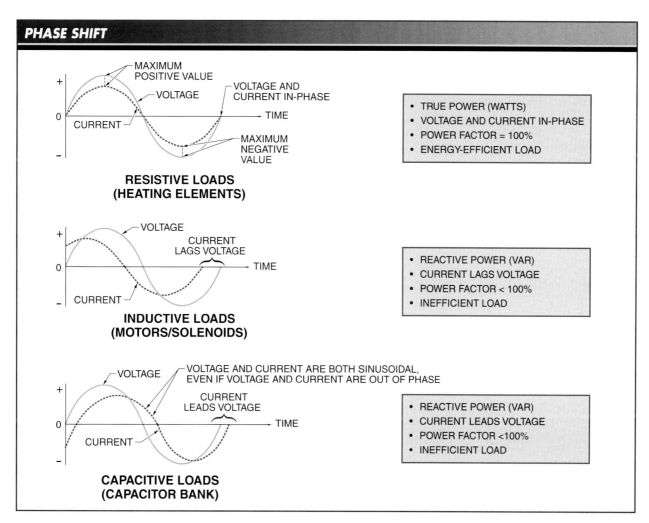

Figure 3-12. The greater the amount that voltage and current are out-of-phase, the greater the phase shift, and the poorer the load or circuit power factor and efficiency.

Because voltage and current are so far out of phase in single-phase motors, single-phase motors are the least energy efficient type of motor. Three-phase motors are more energy efficient than single-phase motors. To improve energy efficiency in motor circuits, three-phase motors are now used in applications where single-phase motors were used because AC variable-speed drives are available that convert single-phase power into three-phase power. **See Figure 3-13.**

POWER FORMULA

The *power formula* is the relationship between power (P), voltage (E), and current (I) in an electrical circuit. Any value in the relationship can be calculated using the power formula when the other two values are known. **See Figure 3-14.**

As with Ohm's law, the power formula is used when troubleshooting and to predict circuit characteristics before power is applied to a circuit or a measurement is taken. The power formula is useful when determining expected current values because most electrical equipment lists a voltage and power rating only. The power rating is listed in watts for most appliances and heating elements or in horsepower for motors.

TECH FACT
In the United States, motors are normally rated in horsepower (HP). In Europe, motors are normally rated in watts (usually kW). There are 746 W per 1 HP. To convert a motor kilowatt rating to the equivalent horsepower rating, multiply the kilowatt rating of the motor by 1.34.

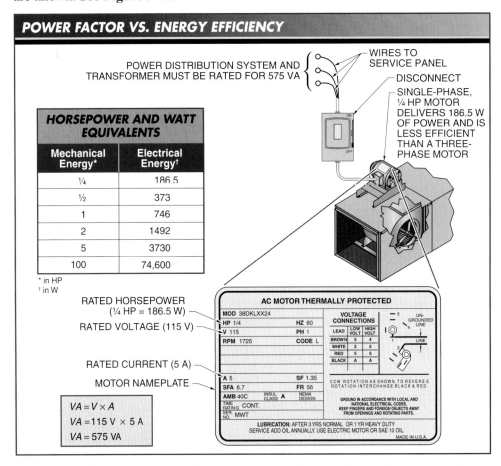

Figure 3-13. Single-phase motors have a poor power factor and are not as energy efficient as three-phase motors.

Chapter 3—Electrical Quantities and Circuits 83

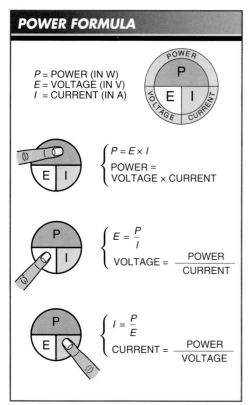

Figure 3-14. The power formula is the relationship between power (P), voltage (E), and current (I) in an electrical circuit.

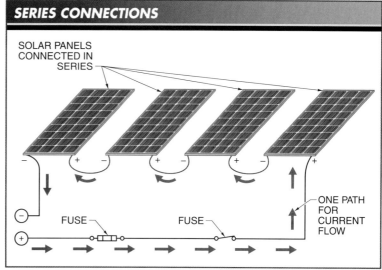

Figure 3-15. There is only one path for current to flow in a series-connected circuit.

TECH FACT

The operation of basic series and parallel circuits obeys certain laws of physics. These laws are referred to as Ohm's law and Kirchhoff's law, named after their originators, Georg Ohm and Gustav Kirchhoff.

SERIES CIRCUITS

Electrical devices and components such as switches, motor windings, transformer coils, batteries, solar panels, and fuses can be connected in series within an electrical circuit. A *series connection* is a connection that has two or more devices connected so there is only one path for current flow. Opening the circuit at any point stops current flow. Current flow stops any time a fuse blows, a circuit breaker trips, or a switch or load opens. **See Figure 3-15.**

For any series-connected circuit, the circuit resistance (R), voltage (E), current (I), and power (P) can be calculated at any point in the circuit. The circuit total resistance, voltage, current, and power are referred to as R_T, E_T, I_T, and P_T. **See Figure 3-16.**

Resistance in Series Circuits

The total resistance in a circuit containing series-connected loads equals the sum of the resistances of all loads. The resistance in the circuit increases if loads are added in series and decreases if loads are removed. The resistance of a series circuit is calculated by using the following formula:

$$R_T = R_1 + R_2 + \ldots + R_N$$

where
R_T = total resistance (in Ω)
R_1 = resistance 1 (in Ω)
R_2 = resistance 2 (in Ω)
R_N = remaining resistances (in Ω)

Example: What is the total resistance of a circuit that has 2 Ω, 4 Ω, and 6 Ω resistors connected in series?

$R_T = R_1 + R_2 + R_3$
$R_T = 2 + 4 + 6$
$R_T = \mathbf{12\ \Omega}$

Series Circuits
Media Clip

SERIES CIRCUITS

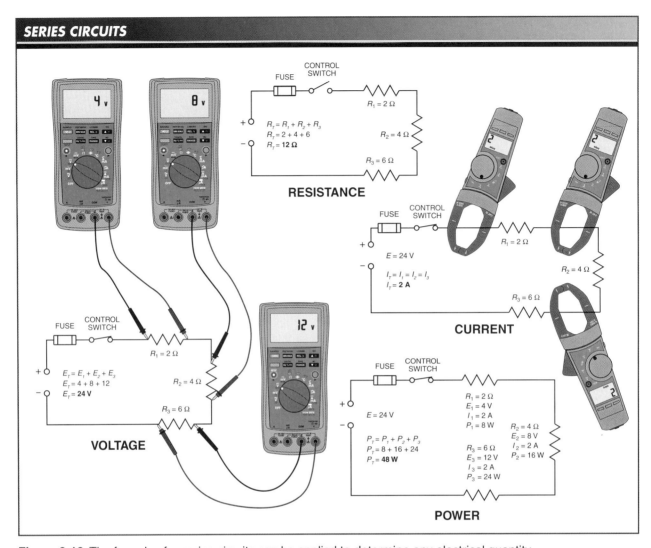

Figure 3-16. The formulas for series circuits can be applied to determine any electrical quantity.

Voltage in Series Circuits

The voltage applied to loads connected in series is divided across the loads. Each load drops a set percentage of the applied voltage. The exact voltage drop across each load depends on the resistance of each load. The voltage drop across each load is the same if the resistance values are the same. The load with the lowest resistance has the lowest voltage drop, and the load with the highest resistance has the highest voltage drop. The voltage of a series circuit when the voltage across each load is known or measured is calculated by using the following formula:

$$E_T = E_1 + E_2 + ... + E_N$$

where

E_T = total applied voltage (in V)//
E_1 = voltage drop across load 1 (in V)//
E_2 = voltage drop across load 2 (in V)//
E_N = voltage drop across all remaining loads (in V)

Example: What is the total applied voltage to a circuit containing 4 V, 8 V, and 12 V drops across three loads?

$$E_T = E_1 + E_2 + E_3$$
$$E_T = 4 + 8 + 12$$
$$E_T = \mathbf{24\ V}$$

Current in Series Circuits

The current in a circuit containing series-connected loads is the same throughout the circuit. The current in the circuit decreases if the circuit resistance increases, and the current increases if the circuit resistance decreases. The total current of a series circuit is calculated by using the following formula:

$I_T = I_1 = I_2 = ... = I_N$

where

I_T = total circuit current (in A)
I_1 = current through load 1 (in A)
I_2 = current through load 2 (in A)
I_N = current through all remaining loads (in A)

Example: What is the total current through a series circuit if the current measured at each load is 2 A?

$I_T = I_1 = I_2 = I_3$
$I_T = 2 = 2 = 2$
$I_T = \mathbf{2\ A}$

Power in Series Circuits

The total power in a series circuit is equal to the sum of the individual power produced by each load. The lower the circuit or load resistance, the higher the current draw and the greater the power. Conversely, the higher the circuit or load resistance, the lower the current draw and the less power produced. Power is produced when voltage is applied to a load and current flows through the load. The power produced is used to make light (lamps), heat (heating elements), rotary motion (motors), or linear motion (solenoids).

The amount of power produced is measured in watts. The total power in a series circuit when the power across each load is known or measured is calculated by using the following formula:

$P_T = P_1 + P_2 + ... + P_N$

where

P_T = total circuit power (in W)
P_1 = power of load 1 (in W)
P_2 = power of load 2 (in W)
P_N = power of all remaining loads (in W)

Example: What is the total power in a series circuit if three loads of 8 W, 16 W, and 24 W are connected in series?

$P_T = P_1 + P_2 + P_3$
$P_T = 8 + 16 + 24$
$P_T = \mathbf{48\ W}$

> **TECH FACT**
> Gustav Kirchhoff stated in 1847 that the sum of all voltage drops across loads in a circuit must equal the applied voltage and that the total current entering a circuit must equal the current leaving the circuit. This became known as both Kirchhoff's voltage law (KVL) and Kirchhoff's current law (KCL) and has been used ever since to help understand how voltage and current act in a circuit. The importance of Kirchhoff's work and subsequent laws is that they proved scientifically through documented experiments that electricity has quantities (voltage and current) that are always predictable and measurable at any point in a circuit.

PARALLEL CIRCUITS

Electrical devices and components such as switches, motor windings, transformer coils, batteries, solar panels, and fuses can be connected in parallel within an electrical circuit. A *parallel connection* is a connection with two or more devices connected so that there is more than one path for current flow. **See Figure 3-17.** Care must be taken when working with parallel circuits because current can be flowing in one part of the circuit even though another part of the circuit is off. Understanding and recognizing parallel-connected components and circuits enables a technician or troubleshooter to take proper measurements, make circuit modifications, and troubleshoot the circuit.

Parallel Circuits Media Clip

For any parallel-connected circuit, the circuit resistance (R), voltage (E), current (I), and power (P) can be calculated at any point in the circuit. As in a parallel circuit, the circuit total resistance, voltage, current, and power are referred to as R_T, E_T, I_T, and P_T. **See Figure 3-18.**

Resistance in Parallel Circuits

The total resistance in a circuit containing parallel-connected loads is less than the smallest resistance value in the circuit. The total resistance decreases if loads are added in parallel and increases if loads are removed. The total resistance in a parallel circuit containing two resistors is calculated by using the following formula:

$$R_T = \frac{R_1 \times R_2}{R_1 + R_2}$$

where
R_T = total resistance (in Ω)
R_1 = resistance 1 (in Ω)
R_2 = resistance 2 (in Ω)

Example: What is the total resistance in a circuit containing resistors of 16 Ω and 24 Ω connected in parallel?

$$R_T = \frac{R_1 \times R_2}{R_1 + R_2}$$

$$R_T = \frac{16 \times 24}{16 + 24}$$

$$R_T = \frac{384}{40}$$

$$R_T = \mathbf{9.6\ \Omega}$$

The total resistance in a parallel circuit containing three or more resistors is calculated by using the following formula:

$$R_T = \frac{1}{\left(\dfrac{1}{R_1}\right) + \left(\dfrac{1}{R_2}\right) + \ldots + \left(\dfrac{1}{R_N}\right)}$$

where
R_T = total resistance (in Ω)
R_1 = resistance 1 (in Ω)
R_2 = resistance 2 (in Ω)
R_N = remaining resistances (in Ω)

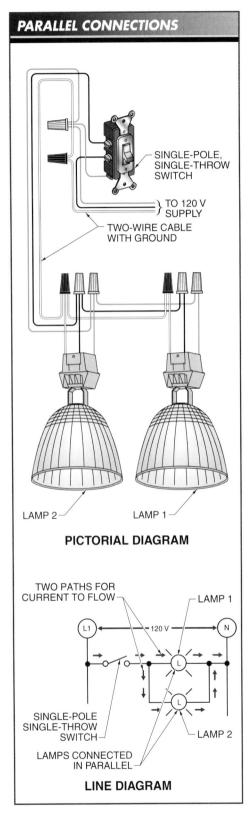

Figure 3-17. A parallel connection has two or more components connected so that there is more than one path of current flow.

Example: What is the total resistance in a circuit containing resistors of 16 Ω, 24 Ω, and 48 Ω connected in parallel?

$$R_T = \frac{1}{\left(\frac{1}{R_1}\right)+\left(\frac{1}{R_2}\right)+\left(\frac{1}{R_3}\right)}$$

$$R_T = \frac{1}{\left(\frac{1}{16}\right)+\left(\frac{1}{24}\right)+\left(\frac{1}{48}\right)}$$

$$R_T = \frac{1}{0.06250 + 0.04166 + 0.02083}$$

$$R_T = \frac{1}{0.12499}$$

$$R_T = \mathbf{8\ \Omega}$$

Voltage in Parallel Circuits

The voltage applied across loads connected in parallel is the same for each load. The voltage across each load remains the same if parallel loads are added or removed. The total voltage in a parallel circuit when the voltage across a load is known or measured is calculated by using the following formula:

$$E_T = E_1 = E_2 = \ldots = E_N$$

where

E_T = total applied voltage (in V)
E_1 = voltage across load 1 (in V)
E_2 = voltage across load 2 (in V)
E_N = voltage across all remaining loads (in V)

Example: What is the total applied voltage if the voltage across three parallel-connected loads is 96 VDC?

$$E_T = E_1 = E_2 = E_3$$
$$E_T = 96 = 96 = 96$$
$$E_T = \mathbf{96\ VDC}$$

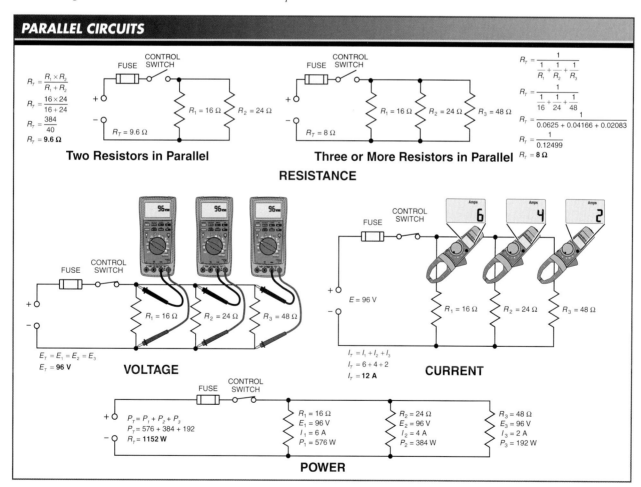

Figure 3-18. The formulas for parallel circuits can be applied to determine any electrical quantity.

Current in Parallel Circuits

Total current in a circuit containing parallel-connected loads equals the sum of the current through all the loads. Total current increases if loads are added in parallel and decreases if loads are removed. The total current in a parallel circuit is calculated by using the following formula:

$$I_T = I_1 + I_2 + ... + I_N$$

where
I_T = total circuit current (in A)
I_1 = current through load 1 (in A)
I_2 = current through load 2 (in A)
I_N = current through all remaining loads (in A)

Example: What is the total current in a circuit containing three loads connected in parallel if the current through the three loads is 6 A, 4 A, and 2 A?

$$I_T = I_1 + I_2 + I_3$$
$$I_T = 6 + 4 + 2$$
$$I_T = \mathbf{12\ A}$$

Series-Parallel Circuits Media Clip

Power in Parallel Circuits

As in a series circuit, the total power in a parallel circuit is equal to the sum of the individual power produced by each load. The lower the circuit or load resistance, the higher the current draw and the greater the power. Conversely, the higher the circuit or load resistance, the lower the current draw and the less power produced. Power is produced when voltage is applied to a load and current flows through the load. The power produced is used to produce light, heat, sound, rotary motion, or linear motion. The higher the applied voltage, the more power produced. The lower the applied voltage, the less power produced. Power produced is measured in watts (W). The total power in a parallel circuit when the power across each load is known or measured is calculated by using the following formula:

$$P_T = P_1 + P_2 + ... + P_N$$

where
P_T = total circuit power (in W)
P_1 = power of load 1 (in W)
P_2 = power of load 2 (in W)
P_N = power of all remaining loads (in W)

Example: What is the total circuit power if four loads are connected in parallel and the loads produce 576 W, 384 W, and 192 W?

$$P_T = P_1 + P_2 + P_3$$
$$P_T = 576 + 384 + 192$$
$$P_T = \mathbf{1152\ W}$$

SERIES-PARALLEL CIRCUITS

A *series-parallel circuit* is a combination of series and parallel-connected components. The majority of electrical applications contain series-parallel-connected components. In a series-parallel-connected circuit, the laws of voltage, current, resistance, and power for series circuits are applied to the series parts of the circuit. The laws of voltage, current, resistance, and power for parallel circuits are applied to the parallel parts of the circuit.

Troubleshooting is not effective unless the laws of series and parallel circuits are known and properly applied. Test instruments are typically used to take voltage and current measurements in electrical circuits, but the measured values are only useful when the troubleshooter understands what the expected voltage and current measurement should be at any point in the circuit.

Series-Parallel Circuit Applications

Every electrical circuit includes series, parallel, or series-parallel-connected components. In some applications, the connections are known. For example, when banks of lamps are connected in parallel, each hot (black) conductor is spliced to the next hot (black) conductor and each neutral (white) conductor is spliced to the next neutral (white) conductor. In other applications, the connection and reason for the connection type is not known.

Electrical loads are used to provide light, heat, sound, motion, and other forms of work. The amount (number) of electrical loads generally increases over time for most circuits and buildings. Increasing loads without changing to energy-efficient loads or downsizing the load increases the total power required and increases the amount of current drawn, even when the voltage remains the same. Increasing current requires larger conductors and higher current-rated switches, fuses/circuit breakers, and electrical panels. For example, in a commercial lighting circuit, the lamps are connected in parallel, while the fuses are connected in series. The current of each lamp adds to the total circuit current flowing to the other lamps. **See Figure 3-19.**

MAGNETISM

Magnetism is a force by which materials exert an attraction or repulsion on other materials. A *ferromagnetic metal* is a material that has extremely high and variable magnetic susceptibility and is strongly attracted to a magnetic field. Ferromagnetic metals are materials, such as soft iron, that are easily magnetized. Magnetism is necessary in the electrical field because it is used to produce electricity in AC and DC generators, develop rotary motion in AC and DC motors, and develop linear motion in solenoids, motor starters, lighting/heating contactors, and relay coils.

Magnets

A *magnet* is a substance that produces a magnetic field and attracts iron. Magnets are either permanent or temporary. A *permanent magnet* is a magnet that holds its magnetism for a long period of time. Permanent magnets are used in electrical applications such as permanent magnet DC motors and magnetically operated switches (reed switches). **See Figure 3-20.**

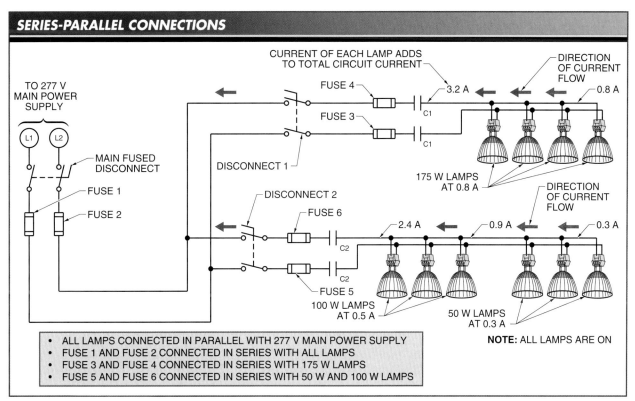

Figure 3-19. In a commercial lamp circuit, the current of each lamp adds to the total circuit current flowing to the other lamps.

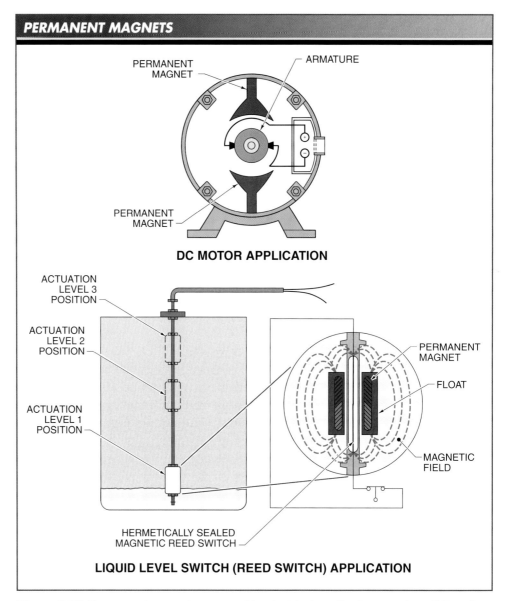

Figure 3-20. Permanent magnets are used in electrical applications such as DC motors and liquid level switches.

Electromagnetism

Electromagnetism is the magnetism produced when electric current passes through a conductor. The magnetic field produced is present only as long as current flows through the conductor. To increase the strength of the magnetic field, the conductor is wrapped into a coil. The strength of a magnetic field is directly proportional to the number of turns in the coil and the amount of current flowing through the conductor. Increasing the voltage increases current. An iron core increases the strength of the magnetic field by concentrating the field. Electromagnetism is used to develop temporary magnets. A *temporary magnet* is a magnet that loses its magnetism as soon as the magnetizing force is removed. Temporary magnets are used in most electrical applications that include a coil, such as motors, transformers, and solenoids. **See Figure 3-21.**

ELECTRICAL CIRCUIT SECTIONS

Electrical circuits in commercial buildings are designed to accomplish specific work in a predetermined manner. An electrical circuit that is properly designed and wired must respond without any changes or inconsistencies. To achieve consistency, each component in the circuit has a function that is based on how the individual components are interconnected.

An electrical circuit can be broken down into four sections. The four sections of an electrical circuit are the output (load), input (switch), interface, and decision sections. Even the most basic electrical circuits have a load and a switch. Electrical circuits can have an interface and a decision device in addition to a load and switch. **See Figure 3-22.**

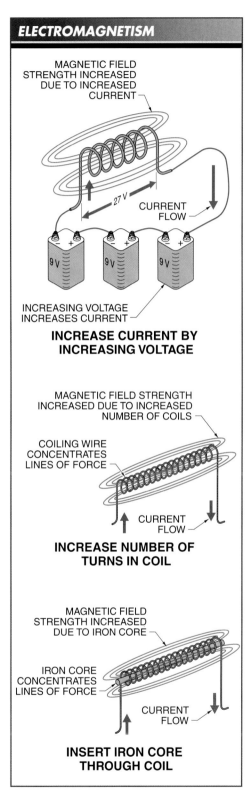

Figure 3-21. The strength of a magnetic field produced by a conductor may be increased by increasing the voltage, increasing the number of coil turns, or inserting an iron core through the coil.

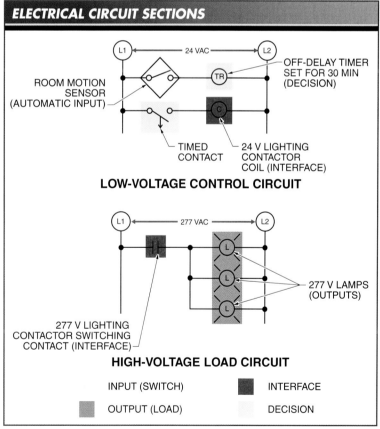

Figure 3-22. The four sections of an electrical circuit are the output (load), input (switch), interface, and decision sections.

The most basic electrical circuit has an input and an output. In a basic circuit, a switch (manual input) is used to control when a lamp (output) is on or off. The switch must carry the current of the lamp and be rated for the same voltage as (or higher voltage than) the lamp.

In many commercial circuits, a switch must control several loads that together draw high current and thus may also be at a higher voltage than what is safest for the switch location and type. For example, a single-pole, single-throw switch may be used to control a 24 V lighting contactor coil. The lighting contactor (interface) high current/power contacts can be used to control high-voltage lamps.

In order to save energy, lights can be automatically turned on and off using motion sensors. Motion sensors can detect warm objects, such as individuals, that move in a room. The sensor could be connected to directly control the ON/OFF function of lamps, but can cause the lights to flicker on and off. To solve this problem, a timer (decision device) can be used. An off-delay timer set for 30 min can keep the lights on for 30 min after the motion sensor no longer detects movement in the room. This can be used in areas where individuals may leave for short periods of time.

TECH FACT
Using motion sensors that detect the movement of heat can have the added advantage of turning on the lights during a fire, which can help individuals leave the area and firefighters enter the area.

Loads
Media Clip

Outputs (Loads)

An *output (load)* is an electrical device that converts electrical energy into some other form of energy, such as light, heat, sound, linear motion, or rotary motion. Outputs are the most necessary part of a circuit because they produce work. When an output that should be energized is not, a problem exists in the system. For example, when outputs are not energized, the following conditions result:

- fan motors do not run (no airflow)
- pump motors do not run (no liquid flow)
- lamps do not turn on (no light)
- heating elements do not operate (no heat)
- refrigeration units do not run (no cooling)
- equipment and machinery do not run (no work accomplished)
- alarms, horns, and sirens do not sound (no audible warning)

An output that is not working is usually the first item reported when there is an electrical problem. Troubleshooting usually starts by first checking to verify which outputs are working and which are not.

Inputs (Switches)

An *input (switch)* is a device or component that controls the flow of electricity or information into a circuit. The four basic groups of inputs are manual, mechanical, automatic, and special.

A *manual input* is a switch that requires an individual to enter information or control a circuit. Typical manual switches include pushbuttons (industrial, keyboard, or keypad types), toggle switches, selector switches, wall switches for lighting circuits, and key-operated switches.

A *mechanical input* is a switch that detects the physical presence of an object. Typical mechanical switches include limit switches. Limit switches are mechanical inputs that require an object to touch the switch actuator (lever, fork, or stick) before the switch is activated.

An *automatic input* is a switch independent of an individual or object that enters information or controls a circuit. Typical automatic switches are level, flow, temperature, motion, and pressure switches.

Inputs start or stop the flow of current in a circuit. Some inputs are designed to stop the flow of current in a circuit, but only when there is a problem. Fuses, circuit breakers, overload contacts on motor starters, electronic overloads on motor drives, and other circuit/component protection devices are switches that react to an input condition such as overcurrent. These types of inputs are considered special because they are switches, but are not used unless there is a major circuit fault. If these types of inputs are never activated, then it indicates that the circuit has operated as designed.

Interfaces

An *interface* is any device that allows two different types of components, voltage levels, voltage types, or systems to be interconnected. In some electrical circuits, the output (load) can be directly connected to the input (switch). For example, two-way, three-way, and four-way switches can be connected directly to most lamps in a lighting circuit. Likewise, the same type of switch can be used to directly control fractional HP single-phase motors, DC fan motors, or heating elements. However, this type of switch cannot be used to directly control three-phase motors, high wattage heaters, or large banks of lights in auditoriums.

An interface is required when a switch cannot directly control a load or two parts of an electrical circuit are not directly compatible. Relays, contactors, motor starters, transformers, and solenoid-operated valves are electrical interfaces. For example, a motor starter allows a basic pushbutton to control a large motor, a solenoid on a valve allows electricity to control the flow of a liquid, and step-down control transformers allow a control circuit to operate at a lower and safer voltage level than the motor, heating element, and lighting bank.

Decisions

A *decision* is an electronic determination of how an electrical circuit is to operate with a predetermined set of known conditions. The decision section of a circuit determines what work is required and the sequential order in which the work is to occur.

Decision devices include timers (on-delay, off-delay, one shot, recycle, etc.), counters (straight and up-down), logic relays (flip-flop, etc.), and other devices that take the circuit input and determine when and how the output will operate. Programmable-decision electrical devices include security keypads, programmable logic relays (PLRs), programmable logic controllers (PLCs), motor drive keypads, and other devices that can be programmed and reprogrammed.

Refer to Chapter 3 Quick Quiz® on CD-ROM.

Case Study—Electric Heating Element Service

The kitchen staff in a commercial building cafeteria reports that some deep fryers are taking much longer to heat up when first turned on and seem to be unable to maintain a high temperature during peak usage.

The maintenance staff is required to check the operating condition of the heating elements and replace the heating elements that are defective. Since there are different size heating elements used in the different fryers, the electrical rating of each fryer must be known to calculate the current draw and resistance of each heating element. The calculated values are then compared to the measured values to determine their operating condition.

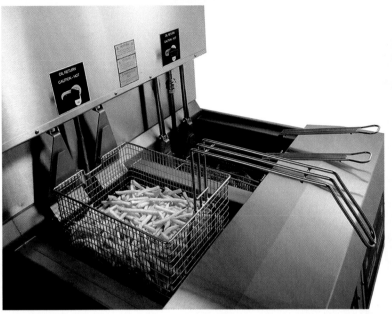

Henny Penny Corporation

Each heating element has a power rating and a voltage rating. Using the power formula, the expected current of each heating element is calculated. Then, using Ohm's law, the expected resistance is calculated.

Heating element 1 has a power rating of 3500 W and an electrical rating of 240 V. Since current equals power divided by voltage ($I = P \div E$), heating element 1 should have a current draw of approximately 14.58 A (3500 ÷ 240 = 14.58).

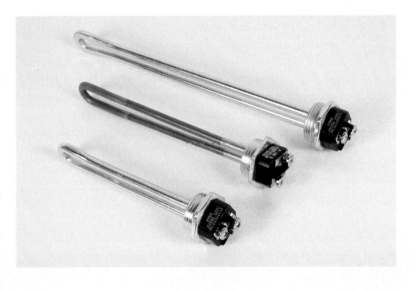

Ohm's law states that resistance equals voltage divided by current ($R = E \div I$). Thus, heating element 1 should have a resistance of approximately 16.46 Ω (240 ÷ 14.58 = 16.46).

Heating element 2 has a power rating of 2000 W and an electrical rating of 120 V. The current draw of heating element 2 should be approximately 16.67 A (2000 ÷ 120 = 16.67). In addition, the resistance of heating element 2 should be approximately 7.2 Ω (120 ÷ 16.67 = 7.2).

Heating element 3 has a power rating of 1500 W and an electrical rating of 120 V. The current draw of heating element 3 should be approximately 12.5 A (1500 ÷ 120 = 12.5). In addition, the resistance of heating element 3 should be approximately 9.6 Ω (120 ÷ 12.5 = 9.6).

Case Study—Electric Heating Element Service (continued)

The heating elements are tested to determine their actual resistance values. When testing heating elements 2 and 3, the measured resistance was much higher than the calculated value. The higher resistance leads to a lower current draw, which reduces the power (heat) output of the heating element. All heating elements that had a resistance reading 20% or more higher than the calculated values were replaced.

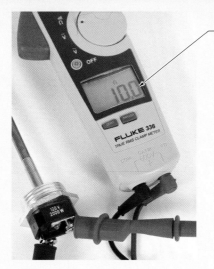

MEASURED RESISTANCE (R) = 10.0 Ω

NOTE: EXPECTED CALCULATED RESISTANCE (R) = 7.2 Ω

Definitions

- *AC voltage* is voltage that reverses its direction of flow at regular intervals.
- An *ampere* is the amount of electrons passing a predetermined point in a circuit in 1 sec.
- *Apparent power (P_A)* is the product of the voltage and current in a circuit calculated without considering the phase shift that can be present between the voltage and current in a circuit.
- An *automatic input* is a switch independent of an individual or object that enters information or controls a circuit.
- *Average voltage (V_{avg})* is the mathematical mean of instantaneous voltage values in the sine wave and is equal to 0.637 of the peak value.
- *Capacitance (C)* is the ability of a component or circuit to store energy in the form of an electric charge.
- *Capacitive reactance (X_C)* is the opposition to current flow by a capacitor.
- A *capacitor* is an electric device that stores electrical energy by means of an electrostatic field.
- *Current (I)* is the amount of electrons flowing through an electrical circuit.
- *DC voltage* is voltage that flows in one direction only.
- A *decision* is an electronic determination of how an electrical circuit is to operate with a predetermined set of known conditions.
- *Electromagnetism* is the magnetism produced when electric current passes through a conductor.
- A *ferromagnetic metal* is a material that has extremely high and variable magnetic susceptibility and is strongly attracted to a magnetic field.
- An *impulse transient voltage* is a short-duration transient voltage commonly caused by a lightning strike, which results in a short, unwanted voltage placed on the power distribution system.
- *Inductance (L)* is the property of a circuit that causes it to oppose a change in current due to energy stored in a magnetic field.
- *Inductive reactance (X_L)* is the opposition to the flow of alternating current in a circuit due to the inductance.

- An *input (switch)* is a device or component that controls the flow of electricity or information into a circuit.
- An *interface* is any device that allows two different types of components, voltage levels, voltage types, or systems to be interconnected.
- A *magnet* is a substance that produces a magnetic field and attracts iron.
- *Magnetism* is a force by which materials exert an attraction or repulsion on other materials.
- A *manual input* is a switch that requires an individual to enter information or control a circuit.
- A *mechanical input* is a switch that detects the physical presence of an object.
- An *oscillatory transient voltage* is a transient voltage that includes both positive and negative polarity values.
- *Ohm's law* is the relationship between voltage (*E*), current (*I*), and resistance (*R*) in a circuit.
- An *output (load)* is an electrical device that converts electrical energy into some other form of energy, such as light, heat, sound, linear motion, or rotary motion.
- *Overvoltage* is a voltage increase more than 10% above the normal rated line voltage for a period of longer than 1 min.
- A *parallel connection* is a connection with two or more devices connected so that there is more than one path for current flow.
- *Peak-to-peak voltage* ($V_{p\text{-}p}$) of a sine wave is the voltage measured from the maximum positive alternation to the maximum negative alternation.
- *Peak voltage* (V_{max}, V_p) of a sine wave is the maximum value of either the positive or negative alternation.
- A *permanent magnet* is a magnet that holds its magnetism for a long period of time.
- *Phase shift* is the state when voltage and current in a circuit do not reach their maximum amplitude and zero level simultaneously.
- *Polarity* is the positive (+) or negative (−) state of an object.
- *Power (P)* is the rate of doing work or using energy.
- *Power factor (PF)* is the ratio of true power used in an AC circuit to apparent power delivered to the circuit.
- The *power formula* is the relationship between power (*P*), voltage (*E*), and current (*I*) in an electrical circuit.
- *Reactive power (VAR)* is power supplied to a reactive load, such as a transformer coil, motor winding, or solenoid coil.
- A *rectifier* is a device that converts AC voltage to DC voltage by allowing the voltage and current to flow in one direction only.
- *Resistance (R)* is the opposition to the flow of electrons in a circuit and is measured in ohms (Ω).
- A *resistive circuit* is a circuit that contains components that produce only resistance, such as heating elements and incandescent lamps.
- *Root-mean-square voltage* (V_{rms}) is the AC voltage value that produces the same amount of heat in a pure resistive circuit as DC voltage of the same value.
- A *series connection* is a connection that has two or more devices connected so there is only one path for current flow.
- A *series-parallel circuit* is a combination of series and parallel-connected components.

- A **short circuit** is an overcurrent that leaves the normal current-carrying path by going around the load and back to the power source or ground.
- A **tap** is a connection point provided along the transformer coil.
- A **temporary magnet** is a magnet that loses its magnetism as soon as the magnetizing force is removed.
- A **transient voltage**, also known as a voltage spike, is a short, temporary, undesirable voltage in an electrical circuit.
- **True power (P_T)** is the actual power used in an electrical circuit to produce work, such as heat, light, sound, and motion.
- **Undervoltage** is a voltage decrease more than 10% below the normal rated line voltage for a period of longer than 1 min.
- **Voltage (E)** is the amount of electrical pressure in a circuit and is measured in volts (V).

Review Questions

1. Explain why DC component polarity must be understood prior to making any installations.
2. List at least one advantage and disadvantage of using AC voltage.
3. What is the acceptable voltage operating range for most motors?
4. Explain the causes of oscillatory transient voltages in electrical circuits.
5. Explain why it is important to measure both voltage and current draw when troubleshooting electrical components such as heating elements.
6. What is the difference between materials that insulate and materials that conduct?
7. Explain the relationship between voltage, current, and resistance in a circuit as defined by Ohm's law.
8. Define true power, reactive power, and apparent power.
9. Why are three-phase motors more energy efficient than single-phase motors?
10. What is the difference between a series connection and a parallel connection?
11. Why is magnetism necessary in an electrical field?
12. Explain how an electrical system must be modified when adding loads or increasing the amount of current.
13. List the four sections of an electrical system.
14. Define output (load) and explain why it is the most necessary part of a circuit.
15. Explain the difference between an interface and a decision.

chapter 4
USING ELECTRICAL TEST INSTRUMENTS

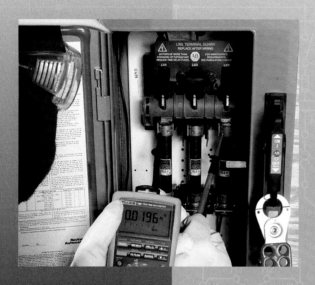

In the electrical field, a dangerous condition cannot be easily seen unless it is obvious, such as a fallen power line or a burning motor. Dangers that cause almost all electrical fires and electrical shocks cannot be seen or heard and are only realized when it is too late. The only safe way to know whether electricity is present in a circuit, the amount of voltage present, the type and amount of current (AC or DC) present, and the condition of wire insulation is to use test instruments to find the necessary information so that safe decisions can be made.

OBJECTIVES

- Explain the usage of test instruments.
- List and describe test instrument terminology.
- Describe voltage tester measurement procedures and applications.
- Describe multimeters and their features.
- Describe resistance measurement test procedures.
- Describe current measurement test procedures.
- Describe voltage measurement test procedures.
- Describe megohmmeter measurement procedures and applications.

TEST INSTRUMENT USAGE

An *electrical test instrument* is a device used to measure electrical quantities such as voltage (V), current (A), resistance (Ω), power (W), capacitance (f), and frequency (Hz). The test instrument used for an application depends on the desired results and the precision required. For example, a voltmeter is used to measure and display the exact amount of voltage present at a switch, load, fuse, or circuit breaker. An ammeter is used to measure the exact amount of current flowing through a switch, load, fuse, or circuit breaker.

When installing, testing, or troubleshooting electrical equipment, and when conducting preventive maintenance tests, various electrical test instruments are used. The best choice of test instrument depends on the desired information to be obtained. **See Figure 4-1.**

- Voltage testers are used to verify the presence of voltage at receptacles, switches, and fuses or circuit breakers.
- A clamp-on meter or digital multimeter (DMM) with a separate current clamp attachment is best for taking AC current measurements above 1 A.
- Advanced multimeters measure electrical quantities such as voltage, current, and resistance and are used to obtain an accurate measurement of the exact electrical quantity being measured.
- Megohmmeters are used to measure insulation deterioration or insulation failure on various wires.

TECH FACT

If an AC voltage seems incorrect because the numbers constantly change during the measurement procedure, then the voltage may actually be DC. Change the meter function switch to measure DC and retest the circuit. If the voltage at the test point is DC, the meter reading will be the same when the meter leads are reversed except for the negative (–) sign in front of one of the readings.

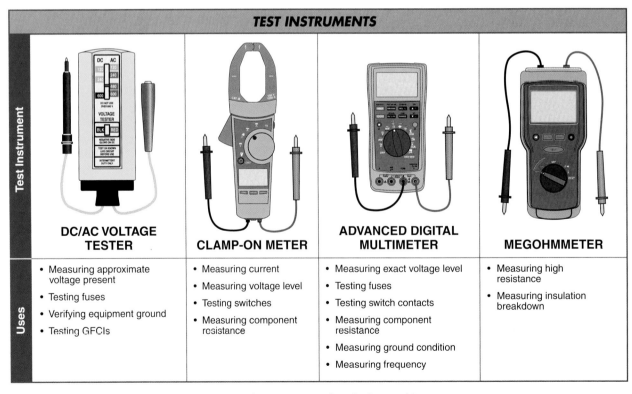

Figure 4-1. Various test instruments are used to measure electrical quantities.

TEST INSTRUMENT TERMINOLOGY

Electrical test instruments normally use electrical terminology in the form of abbreviations such as V, A, and W, symbols such as Ω and ϕ, and metric prefixes such as k, M, and m to display the measured value. As the number of test instrument measurement capabilities increases, the amount and/or type of abbreviations, symbols, and prefixes used also increases. Understanding the meaning of each abbreviation, symbol, and prefix is required to understand the displayed measurement and prevent an unsafe condition. For example, safety considerations change considerably when measuring 1 mV (millivolt), which equals 0.001 V, and 1 MV (megavolt), which equals 1,000,000 V.

Test Instrument Abbreviations

Test instruments use abbreviations to display information. An *abbreviation* is a letter or combination of letters that represents a word or words. For example, "V" represents voltage and "mV" represents millivolts. Abbreviations can be used individually or in combination with prefixes. **See Figure 4-2.**

Test Instrument Symbols

Test instruments also use symbols to display information. A *symbol* is a graphic element that represents a quantity, unit, or component. Symbols provide quick recognition and are independent of any particular language. For example, electrical resistance is measured in ohms, which can be shown using the symbol "Ω." Symbols can be shown individually or along with abbreviations. For example, a measurement of 1000 Ω can be displayed on a test instrument as 1 kΩ. **See Figure 4-3.**

VOLTAGE TESTERS

A *voltage tester* is an electrical test instrument that indicates the approximate voltage amount and type (AC or DC) in a circuit using a movable pointer or vibration. When a voltage tester includes a solenoid, the solenoid vibrates when the tester is connected to AC voltage. Some voltage testers include a colored plunger or other indicators, such as a light that indicates the polarity of the test leads as positive or negative, when measuring a DC circuit.

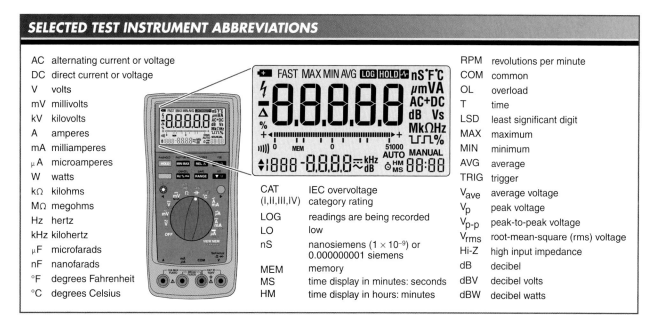

Figure 4-2. Abbreviations are used individually or in combination with prefixes to represent a word or words.

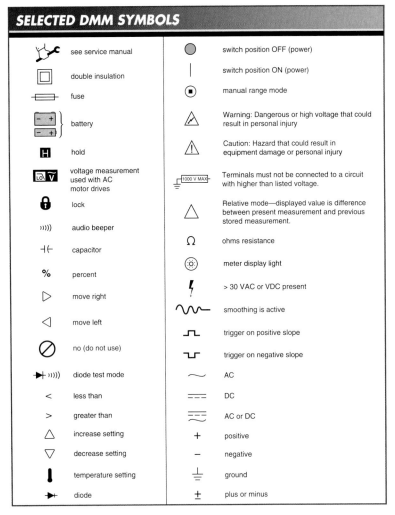

Figure 4-3. Test instrument symbols represent a quantity, unit, or component and can be recognized regardless of the language an individual speaks.

There are two main advantages to using a voltage tester over other types of test instruments. First, when using a vibrating voltage tester, an individual can concentrate on the placement of the test leads instead of the meter reading and still feel the meter vibrating when voltage is present. Second, a voltage tester is designed to draw more than 8 mA when taking a voltage measurement. This design is required when testing ground-fault circuit interrupters (GFCIs). GFCIs are designed to trip at approximately 6 mA (0.006 A) of current flow from the hot line to ground through a path that is not intended for current flow. Since standard multimeters draw much less current when taking a voltage measurement, they cannot be used to test GFCIs.

Voltage testers are used to take voltage measurements any time the voltage of a circuit is within the rating of the tester and an exact voltage measurement is not required. Exact voltage measurements are not required to determine when a receptacle is hot (energized), a system is grounded, fuses or circuit breakers are good or bad, or when a circuit has 115 VAC, 230 VAC, or 460 VAC.

Voltage Tester Measurement Procedure

Before any voltage measurements are taken using a voltage tester, the tester must be checked to ensure that it is designed to take measurements on the circuit being tested. Voltage testers must always be checked before use on a known energized circuit that is within the voltage rating of the tester to verify proper operation. The operator's manual should be referred to for all measurement precautions, limitations, and procedures. Required personal protective equipment (PPE) must always be worn and all safety precautions followed when taking the measurement. **See Figure 4-4.** Voltage measurements are taken with a voltage tester using the following procedure:

1. Verify that the voltage tester is rated for the voltage level and type to be tested.
2. Wear all required/recommended PPE for the area and procedure.
3. Verify that the voltage tester is working by testing it on a known energized voltage source.
4. Connect the negative test lead to the neutral or ground side.
5. Connect the positive test lead to the hot voltage source.
6. Read the displayed voltage.
7. Retest the voltage tester on a known energized source to verify the tester still operates properly.

Chapter 4—Using Electrical Test Instruments 103

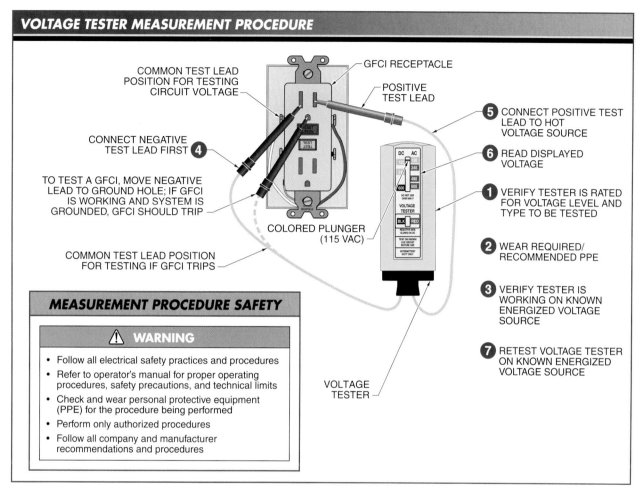

Figure 4-4. Voltage testers are used to take voltage measurements any time the voltage of a circuit is within the rating of the tester and an exact voltage measurement is not required.

A GFCI is tested by moving the negative lead to the ground hole while the positive lead is still connected to the hot slot. If the GFCI trips, the GFCI is working and the system is grounded. If the GFCI does not trip but the tester displays the same voltage as it did from hot to neutral, the system is grounded, but the GFCI is not working and/or not wired correctly. If the GFCI does not trip and there is no voltage reading, then the GFCI is not grounded.

MULTIMETERS

A *multimeter* is a portable test instrument that is capable of measuring two or more electrical quantities. Multimeters are either analog or digital. **See Figure 4-5.**

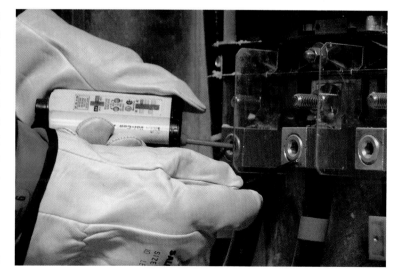

Voltage testers are used when exact voltage measurement is not required, such as when measuring a circuit to determine if it is a 120 V, 240 V or 480 V circuit.

104 ELECTRICAL SYSTEMS for FACILITIES MAINTENANCE PERSONNEL

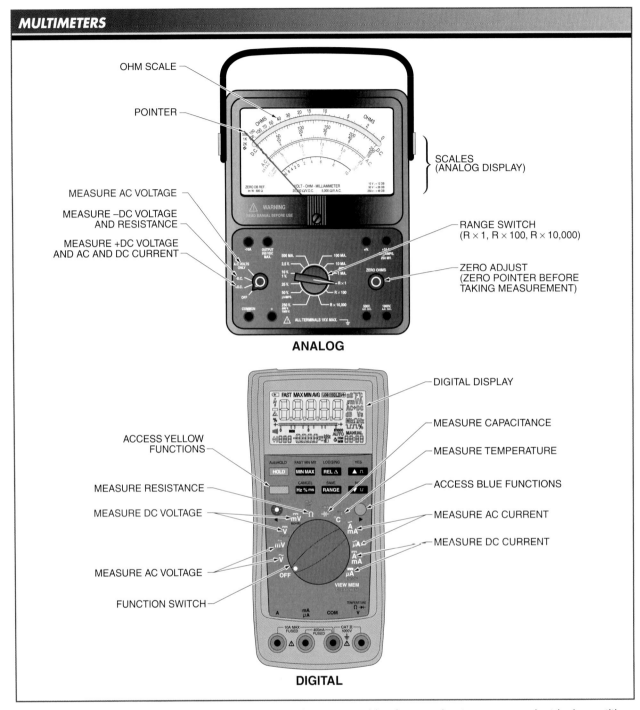

Figure 4-5. Multimeters are portable test instruments that are capable of measuring two or more electrical quantities.

Multimeters
Media Clip

Analog Multimeters

An *analog multimeter* is a meter that can measure two or more electrical properties and displays the measured properties using a pointer that moves along several calibrated scales. These scales correspond to the different function switch settings (AC, DC, and R) and jacks (mA jack and 10 A jack) on the meter. Properly reading an analog multimeter requires reading the correct scale. The most common measurements made with analog multimeters are voltage, resistance, and current.

Reading Analog Displays. An *analog display* is an electromechanical device that displays measurements using the mechanical motion of a pointer. Analog displays use linear and nonlinear scales to display measured values. A *linear scale* is a scale that is divided into equally spaced segments. A *nonlinear scale* is a scale that is divided into unequally spaced segments. A mirror on the analog display helps to reduce measurement errors due to parallax (distortion caused by reading the display from an angle). **See Figure 4-6.**

Analog scales are divided using primary divisions, secondary divisions, and subdivisions. A *primary division* is a division with a listed value. A *secondary division* is a division that divides primary divisions into halves, thirds, fourths, fifths, etc. A *subdivision* is a division that divides secondary divisions into halves, thirds, fourths, fifths, etc. Secondary divisions and subdivisions do not have listed numerical values. When reading an analog scale, the primary, secondary, and subdivision readings are added together. **See Figure 4-7.** An analog scale is read using the following procedure:

1. Read the primary division.
2. Read the last secondary division the pointer moves past. *Note:* This may not occur with very low readings.
3. Read the subdivision when the pointer is not directly on a primary or secondary division. Round the reading to the nearest subdivision when the pointer is not directly on a subdivision. Round the reading to the next highest subdivision when rounding to the nearest subdivision is unclear.
4. Add the primary division, secondary division, and subdivision readings to obtain the analog reading.

Digital Multimeters

A *digital multimeter (DMM)* is a meter that can measure two or more electrical properties and display the measured properties as numerical values. Basic DMMs measure voltage, current, and resistance. Advanced DMMs can measure capacitance, temperature, and frequency. The main advantages of using a DMM over an analog multimeter are that DMMs are easier to read and can also record measurements. Because of this, DMMs are the most commonly used meters in industry.

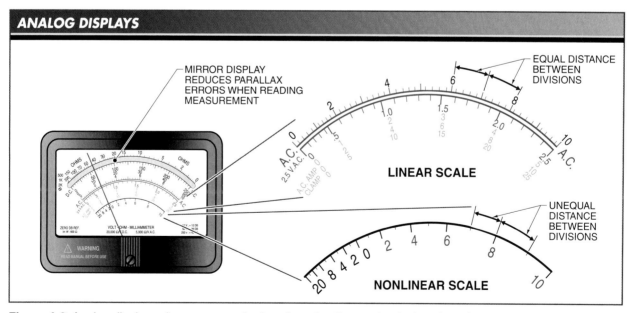

Figure 4-6. Analog displays show measured values by using the mechanical motion of a pointer on a linear or nonlinear scale.

READING ANALOG SCALES

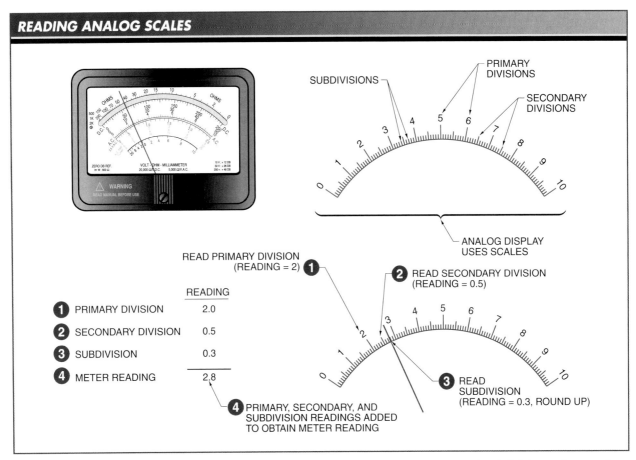

Figure 4-7. Analog scales are divided into primary divisions, secondary divisions, and subdivisions whose values are added together to obtain a reading.

Reading Digital Displays. A *digital display* is an electronic device that displays measurements as numerical values. Digital displays help eliminate human error by displaying the exact values measured. However, errors can still occur when reading a digital display if displayed abbreviations, prefixes, symbols, and decimal points are not properly applied to the measurement.

Digital displays show values using either light-emitting diodes (LEDs) or liquid crystal displays (LCDs). LED displays are easier to read but use more power than LCDs. Most portable DMMs use LCDs. The exact value on a digital display is determined from the numbers displayed and the position of the decimal point. Range switches determine the placement of the decimal point on a digital display.

Typical voltage ranges on a digital display are 0 V to 3 V, 0 V to 30 V, and 0 V to 300 V. The highest possible reading with the range switch set on 3 V is 2.99 V. The highest possible reading with the range switch set on 30 V is 29.99 V, and so on. **See Figure 4-8.**

Bar Graphs. Most digital displays include a bar graph to show changes and trends in a circuit. A *bar graph* is a graph composed of segments that function similar to an analog pointer. The displayed bar graph segments increase as the measured value increases and decrease as the measured value decreases. The polarity of the test leads should be reversed when a negative sign is displayed at the beginning of the bar graph. **See Figure 4-9.**

Chapter 4—Using Electrical Test Instruments 107

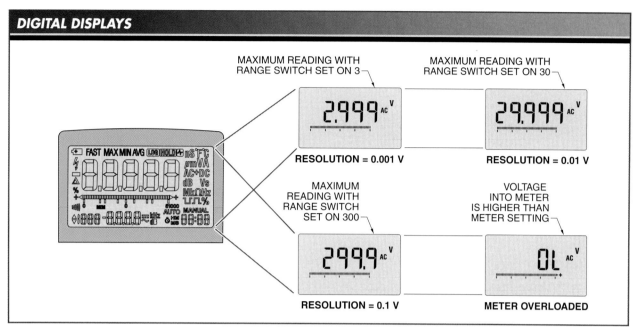

Figure 4-8. Digital displays use numbers to give exact values.

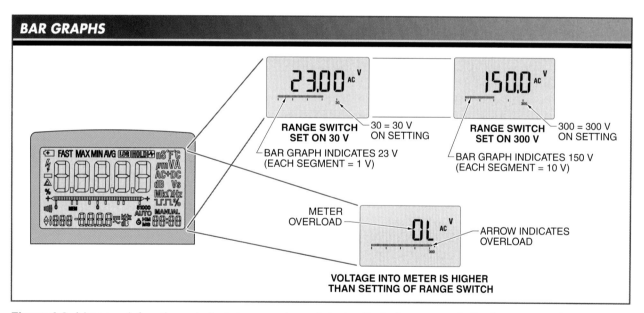

Figure 4-9. A bar graph functions similarly to an analog pointer by displaying segments that increase as the measured value increases and that decrease as the measured value decreases.

A bar graph reading is normally updated 30 times per second. A digital display is normally updated four times per second. A bar graph is used when quick-changing signals cause a digital display to flash or when there is a change in the circuit that is too rapid for the digital display to detect. For example, mechanical relay contacts may bounce open when exposed to vibration. Contact bounce causes intermittent problems in electrical equipment. The frequency and severity of contact bounce increase as a relay ages. A bar graph is updated at sufficient speed to detect relay contact bounce even if the numerical display cannot.

Multimeter Resistance Measurement

Resistance (R) is the opposition to the flow of electrons in a circuit and is measured in ohms (Ω). Since resistance is measured in ohms, meters that measure only resistance are often referred to as ohmmeters. Resistance measurements are normally taken to indicate the condition of a circuit or component. The higher the resistance in a circuit, the lower the current flow through the circuit. Conversely, the lower the resistance in a circuit, the higher the current flow through the circuit. Insulators, such as rubber and plastic, have very high resistance. Conductors, such as wires or switch contacts, have very low resistance.

When insulators are damaged their resistance decreases. When conductors are damaged their resistance increases. Components such as heating elements and resistors have a fixed resistance value. Any significant change in the fixed resistance value normally indicates a problem. Components such as resistors include a tolerance along with a resistance value. When a tolerance is indicated, the measured resistance value must be within the specified tolerance range.

WARNING: Circuit power must be deenergized before taking resistance measurements. A DMM set to measure resistance uses an internal battery to supply voltage for resistance measurements and should never be connected to an energized circuit. When a circuit includes a capacitor, the capacitor must be discharged before taking any resistance measurements.

A good capacitor stores an electrical charge and may be energized when power is removed. Before touching a capacitor, all power must be off and the capacitor discharged. A DMM is used to ensure the power is off. To safely discharge a capacitor after power is removed, a 20,000 Ω (20 kΩ), 5 W resistor is connected across the capacitor terminals for 5 sec. On very large capacitors, the process may have to be repeated as the capacitor can build up a charge again after discharging. A DMM is also used to confirm the capacitor is fully discharged.

Resistance measurements are used to determine the resistance of a circuit or component and how the circuit is wired. For example, a DMM set to measure resistance can be used to determine if a nine-lead, 3ϕ motor is wye or delta connected and if the windings of the motor are in good condition. **See Figure 4-10.** A DMM set to measure resistance is used to test for open or short motor windings and/or to test the resistance of the windings. An open motor winding is displayed as an overload (OL). A shorted motor winding is displayed as 0 Ω or as a lower-than-normal winding resistance.

DMMs have a function switch that must be set to the resistance measurement (Ω) mode. Some DMMs include a continuity test function as part of the resistance measurement mode. The continuity test function is used to indicate very low or no resistance (closed switch) or infinite or very high resistance (open switch). In the continuity mode, the

Fluke Corporation

Circuits must be deenergized when taking resistance measurements with a DMM. The DMM's power source supplies the required power to test the resistance of the circuit.

meter sounds an audible "beep" when there is very low or no resistance. This mode is helpful when checking switches that have only two circuit conditions, such as closed or open, or when checking fuses that may be good or bad. **See Figure 4-11.**

> ### TECH FACT
> Analog ohmmeters require calibration before use and are calibrated using an adjustment knob. To calibrate an analog ohmmeter, set the meter to resistance, touch the red and black leads together, and rotate the knob until the needle is at zero. This should be repeated with each use and with each scale change.

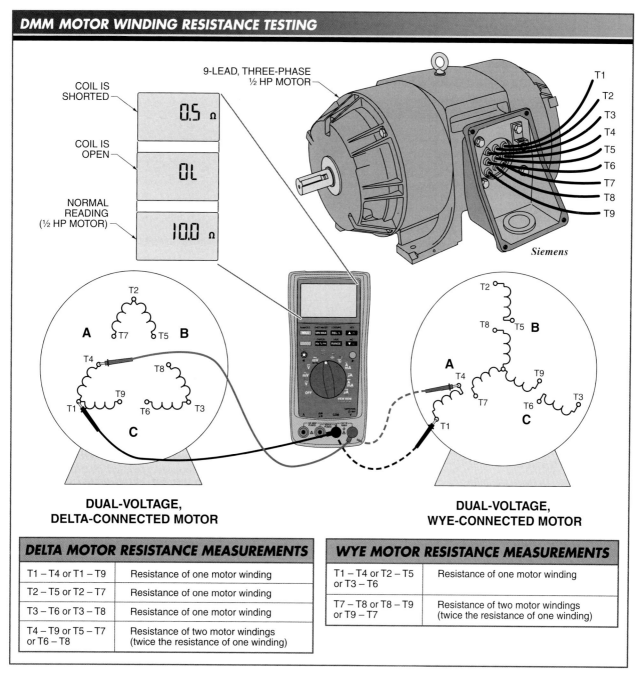

Figure 4-10. A DMM set to measure resistance can be used to determine if a nine-lead, three-phase motor is wye or delta connected and if the windings are open or shorted.

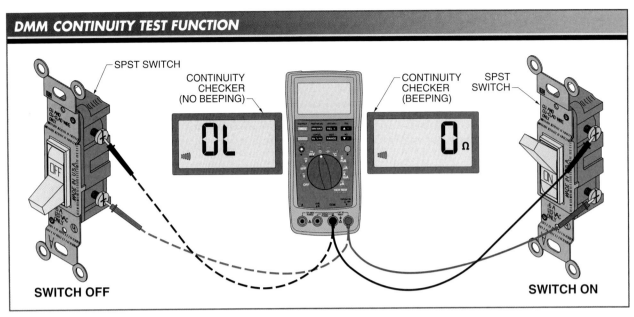

Figure 4-11. Some DMMs include a continuity test function that can be used to indicate very low or no resistance (closed switch) or high/infinite resistance (open switch).

Multimeter Resistance Measurement Procedure. The DMM resistance mode is used to measure resistance values from 1 Ω to millions of ohms (MΩ). When reading a resistance measurement, the symbol and metric prefix (kΩ and MΩ) associated with the displayed value must be noted. Resistance measurements are always taken when the circuit power is off and the DMM set to the resistance (Ω) mode. Therefore, PPE such as electrical insulated gloves are not required when taking resistance measurements. However, when using a DMM to verify that all power is off, electrical insulated gloves must be worn. In addition, other PPE such as safety glasses, head protection, and foot protection should be worn as required for the specific environment in which the measurements are taken.

A DMM set to the resistance mode must be checked to ensure the meter can safely take measurements on the circuit being tested. The operator's manual should be referred to for all measurement precautions, limitations, and procedures. Required PPE must always be worn and all safety rules followed when taking the measurement. **See Figure 4-12.** Resistance measurements taken using a DMM are obtained using the following procedure:

1. Verify that all power to the circuit is off and remove the component being tested from the circuit.
2. Set the DMM function switch to resistance mode. *Note:* DMMs display OL and Ω when set to measure resistance.
3. Plug the black test lead into the common jack.
4. Plug the red test lead into the resistance jack.
5. Ensure that the batteries of the test instrument are in good condition. *Note:* Most DMMs display a battery symbol when the batteries are low.
6. Connect the test leads across the component or circuit being tested. Ensure that the contact between the test leads and the circuit or component is correct. *Note:* Dirt, solder flux, oil, and other foreign substances greatly affect resistance readings. Also, any contact with the metal ends of the test leads by the fingers of the electrician during the resistance test affects the resistance measurement.

7. Read the resistance measurement displayed.
8. Remove the DMM from the component or circuit after completing all resistance measurements and turn the meter off to preserve battery life.
9. Retest the DMM by touching the leads together to verify the meter still operates properly.

Multimeter Current Measurement

DMMs include a current measurement function. Current flows through a closed circuit when a power source is connected to a load that uses electricity. Every load that converts electrical energy (such as a lamp, motor, or heating element) into some other form of energy (such as light, rotating motion, or heat) uses current. The more electrical energy required, the higher the current usage. Every time a load is switched on or a new load is added to the circuit, the power source must provide more current.

Current measurements are used to determine the amount of circuit loading or the condition of a load. For example, a motor with a nameplate rating of 10 A draws 10 A when it is fully loaded, less than 10 A when the motor is not fully loaded, and over 10 A when the motor is overloaded. Since current is measured in amperes, meters designed to measure current are often referred to as ammeters or clamp-on ammeters.

Current is normally measured using clamp-on ammeters or DMMs with clamp-on current probe accessories. A *clamp-on ammeter* is a test instrument that measures current in a circuit by measuring the strength of the magnetic field around a conductor. To do this, the jaws of the clamp-on ammeter are placed around the conductor.

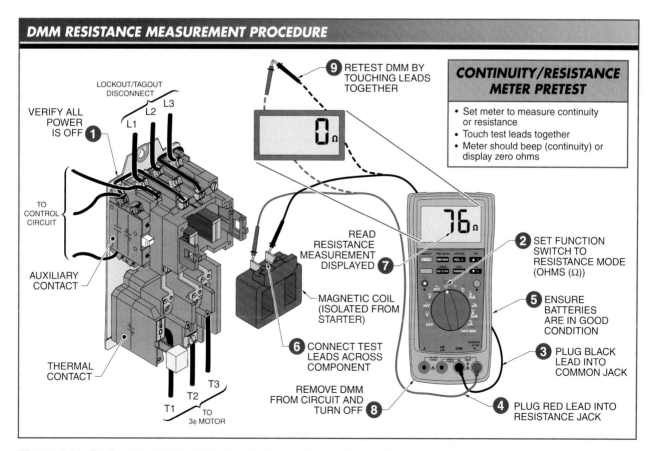

Figure 4-12. Resistance measurements are always taken with circuit power off.

Clamp-on ammeters are available for measuring AC only or for measuring both AC and DC. When a clamp-on current probe accessory is used with a DMM, the meter functions as a clamp-on ammeter. Clamp-on current probe accessories are available that measure AC only or AC and DC.

Small amounts of current can be measured safely using a DMM connected as an in-line ammeter. A DMM connected as an in-line ammeter measures small amounts of current in a circuit when the test leads of the meter are inserted in series with the components being tested. In-line ammeters require that the circuit be opened so the meter can be inserted into the circuit to acquire a reading. In-line ammeters are available for DC only or AC and DC. **See Figure 4-13.**

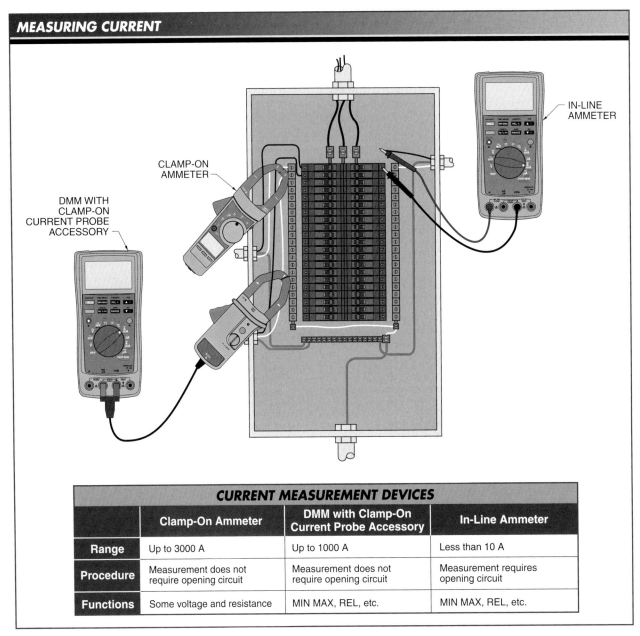

Figure 4-13. Current is normally measured using a clamp-on ammeter, a DMM with a clamp-on current probe accessory, or an in-line ammeter.

The advantage of using a clamp-on ammeter or DMM with a clamp-on current probe accessory is that current measurements are taken without opening the circuit. Clamp-on ammeters and DMMs with current probes are commonly used to measure currents up to 3000 A. In-line ammeters are used to measure small amounts of current (normally less than 1 A).

The maximum current that can be measured by an in-line ammeter is determined by the range switch and the placement of the test leads in the current jacks of the meter. A DMM normally has two current jacks. One jack is used for measuring small amounts of current (normally up to 300 mA or 400 mA). The other jack is used for measuring larger amounts of current (normally up to 1 A). In-line current measurements must be limited to circuits that can be easily opened and are known to have currents less than 1 A.

Clamp-On Ammeters. A clamp-on ammeter is most commonly used to measure current over 1 A and in applications in which current can be measured easily by placing the jaws of the ammeter around one of the conductors. Most clamp-on ammeters can also measure voltage and resistance. To measure voltage and resistance, the clamp-on ammeter must include test leads, a voltage mode, and a resistance mode. **See Figure 4-14.**

Current is a common troubleshooting measurement because it can be used to determine the degree to which a circuit is loaded. Current in a series circuit is constant throughout the circuit. When loads are added to parallel circuits, the supply voltage remains the same, but the current increases with each load added. The highest amount of current in a parallel circuit is at the point closest to the power source because current decreases as the system distributes current to each of the parallel loads.

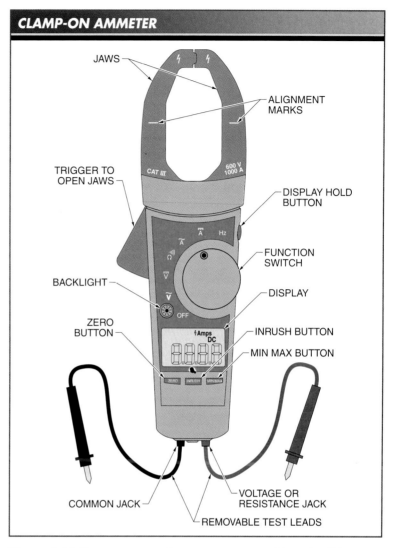

Figure 4-14. Clamp-on ammeters measure current in a circuit by measuring the strength of the magnetic field around a single conductor.

On an individual leg of a combination series-parallel circuit, the current flowing to the load is the same as the current on the return line from the load. A slight difference of up to 10% between the supply and return lines is possible when using a clamp-on ammeter because of the placement of the meter, or when low current measurements are taken with the meter set to a high current range (when measuring 2 A on a 400 A range setting, for example). Any variation that is excessive must be investigated because the current measurement may indicate that a partial short exists on one of the lines and current is flowing to ground.

Clamp-On Ammeter Measurement Procedure. Clamp-on ammeters measure current in a circuit by measuring the strength of the magnetic field around a single conductor. An individual must ensure that a clamp-on ammeter does not pick up stray magnetic fields by separating the conductors being tested from other conductors as much as possible. If stray magnetic fields are affecting a measurement, various locations along the same conductor must be tested.

Before any current measurements are taken using a clamp-on ammeter, the meter must be checked to ensure it is designed to take measurements on the circuit being tested. The operator's manual for the test instrument should be referred to for all measurement precautions, limitations, and procedures. Required PPE must always be worn and all safety rules followed when taking the measurement.

AC or DC current measurements performed with clamp-on ammeters or with DMMs with clamp-on current probe accessories must be performed following standard procedures. **See Figure 4-15.** Current measurements are taken using a clamp-on ammeter or a DMM with a clamp-on current probe accessory using the following procedure:

1. Determine if the current in the circuit is AC or DC.
2. Select a clamp-on ammeter that can measure the circuit current (AC only or AC and DC).
3. Determine if the ammeter range is high enough to measure the maximum current that exists in the circuit.
4. Set the function switch to the proper current setting. Set the function switch to as high as, or higher than, the highest possible circuit current if there is more than one test position or if the circuit current is unknown. When using a DMM with a clamp-on current probe accessory that produces current output, plug the black test lead of the current probe into the common jack and plug the red test lead into the mA jack. When using a DMM with a clamp-on current probe accessory that measures voltage output, plug the black test lead into the common jack and the red test lead into the voltage (V) jack. *Note:* The current measurement accessories that produce current output are designed to measure AC only and generally deliver 1 mA to the meter for every 1 A of measured current (1 mA/A). Current probe accessories that produce a voltage output are designed to measure AC or DC current and deliver 1 mV to the meter for every 1 A of measured current (1 mV/A).
5. Open the clamp-on ammeter or current probe accessory jaws by pressing against the trigger.
6. Clamp jaws around one conductor. Ensure that the jaws are completely closed before taking any current measurements.
7. Read the current measurement displayed.
8. Remove the clamp-on ammeter or clamp-on current probe accessory from the circuit.
9. Retest the clamp-on ammeter on a known energized source to verify the meter still operates properly.

In-Line Ammeters. In-line ammeters provide the most accurate current measurements when measuring currents less than 1 A. In-line ammeters are the most accurate because the meter is connected in series with the load and every milliamp of current that flows through the load must flow through the meter. Some applications require that in-line current measurements be used because they are the best way, and sometimes the only way, to test a circuit or component. In-line current measurement is normally used only for applications requiring very accurate current measurements and having less than 1 A of current.

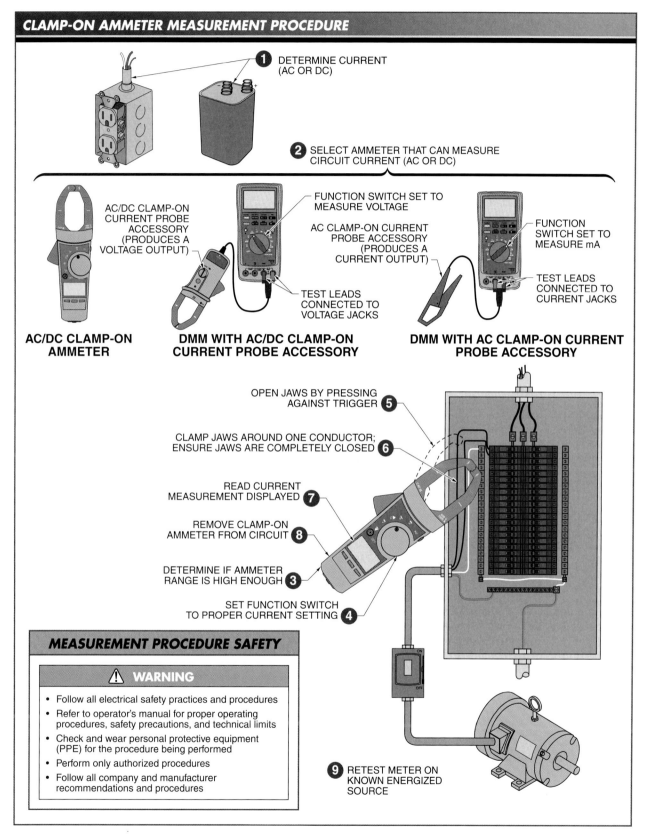

Figure 4-15. AC or DC current measurements performed with clamp-on ammeters or DMMs with clamp-on current probe accessories must be performed following standard procedures.

In-Line Ammeter Measurement Procedure. Care is required to protect the in-line ammeter, circuit, and individual when measuring AC or DC with an in-line ammeter. Before taking any current measurements using an in-line ammeter, the meter is checked to ensure it is designed to take measurements on the circuit being tested. The meter's operator manual should be referred to for all measurement precautions, limitations, and procedures. Required PPE must be worn and all safety rules followed when taking the measurement. **See Figure 4-16.** Current measurements are taken using an in-line ammeter using the following procedure:

1. Set the function switch of the in-line ammeter to the proper position for measuring current (A, mA, or µA). Select a setting with a high enough rating to measure the highest possible circuit current when the ammeter has more than one position.
2. Plug the black test lead into the common jack.
3. Plug the red test lead into the current jack. *Note:* The current jack may be marked A or mA/µA.

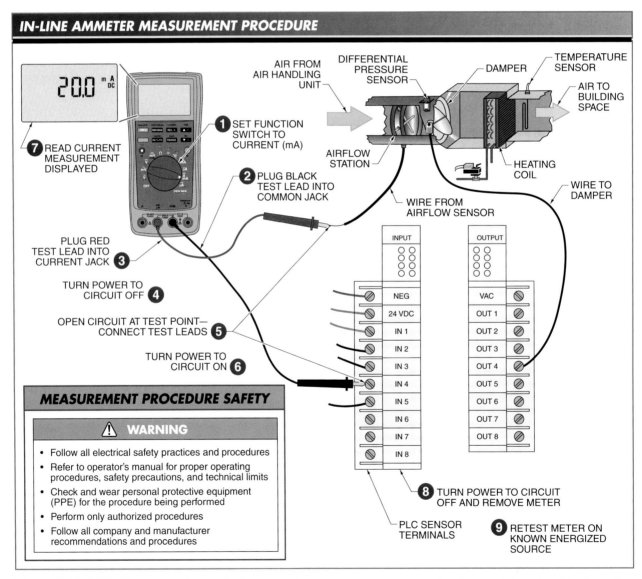

Figure 4-16. A DMM connected as an in-line ammeter measures small amounts of current in a circuit by inserting the test leads of the meter in series with the components being tested.

4. Turn the power to the circuit or component off and discharge all capacitors when possible. Verify that the power to the circuit being tested is off before connecting and disconnecting the in-line ammeter test leads. When necessary, take a voltage measurement to ensure that the circuit is deenergized.
5. Open the circuit at the test point and connect the test leads to each side of the opening.
6. Turn the power to the circuit being tested on.
7. Read the current measurement displayed.
8. Turn the power to the circuit off and remove the meter from the circuit.
9. Retest the meter on a known energized source to verify the meter still operates properly.

Many test instruments include a fuse in the current-measuring circuit to prevent damage caused by excessive current. Before using an in-line ammeter, it must be verified that the meter is fused on the current range being used. Most in-line ammeters are marked as fused or not fused at the test lead current terminals. When an in-line ammeter is not marked as fused, the meter is considered unusable for measuring current. A clamp-on ammeter or a clamp-on current probe accessory is used instead of an unfused meter.

Multimeter Voltage Measurement

All electric devices are designed to operate at a certain voltage level or range. Most manufacturers list the voltage range at which their equipment operates properly. Problems occur if the supply voltage is not within the equipment's acceptable voltage range. Low voltage occurs due to loose connections, corrosion at a connection, undersized feeder conductors, damaged conductors, or when additional loads are added to a circuit.

A DMM set to measure voltage is used when exact voltage values are required. The meter is connected across the part of the circuit requiring measurement. Voltage measurements are taken while equipment is operating. **See Figure 4-17.** Voltage measurements are taken with a DMM set to measure voltage using the following procedure:

WARNING: Ensure that no body part contacts any part of a live circuit, including the metal contact points at the tip of the test leads. Exercise caution when measuring AC voltages over 24 V.

1. Set the function switch to the correct voltage type (AC or DC). Select the highest setting if the meter has more than one voltage position or if the voltage in the circuit is unknown.
2. Plug the black test lead into the common jack. *Note:* The common jack may be marked com (common), – (negative), or lo (low) and have a black collar ring.
3. Plug the red test lead into the voltage jack. The voltage jack may be marked V (voltage), + (positive), or hi (high) and have a red collar ring.
4. Turn the power to the circuit or device off. Discharge all capacitors if possible.
5. Connect the test leads in the circuit. *Note:* The position of the test leads is arbitrary. A common practice is to connect the black test lead to the grounded (neutral) side of the AC voltage first and then the red lead. However, the red lead must be removed first and the black lead removed second.
6. Turn the power to the circuit on.
7. Read the voltage displayed on the meter.
8. Turn the power to the circuit off and remove the meter from the circuit.
9. Retest meter on a known energized source to verify meter still operates properly.

118 ELECTRICAL SYSTEMS for FACILITIES MAINTENANCE PERSONNEL

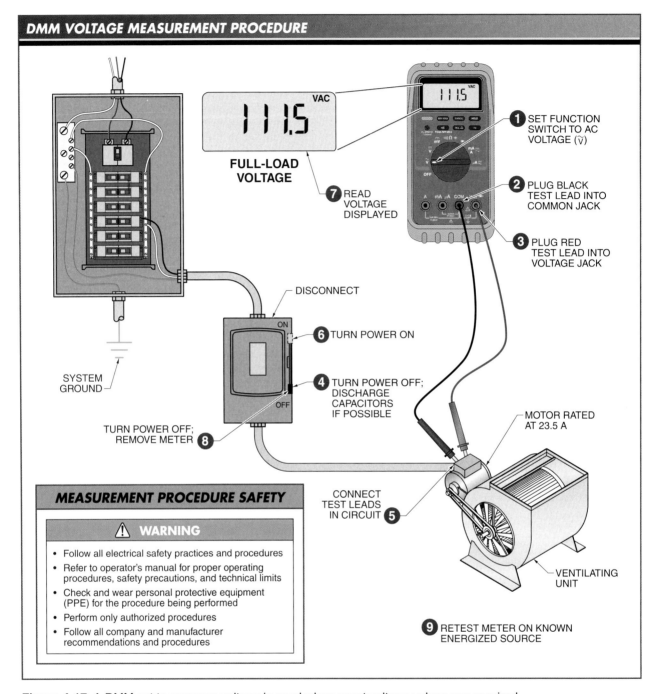

Figure 4-17. A DMM set to measure voltage is used when exact voltage values are required.

MEGOHMMETERS

A *megohmmeter* is a high-resistance ohmmeter used to measure insulation deterioration on conductors by measuring high resistance values during high-voltage test conditions. Megohmmeter test voltages range from 50 V to 5000 V. Megohmmeters detect insulation failure or potential insulation failure in long wire runs, motor windings, and transformers. The minimum resistance of motor windings depends on the voltage rating of the motor. The minimum acceptable resistance measurements are typically found in manufacturer recommended resistance tables. **See Figure 4-18.** Insulation resistance failures may be caused by excessive moisture, dirt, heat, cold, corrosive substances, vibration, or aging.

Insulation allows conductors to stay separated from each other and from earth ground. Insulation must have a high resistance to prevent current from leaking through it. All insulation allows some leakage to occur. *Leakage current* is current that leaves the normal path of current and flows through a ground wire. Under normal operating conditions, the amount of leakage current is so small that the leakage current has no effect on the operation or safety of the circuit. The total leakage current through insulation is the combination of conductive leakage current, capacitive leakage current, and surface leakage current. **See Figure 4-19.**

> **TECH FACT**
>
> Insulation resistance is inversely related to temperature. When measuring insulation resistance, the temperature of the material being measured should be at the same base temperature as previous measurements.

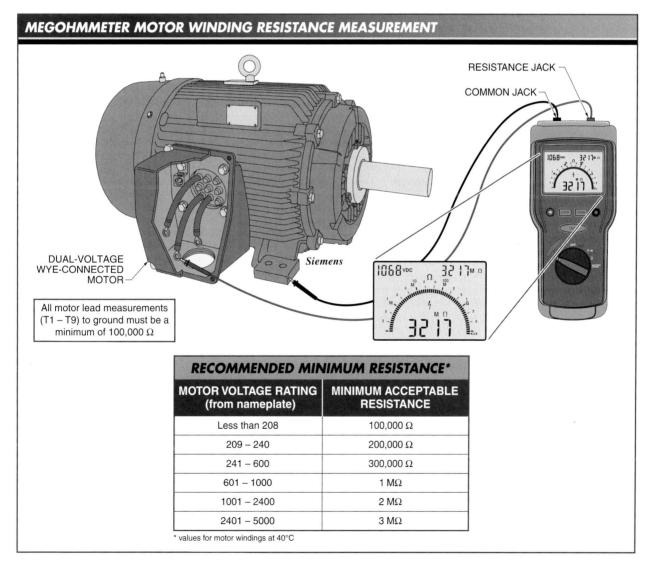

Figure 4-18. Megohmmeters are used to test for insulation breakdown in long wire runs, motor windings, and transformers.

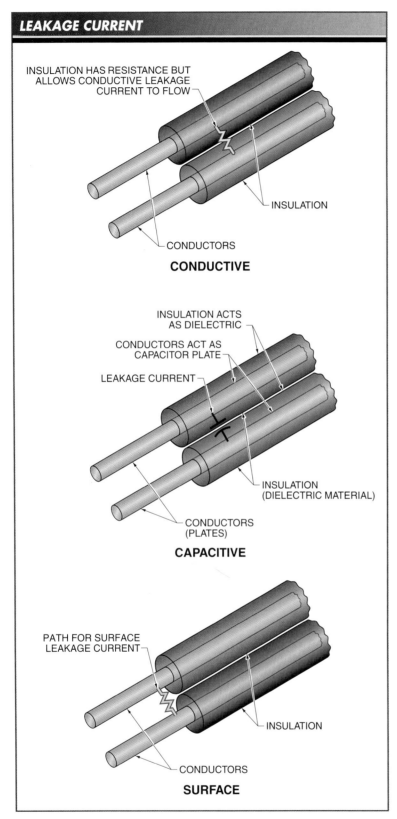

Figure 4-19. The total leakage current through insulation is the combination of conductive leakage current, capacitive leakage current, and surface leakage current.

Conductive Leakage Current

Conductive leakage current is the small amount of current that normally flows through the insulation of a conductor. Conductive leakage current flows from conductor to conductor or from a hot conductor to ground.

Insulation decreases in resistance as it ages and is exposed to damaging elements. The resistance of conductor insulation decreases as conductive leakage current increases. An increase in conductive leakage current results in additional insulation deterioration. Conductive leakage current is kept to a minimum by keeping insulation clean and dry.

Capacitive Leakage Current

Capacitive leakage current is leakage current that flows through conductor insulation due to a capacitive effect. A *capacitor* is an electric device that stores electrical energy by means of an electrostatic field. A capacitor is created by separating two plates with a dielectric material. Two conductors that run next to each other act as a low-level capacitor. The insulation between the conductors acts as the dielectric material, and the conductors act as the plates.

Conductors carrying DC voltage typically produce little capacitive leakage current because the leakage current flows for only a few seconds and then stops. AC voltages produce continuous capacitive leakage current, but the leakage can be kept to a minimum by separating or twisting the conductors along the run.

Surface Leakage Current

Surface leakage current is current that flows from areas on conductors where insulation has been removed to allow electrical connections. Conductors are terminated with wire nuts, splices, spade lugs, terminal posts, and other fastening devices at different points along an electrical circuit.

The point at which insulation is removed from a wire provides a low-resistance path for surface leakage current, with dirt and moisture allowing additional leakage to occur. Surface leakage current results in increased heat at the point of connection. Increased heat contributes to an increase in insulation deterioration, which makes the conductor brittle. Surface leakage current is kept to a minimum by making all connections clean and tight.

AC voltage in conductors varies between 0 V and peak voltage due to the alternating characteristic of AC voltage. Surface leakage current flows continuously as AC voltage alternates.

Megohmmeter Measurement Procedure

Megohmmeters send high voltages into conductors to test the insulation. Before any resistance measurements are taken using a megohmmeter, the meter is checked to ensure it is designed to take measurements on the circuit being tested. The operator's manual should be referred to for all measurement precautions, limitations, and procedures. Required PPE must always be worn and all safety rules followed when taking measurements. **See Figure 4-20.** Resistance measurements are taken with a megohmmeter using the following procedure:

1. Ensure that all power is off in the circuit or component being tested. Test the circuit for voltage using a DMM set to measure voltage.
2. Set the function switch to the proper voltage range for the circuit or component being tested. *Note:* The test voltage used must be as high as, or higher than, the highest voltage to which the circuit or component being tested is exposed.
3. Plug the black test lead into the negative jack.
4. Plug the red test lead into the positive jack.

WARNING: Verify that voltage is not present in a circuit or component being tested before taking any resistance measurements with a megohmmeter. Ensure that no body part contacts the high-voltage section of the megohmmeter test leads.

5. Ensure that the batteries are in good condition. *Note:* A megohmmeter does not contain batteries if the meter includes a hand crank or if the meter plugs into a standard 115 V receptacle.
6. Connect the red test lead to the conductor being tested.
7. Connect the black test lead to a second conductor in the circuit or to earth ground.
8. Press the test button or turn the hand crank and read the resistance displayed. (This may take 10 sec to 20 sec.) Change the megohmmeter resistance or voltage range if necessary.
9. Remove the megohmmeter from the conductors.
10. Discharge the circuit or conductors being tested.
11. Turn the megohmmeter off.
12. Consult the equipment or meter manufacturer for the minimum recommended resistance values. *Note:* The insulation is in good condition when the megohmmeter reading is equal to or higher than the minimum values indicated by the manufacturer.

Megger Group Limited

Megohmmeters can be used to test the level of deterioration of motor insulation.

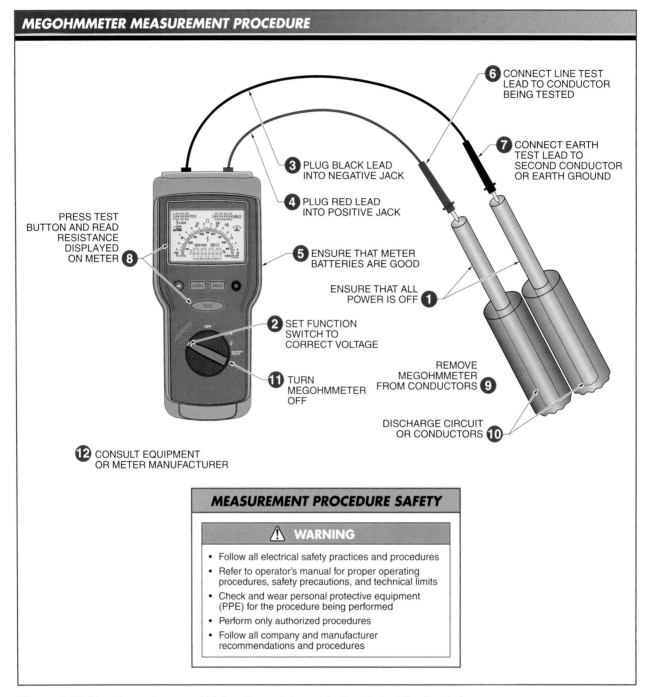

Figure 4-20. Megohmmeters send high voltages into conductors to test the insulation.

Measurement Categories

IEC standard 61010 classifies the applications in which DMMs may be used into four measurement categories, Category I through Category IV. The four categories are typically abbreviated as CAT I, CAT II, CAT III, and CAT IV. The CAT ratings indicate what magnitude of transient voltage a DMM or other electrical appliance can withstand when used on a power distribution system.

For example, a DMM specified for use in a CAT III installation must withstand a 6000 V transient (2 ms rise time and a

50 ms, 50% duration) voltage without resulting in a hazard. When a DMM is operated on voltages above 600 V, the DMM must be capable of withstanding an 8000 V transient voltage. The meters must withstand a total of 20 transient voltages—10 positive transients and 10 negative transients for the respective transient level. A DMM that can withstand these transient voltages may be damaged, but the transient will not result in a hazard to the technician or the facility. To protect technicians from transient voltages, protection must be built into all test equipment.

TECH FACT
Most DMMs include a continuity test function. The continuity test function may be used on an electrical component to check a short wire to determine if the wire is continuous or broken. If the circuit is complete, the meter beeps when the test leads touch both ends of a continuous (unbroken) wire. The wire may be broken if the meter does not beep.

Safety standards such as IEC 61010-1 and the UL standard 61010B have been harmonized per the North America standard. The requirements of the standards are used to rate test equipment in order to minimize hazards such as shock, fire, and arc blast, among other concerns. A DMM that meets these standards offers a high level of protection.

A measurement category rating such as CAT III or CAT IV indicates acceptable usage on three-phase permanently installed loads and three-phase distribution panels in a building or facility. All exposed electrical installations and the power panels of a facility are considered high-voltage areas. Measurement categories such as CAT III and CAT IV are important for DMMs used in commercial applications. A higher CAT number indicates an electrical environment with higher power available, larger short circuit current available, and higher energy transients. For example, a DMM that meets the CAT III standard is resistant to higher energy transients than a DMM that meets the CAT II standard. **See Figure 4-21.**

Power distribution systems are divided into categories based on the magnitude of transient voltage. DMMs must be rated for the category in which they are used. Dangerous high-energy transient voltages, such as a lightning strike, are attenuated (lessened) or dampened as the transient travels through the impedance (AC resistance) of the system and system grounds.

Within an IEC 61010 standard category, a higher voltage rating denotes a higher transient voltage withstanding rating. For example, a CAT III-1000 V (steady-state) rated DMM has better protection compared to a CAT III-600 V (steady-state) rated DMM. But, a CAT III-600 V DMM has an equivalent withstanding capability of a CAT II-1000 V DMM. A DMM must be chosen based on the IEC measurement category first and voltage second. *Note:* IEC standards for test instruments are not the same as the hazard categories defined by NFPA 70E.

Refer to Chapter 4 Quick Quiz® on CD-ROM.

Fluke Corporation
Test instruments used for taking measurements in commercial applications should have a minimum measurement category rating of CAT III.

124 ELECTRICAL SYSTEMS for FACILITIES MAINTENANCE PERSONNEL

IEC 61010 MEASUREMENT CATEGORIES

Category	In Brief	Examples
CAT I	Electronic	• Protected electronic equipment • Equipment connected to (source) circuits in which measures are taken to limit transient voltages to an appropriately low level • Any high-voltage, low-energy source derived from a high-winding resistance transformer, such as the high-voltage section of a copier
CAT II	Single-phase receptacle-connected loads	• Appliances, portable tools, and other household and similar loads • Outlets and long branch circuits • Outlets at more than 30′ (10 m) from CAT III source • Outlets at more than 60′ (20 m) from CAT IV source
CAT III	Three-phase distribution, including single-phase commercial lighting	• Equipment in fixed installations, such as switchgear and polyphase motors • Bus and feeder in industrial plants • Feeders and short branch circuits and distribution panel devices • Lighting systems in larger buildings • Appliance outlets with short connections to service entrance
CAT IV	Three-phase at utility connection; any outdoors conductors	• Refers to the origin of installation, where low-voltage connection is made to utility power • Electric meters; primary overcurrent protection equipment • Outside and service entrance; service drop from pole to building; run between meter and panel • Overhead line to detached building

Figure 4-21. The IEC 61010 standard classifies the application in which a DMM may be used into four measurement categories.

Case Study—Testing Office Receptacles

Different types of receptacles, such as standard, GFCI, and isolated ground, that have different current ratings, such as 15 A and 20 A, are used in commercial buildings. Receptacles must be tested for proper wiring, voltage, and grounding. If a receptacle is a GFCI receptacle, the receptacle must also be tested for proper operation during a current-to-ground fault.

A DMM set to measure AC voltage is used when testing a standard 15 A or 20 A receptacle. The DMM is first connected between the hot (short) slot and the neutral (long) slot to measure the voltage at the receptacle. The black meter lead is connected to the neutral slot, and the red meter lead is connected to the hot slot. The voltage should be within +5% to –10% of the 120 VAC rating.

If there is a measured voltage between the hot and neutral slots, a measurement is taken to determine if the receptacle is grounded. The black meter lead is moved from the neutral slot to the ground slot, and the red meter lead remains connected to the hot slot. If the receptacle is grounded, the voltage measured between the ground and hot slot should be the same as the voltage measured between the hot and neutral slots.

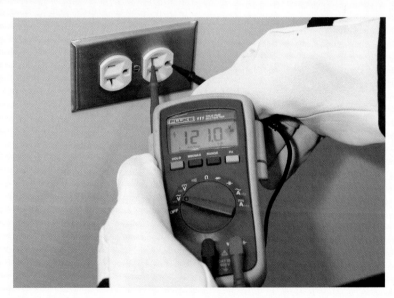

Case Study—Testing Office Receptacles (continued)

If the measured voltage is not the same, the receptacle is either wired backwards (hot wire connected to neutral slot and neutral wire connected to hot slot) or there is no ground. To ensure the receptacle is not wired backwards, the meter leads are connected to the neutral slot and the ground slot. If there is a voltage reading, the receptacle is wired backwards and must be rewired correctly.

To test a GFCI, the electrical part of the GFCI must be tested using a meter. Pressing the button on the GFCI only tests the mechanical part of the GFCI and does not ensure it will trip during an electrical fault. To test a GFCI, a meter that includes a GFCI test button must be used. Standard DMMs cannot be used because they do not draw enough current to trip a GFCI when connected between the hot slot and the ground slot.

To test a GFCI, the meter leads are connected between the hot slot and the ground slot. The circuit voltage should be measured. During the measurement, the GFCI test button on the meter is pressed. If the GFCI is wired correctly and operating properly, the GFCI should trip.

Definitions

- An *abbreviation* is a letter or combination of letters that represents a word or words.

- An *analog display* is an electromechanical device that displays measurements using the mechanical motion of a pointer.

- An *analog multimeter* is a meter that can measure two or more electrical properties and displays the measured properties using a pointer that moves along several calibrated scales.

- A *bar graph* is a graph composed of segments that function similar to an analog pointer.

- *Capacitive leakage current* is leakage current that flows through conductor insulation due to a capacitive effect.

- A *capacitor* is an electric device that stores electrical energy by means of an electrostatic field.

- A *clamp-on ammeter* is a test instrument that measures current in a circuit by measuring the strength of the magnetic field around a conductor.

- *Conductive leakage current* is the small amount of current that normally flows through the insulation of a conductor.

- A *digital display* is an electronic device that displays measurements as numerical values.

- A *digital multimeter (DMM)* is a meter that can measure two or more electrical properties and displays the measured properties as numerical values.

- An *electrical test instrument* is a device used to measure electrical quantities such as voltage (V), current (A), resistance (Ω), power (W), capacitance (f), and frequency (Hz).

- *Leakage current* is current that leaves the normal path of current and flows through a ground wire.

- A *linear scale* is a scale that is divided into equally spaced segments.

- A *megohmmeter* is a high-resistance ohmmeter used to measure insulation deterioration on conductors by measuring high resistance values during high-voltage test conditions.

- A *multimeter* is a portable test instrument that is capable of measuring two or more electrical quantities.

- A *nonlinear scale* is a scale that is divided into unequally spaced segments.

- A *primary division* is a division with a listed value.

- *Resistance (R)* is the opposition to the flow of electrons in a circuit and is measured in ohms (W).

- A *secondary division* is a division that divides primary divisions into halves, thirds, fourths, fifths, etc.

- A *subdivision* is a division that divides secondary divisions into halves, thirds, fourths, fifths, etc.

- *Surface leakage current* is current that flows from areas on conductors where insulation has been removed to allow electrical connections.

- A *symbol* is a graphic element that represents a quantity, unit, or component.

- A *voltage tester* is an electrical test instrument that indicates the approximate voltage amount and type (AC or DC) in a circuit using a movable pointer or vibration.

Review Questions

1. What are the advantages of using a voltage tester to measure the voltage in a circuit connected to a GFCI receptacle?

2. List the procedural steps that should be followed when using a voltage tester to ensure the proper operation of a GFCI.

3. What are the different types of scales found on an analog display and what are the differences between them?

4. List the procedural steps that should be followed when reading a linear analog scale.

5. What is the purpose of a bar graph on a digital multimeter display and why is it useful when used in conjunction with a digital display?

6. List the tests that can be performed on a motor using a DMM set to measure resistance.

7. What are the precautions for measuring the resistance of a circuit using a DMM?

8. What devices are typically used to measure current and how do they function?

9. What device and procedural steps would be used to test the current of a circuit containing a 480 V duct heating element if the element were suspected of being faulty?

10. Why are in-line ammeters the most accurate current measurement devices and what are their limitations?

11. What is the most common device for measuring the current in a circuit containing a door motion sensor with an expected current of 1000 mA, and what are the procedural steps used?

12. List the procedural steps for testing voltage in a circuit with a DMM.

13. Why are megohmmeters used to measure resistance in long wire runs, motor windings, and transformers, and where are the minimum resistance values found?

14. List and describe the three types of leakage current.

15. List the procedural steps for testing the insulation resistance between a motor circuit and earth ground.

chapter 5
ELECTRICAL DISTRIBUTION SYSTEMS

Electrical distribution system troubleshooting and maintenance begins by understanding how electrical power is routed into a facility from an electrical service (utility) provider. Electrical power is delivered to electrical devices through receptacles. Receptacles must be wired correctly, and the devices must be connected to the correct receptacle based on required voltage level and phase requirements. Working with distribution system components and devices requires specialized training and equipment. Power quality problems can cause poor device operation and damage to wiring and distribution system components.

OBJECTIVES

- Identify types of receptacles and describe how to test each.
- Describe how to troubleshoot electric power from a starter or fuse box to isolate electrical problems.
- Identify fuses and circuit breakers and describe how to test each.
- List the steps for measuring circuit loading and calculating percent voltage drop, voltage unbalance, and current unbalance.
- Define and give an example of linear and nonlinear loads.
- Describe the effect harmonic distortion has on an electrical system.
- Explain how transformers operate and are tested for proper operation.

RECEPTACLES

A *receptacle,* also known as an outlet, is a device used to connect equipment with a cord and plug to an electrical system. Receptacles are used to deliver power to electrical devices such as computers, copy machines, portable lamps, and entertainment systems. Receptacles are available in different designs to meet the requirements of different loads, locations, and usages. Types of receptacles include standard and isolated-ground receptacles, GFCI receptacles, 120 V, single-phase locking and nonlocking receptacles, special-use receptacles, 208 V, single-phase receptacles, 120/208 V, single-phase receptacles, and three-phase receptacles.

Receptacles and neutral-to-ground connections are tested for proper wiring. The type, location, and usage of receptacles are identified by a schematic symbol on a schematic drawing or electrical print. **See Figure 5-1.** The most common type of receptacle installed in a commercial building is a duplex receptacle.

Testing Standard and Isolated-Ground Receptacles

Standard and isolated-ground receptacles can be tested using a DMM or voltage tester. Isolated-ground receptacles are used for protection for computers and other sensitive electronic equipment. They protect equipment from electromagnetic interference and electrical noise caused by harmonics and undesirable current flow on the equipment grounding conductor (EGC). **See Figure 5-2.** A standard or isolated-ground receptacle is tested using the following procedure:

Note: Always wear proper PPE for the location and type of test.
1. Set the DMM or voltage tester to measure AC voltage. Test the DMM or voltage tester on a known energized source before taking a measurement to verify that the test instrument is in proper working condition.
2. Use a DMM or voltage tester to measure the voltage between the hot slot and neutral slot of the receptacle. *Note:* The voltage measurement should be within +5% to –10% of the rating of loads that are to be connected to the receptacle.
3. Use a DMM or voltage tester to measure the voltage between the hot slot and ground slot to determine if the ground slot and the hot slot are connected correctly. *Note:* The voltage measured should be the same as in step 2. If there is no measured voltage, it is possible the ground slot is properly connected, but the hot and neutral connections are reversed.
4. Use a DMM or voltage tester to measure the voltage between the neutral slot and ground slot. *Note:* The voltage should be near 0 V (no load) to no more than 3.5 V when the circuit is fully loaded (80% of circuit breaker rating). The greater the load on the branch circuit, the greater the voltage drop between the neutral and ground conductors. This is because of the voltage drop on the neutral conductor due to the resistance of the conductor and current flow through the conductor. If the measured voltage is the same as in step 2, the hot and neutral connections are reversed.
5. Retest the DMM or voltage tester on the known energized source to verify that the meter is still working properly.

TECH FACT

In order to reduce power quality problems such as voltage sags or transient voltages, the amount of installed standard or isolated-ground receptacles should be reduced by half of the NEC® permissible maximum in any area in which sensitive equipment such as computers, printers, and monitors are connected.

Chapter 5—Electrical Distribution Systems 131

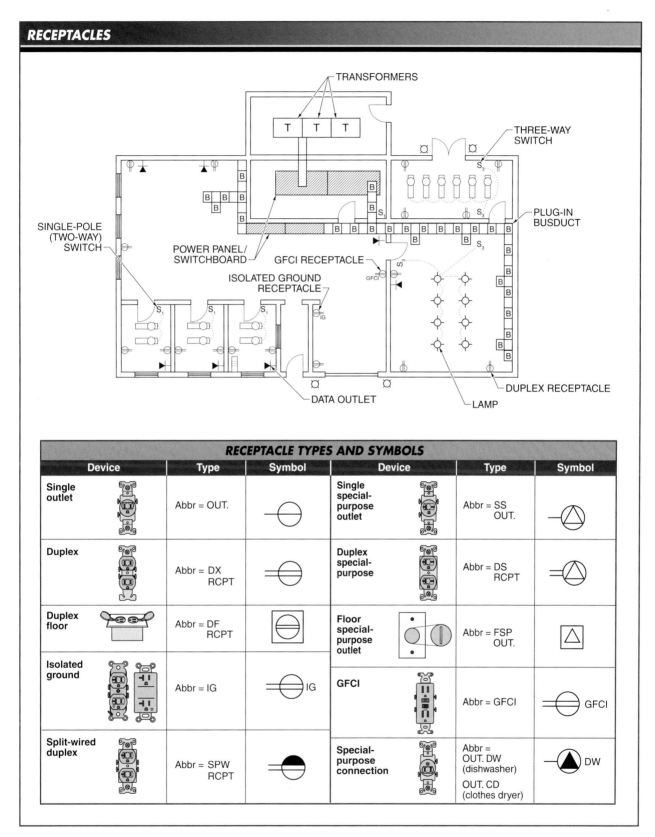

Figure 5-1. The type, location, and usage of receptacles are identified by a schematic symbol on a schematic drawing or electrical print.

132 ELECTRICAL SYSTEMS for FACILITIES MAINTENANCE PERSONNEL

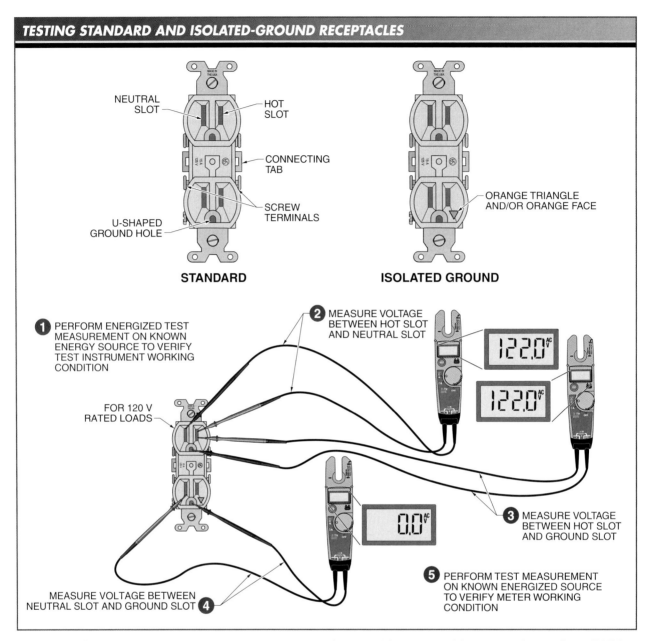

Figure 5-2. Standard and isolated-ground receptacles can be tested for proper wiring connections using a DMM or voltage tester.

Testing GFCI Receptacles

Ground-fault circuit interrupter (GFCI) receptacles are tested in the same manner as standard and isolated-ground receptacles in order to measure circuit voltage and verify that the hot, neutral, and ground slots are properly connected. However, one additional test is performed when testing a GFCI receptacle to verify that the electronic fault detection and trip circuit is functioning properly. The test instrument used must have a GFCI test function. A GFCI test function verifies that the test instrument is able to draw enough current (approximately 8 mA to 10 mA) to trip the GFCI when the test instrument or meter is connected between the hot slot and the ground slot. **See Figure 5-3.**

A GFCI receptacle is tested using the following procedure:

Note: Always wear proper PPE for the location and type of test. Always test a GFCI receptacle with a voltage tester or GFCI tester because the test button on a GFCI receptacle only tests the mechanical capabilities of the GFCI.

1. Use only a DMM or voltage tester that includes a GFCI test function. Set the DMM or voltage tester to measure AC voltage. Test the DMM or voltage tester on a known energized source before taking a measurement to verify that the test instrument is in proper working condition.

2. Measure the circuit voltage between the hot slot and neutral slot of the receptacle. *Note:* The voltage should be within +5% to –10% of the rating of load(s) that are to be connected to the receptacle.

3. Measure the voltage between the hot slot and ground slot to verify that the ground and the hot slot are connected correctly. *Note:* The voltage measured should be the same as in step 2. When using a voltage tester or DMM that has a GFCI test function included but does not have a separate GFCI test button, the GFCI should automatically trip, indicating that the GFCI is functioning.

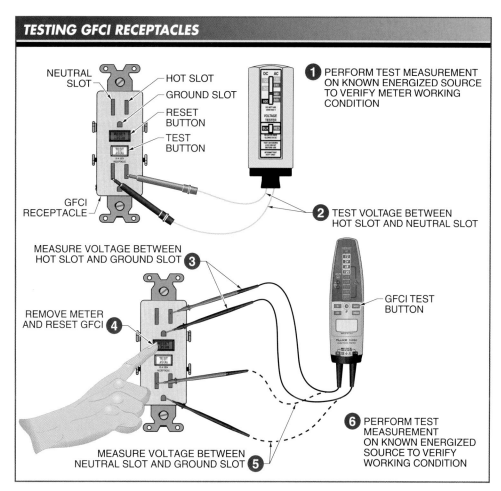

Figure 5-3. GFCI receptacles are tested to verify that the hot, neutral, and ground slots are properly wired and to verify that the electronic fault detection and trip circuit is functioning properly.

4. Remove the meter and reset the GFCI using the reset button on the GFCI. When using a voltage tester or DMM that has a GFCI test function included, measure the voltage between the hot slot and ground slot and press the GFCI test button on the meter. *Note:* The GFCI should automatically trip, indicating the GFCI is functioning.
5. Use a voltage tester or DMM that has a GFCI test function included to measure the voltage between the neutral slot and ground slot. *Note:* The voltage should be near 0 V (no load) to no more than 3.5 V when the circuit is fully loaded (80% of circuit breaker rating). The greater the load on the branch circuit, the greater the voltage drop between the neutral and ground conductors. This is because of the voltage drop on the neutral conductor due to the resistance of the conductor and current flow through the conductor. If the measured voltage is the same as in step 2, the hot and neutral connections are reversed.
6. Retest the DMM or voltage tester by performing a voltage measurement on a known energized source to verify that the meter is still working properly.

TECH FACT

Test instruments and meters that take voltage measurements use so little current when taking the measurement that they cannot trip a GFCI when connected between the hot and neutral receptacle slots. Only DMMs or voltage testers that are designed for testing GFCIs can be used to test GFCI operation.

Testing 120 V, Single-Phase, Locking and Nonlocking Receptacles

Receptacles and plugs are configured so that only the correct size (15 A, 20 A, 30 A, etc.), voltage level (120 V, 240 V, etc.), and type (locking or nonlocking) can fit together. This ensures that the plug on a piece of equipment can only be plugged into a receptacle that is the right type and has the right amperage rating. For example, a piece of equipment with a 20 A/120 V rated plug cannot be plugged into a 15 A/120 V rated receptacle. However, a piece of equipment with a 15 A/120 V rated plug can be plugged into a 20 A/120 V rated receptacle. **See Appendix.**

When testing a 120 V, single-phase receptacle, the receptacle is tested to verify that the voltage level is within an acceptable range for the equipment and circuit (usually +5% to –10%), and the receptacle is connected properly. **See Figure 5-4.** A 120 V, single-phase receptacle is tested using the following procedure:

Note: Always wear proper PPE for the location and type of test.

1. Set the meter to measure AC voltage. Test the meter on a known energized source before taking a measurement to verify that the meter is in proper working condition.
2. Perform a circuit voltage measurement using a DMM or voltage tester to measure the voltage between the hot slot and neutral slot of the receptacle. *Note:* The voltage measurement should be within +5% to –10% of the rating of load(s) that are to be connected to the receptacle.
3. Perform a circuit ground test using a DMM or voltage tester to measure the voltage between the hot slot and ground slot to determine if the ground slot and the hot slot are connected correctly. *Note:* The voltage measured should be the same as in step 2. If there is no measured voltage, it is possible the ground slot is properly connected, but the hot and neutral connections are reversed.
4. Perform a neutral to ground test using a DMM or voltage tester to measure the voltage between the neutral slot and ground slot. *Note:* The voltage should be near 0 V (no load) to no more than 3.5 V when the circuit is fully loaded (80% of circuit breaker rating).

5. Retest the DMM or voltage tester on a known energized source to verify that the meter is still working properly.

TECH FACT

Receptacles and plugs are rated at a slightly higher voltage rating than the piece of electrical equipment is rated for and/or the actual circuit voltage. For example, a receptacle that is rated at 125 V can be used with equipment or circuits rated at 110 V, 115 V, or 120 V. The 125 V rating signifies that the receptacle can be used with any voltage up to a maximum voltage of 125 V.

Test measurements are taken on hot electrical circuits by inserting a test instrument's probes directly into the receptacle slots.

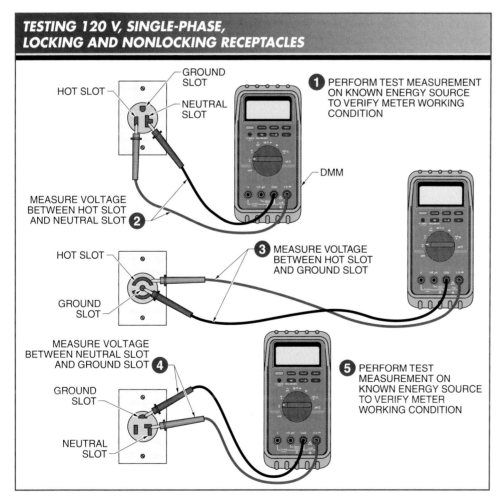

Figure 5-4. When testing a 120 V, single-phase receptacle, the receptacle is tested to verify that the voltage level is within an acceptable range for the equipment and circuit and that the receptacle is wired properly.

TECH FACT

Nonlocking plugs and receptacles are the most common types of plugs and receptacles used in commercial environments and are suitable for most applications. Locking plugs and receptacles require the plug to be inserted and then twisted in place. Locking plugs cannot be pulled out without first twisting them back into a specific position and should only be used in applications in which the accidental removal of the plug or power can cause significant problems.

Special-Use Receptacles

Electrical loads and equipment are designed to operate on different voltages, different voltage types, and different amperages. Single-phase, high-power loads such as electric heating elements and hot water heaters are designed to operate at 208 V instead of 120 V. As with 120 V plugs, 208 V receptacles also have a configuration that matches a 208 V current rating and are designed so smaller and larger current-rated plugs cannot fit into the 208 V configuration. Some 208 V receptacles are connected to provide only 208 V and a ground, and other receptacles are connected to provide both 208 V and 120 V. Because there are different requirements for different applications, special-use receptacles are available in a variety of sizes and configurations. **See Figure 5-5.**

Testing 208 V, Single-Phase Receptacles

A 208 V, single-phase receptacle is used to provide more power to single-phase loads than a 120 V, single-phase receptacle of the same amperage rating because power (in W or HP) is the product of voltage times current ($P = E \times I$). When testing a 208 V, single-phase receptacle, the receptacle is tested to ensure that the voltage level is within an acceptable range for the equipment and circuit (usually +5% to –10%), and the receptacle is properly grounded. **See Figure 5-6.** A 208 V, single-phase receptacle is tested using the following procedure:

Note: Always wear proper PPE for the location and type of test.

1. Set the test instrument to measure AC voltage. Test the test instrument on a known energized source before taking a measurement to verify that the test instrument is in proper working condition.
2. Perform a circuit voltage measurement using a DMM or voltage tester to measure the voltage between the two hot slots of the receptacle. *Note:* The voltage measurement should be within +5% to –10% of the rating of loads that are to be connected to the receptacle.

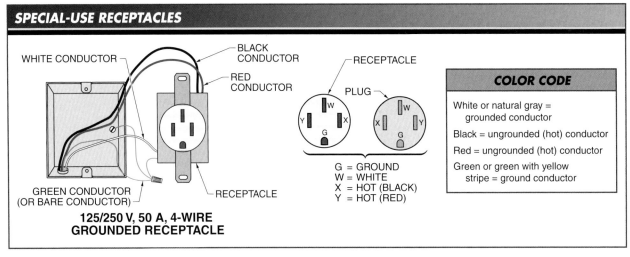

Figure 5-5. Single-phase high-power loads are designed to operate on 208 V, while some loads can operate on either 208 V or 120 V.

3. Perform a circuit ground test using a DMM or voltage tester to measure the voltage between the two hot slots and the ground slot to determine if the ground slot and the two hot slots are connected correctly. *Note:* The voltage between either of the hot slots and the ground slot should be the same.
4. Retest the DMM or voltage tester on the known energized source to verify that the meter is still working properly.

Testing 120/208 V, Single-Phase Receptacles

A 120/208 V, single-phase receptacle is used to provide a combination of different voltages to different types of equipment. Some equipment uses the higher 208 V for high power consumption and the 120 V for low power consumption. For example, an electric heater would apply 208 V to the high power heating elements and 120 V to a low power fan motor used to move air over the heating elements. A 120/208 V receptacle is also used to provide different amounts of power to the same load to allow flexibility. For example, in an electric heater, a heating element rated at 208 V can be connected to either 208 V or 120 V. The amount of heat produced when the heating element is connected to 208 V is four times more than when the heating element is connected to 120 V.

TECH FACT

Doubling the voltage on a heating element quadruples the power output.

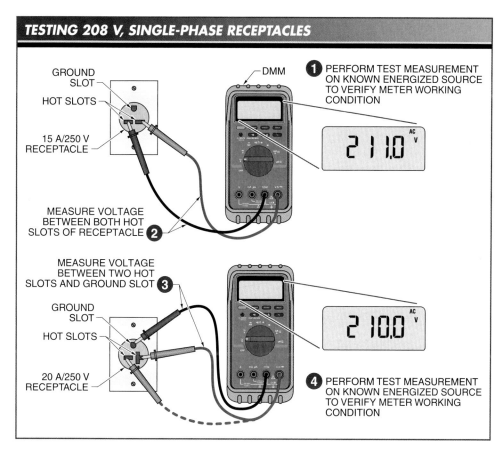

Figure 5-6. A 208 V, single-phase receptacle is tested to ensure that the voltage level is within an acceptable range for the equipment and circuit and that the receptacle is properly grounded.

When testing a 120/208 V, single-phase receptacle, the receptacle is tested to verify that both voltage levels are within an acceptable range for the equipment and circuit (usually +5% to –10%) and the receptacle is connected properly. **See Figure 5-7.** A 120/208 V, single-phase receptacle is tested using the following procedure:

Note: Always wear proper PPE for the location and type of test.
1. Set the test instrument to measure AC voltage. Test the test instrument on a known energized source before taking a measurement to verify that the test instrument is in proper working condition.

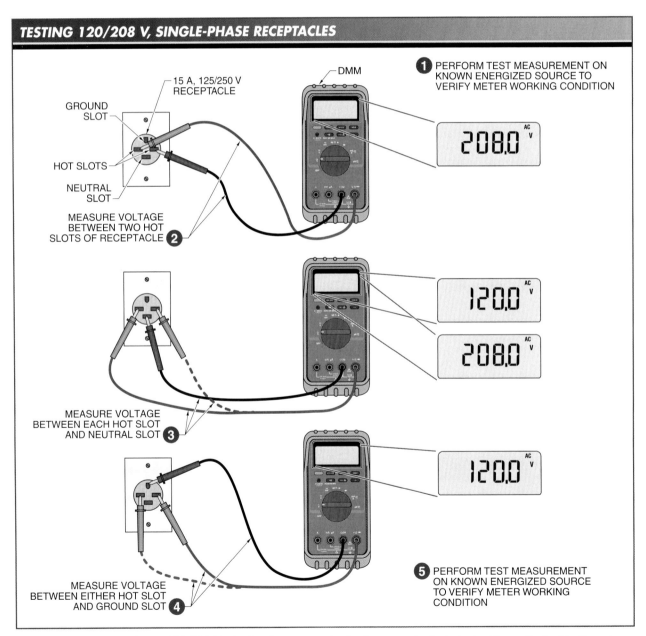

Figure 5-7. When testing a 120/208 V, single-phase receptacle, the receptacle is tested to verify that both voltage levels are within an acceptable range for the equipment and circuit and that the receptacle is wired properly.

2. Perform a high-voltage measurement using a DMM or voltage tester to measure the voltage between the two hot slots of the receptacle. *Note:* The voltage measurement should be within +5% to –10% of the rating of loads that are to be connected to the receptacle.
3. Perform a low-voltage measurement using a DMM or voltage tester to measure the voltage between each of the two hot slots and the neutral slot of the receptacle. *Note:* The voltage should be within +5% to –10% of the rating of loads that are to be connected to the receptacle.
4. Perform a circuit ground test using a DMM or voltage tester to measure the voltage between either hot slot and ground slot to determine if the ground slot and the hot slot are connected correctly. *Note:* The voltage between either of the hot slots and the ground slot should be the same.
5. Retest the DMM or voltage tester on the known energized source to verify that the meter is still working properly.

Testing Three-Phase Receptacles

Three-phase power is used to provide power to large electrical equipment because it is more energy efficient than single-phase power. Three-phase receptacles can be connected to either a wye or a delta system and can be powered by different voltages. For this reason, three-phase receptacles require that the voltage between every slot on the receptacle be tested. When testing three-phase receptacles, the receptacle is tested to ensure the voltage level is within an acceptable range for the equipment and circuit (usually +5% to –10%), and the receptacle is connected properly (ground slot is grounded, etc.). **See Figure 5-8.** A three-phase power receptacle is tested using the following procedure:

Note: Always wear proper PPE for the location and type of test.
1. Set the test instrument to measure AC voltage. Test the test instrument on a known energized source before taking a measurement to verify that the test instrument is in proper working condition.
2. Test for three-phase circuit voltage using a DMM or voltage tester to measure the voltage between the three hot slots of the receptacle. *Note:* The voltage should be within +5% to –10% of the rating of loads that are to be connected to the receptacle.
3. If a three-phase with a neutral is used, test for low hot to neutral circuit voltage using a DMM or voltage tester to measure the voltage between each of the three hot slots and the neutral slot of the receptacle. *Note:* The voltage should be within +5% to –10% of the rating of loads that are to be connected to the receptacle.
4. Perform a circuit ground test using a DMM or voltage tester to measure the voltage between each of the hot slots and ground slot of the receptacle to determine if the ground slot and the hot slots are connected correctly. The voltage between each of the hot slots and the ground slot should be the same.
5. Retest the DMM or voltage tester on the known energized source to verify that the meter is still working properly.

TECH FACT

The neutral conductor and ground conductor are not the same, even though the neutral conductor is grounded in the main service panel. A neutral conductor is intended to carry the load current the same as the hot conductor. A ground conductor should never carry current unless there is a fault, at which time the ground conductor should carry the fault current.

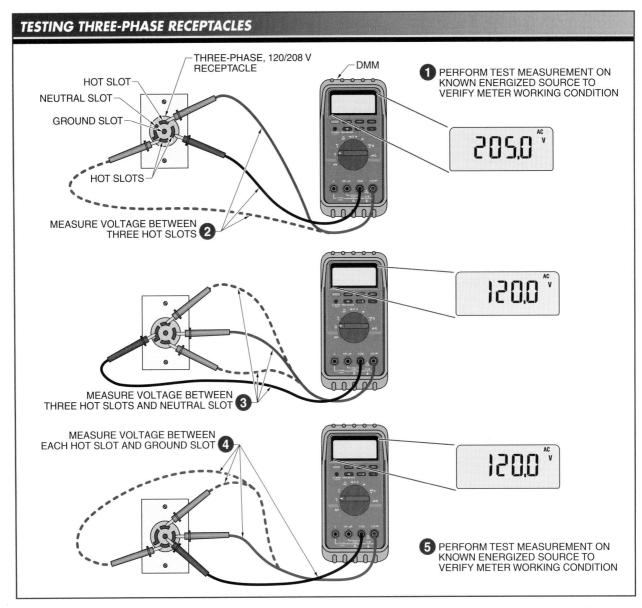

Figure 5-8. Three-phase receptacles require that the voltage between every slot on the receptacle be tested because they can be powered by different voltages.

GREEN TECH FACT

Whenever possible, use a 3ϕ motor rather than a 1ϕ motor because a 3ϕ motor is more energy efficient than a 1ϕ motor.

Neutral-to-Ground Voltage Troubleshooting

A *ground* is an intentional or accidental electrical connection between a circuit or equipment and the earth or other conducting member. A ground is provided at the main service equipment or the source of a separately derived system. A *separately derived system (SDS)* is a system that supplies a facility with electrical power derived, or taken, from a transformer, storage batteries, a photovoltaic system, or a generator. Because an SDS does not have electrical connections to any other part of the distribution system, a new ground reference is required.

A neutral and ground connection is made at the main service entrance panel by connecting the neutral bus to the ground bus with a main bonding jumper. A *main bonding jumper (MBJ)* is a connection at the service equipment that connects the equipment grounding conductor, grounding electrode conductor, and grounded conductor (typically the neutral conductor).

An *equipment grounding conductor (EGC)* is an electrical conductor that provides a low-impedance ground path between electrical equipment and enclosures within a distribution system. A *grounding electrode conductor (GEC)* is a conductor that connects grounded parts of a power distribution system to an NEC® approved earth grounding system. A *grounded conductor* is a conductor that has been intentionally grounded and is typically known as the neutral conductor.

Neutral-to-ground voltage connections are made at the main panel and should not be made in any subpanels, receptacles, or equipment. If a neutral-to-ground connection is made, a parallel path for the normal return current from system loads is created. The parallel path allows current to flow through metal parts of the system and cause electrical shocks.

Measuring the voltage between the neutral and ground can help determine if there are any illegal neutral-to-ground connections. There should be a voltage difference caused by a voltage drop across the neutral conductor because the conductor has resistance and is carrying current. The farther away the neutral conductor is located from the power panel, the higher the voltage drop, power rating of the loads, and measured voltage difference. A measured voltage difference of 5 V or higher indicates that the branch circuit is overloaded. A low or no voltage difference measurement may indicate an illegal neutral-to-ground connection and should be checked. **See Figure 5-9.**

POWER AND DISTRIBUTION PANELS

Power and distribution panels can be a starting point for taking measurements when troubleshooting electrical problems and for understanding how a commercial power distribution system operates (amount each circuit and main power lines are loaded, etc.). Common electrical measurements taken for circuit loading include current and voltage measurements.

Additional electrical measurements taken to isolate specific problems include voltage and current (minimum and maximum) over time, peak voltage, peak current, temperature measurements, harmonic measurements, power, power factor, waveform analysis, and transient voltage measurements. In addition to clamp-on ammeters and DMMs, electrical measurements can be taken at a power or distribution panel with power quality meters and noncontact thermometers. **See Figure 5-10.**

Measuring Circuit Loading

Electrical loads convert electrical energy into other forms of energy such as heat, light, motion, and sound. When converting energy, electrical loads use power. Since power (P) is the product of the applied voltage (E) times the current (I), the higher the power, the greater the current draw. To safely deliver the required power to a load, the conductor connecting the load must be rated high enough to carry the amount of current draw from the power panel. To ensure the conductors do not overheat and cause a problem such as a fire or short circuit, fuses and circuit breakers are connected in series between the loads and the main power lines feeding the panel. The fuses and circuit breakers become the weakest point in the circuit and must be properly rated so that they can open an overloaded circuit before it becomes a safety hazard.

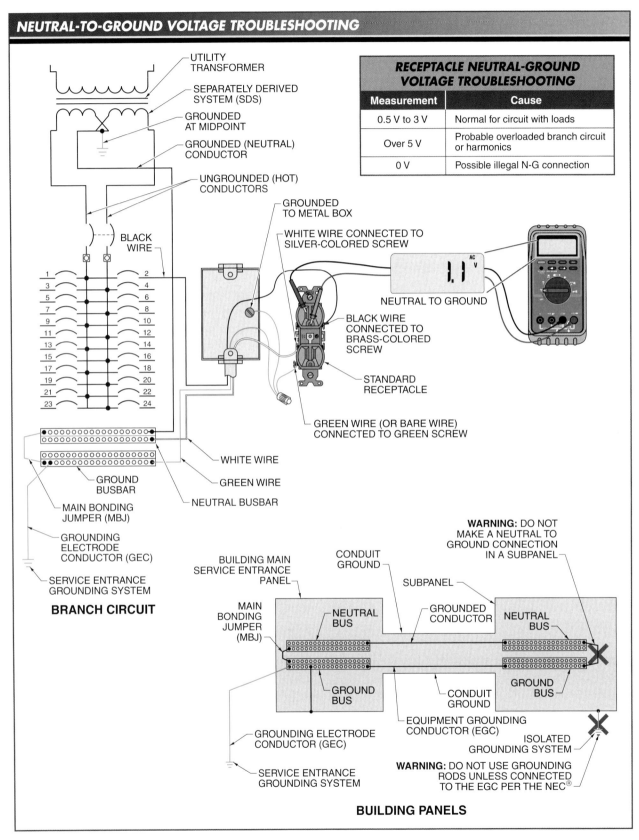

Figure 5-9. Measuring the voltage between the neutral and ground can help determine if there are any illegal neutral-to-ground connections.

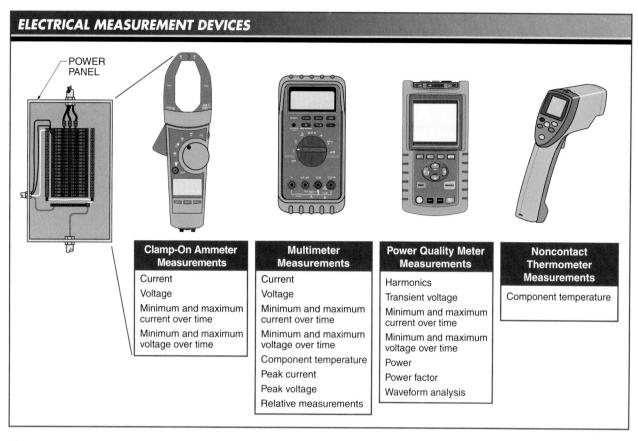

Figure 5-10. In addition to clamp-on ammeters and DMMs, electrical measurements can be taken at a power panel with power quality meters and noncontact thermometers.

Measuring the current at fuses and circuit breakers indicates the amount of load on a circuit (current draw). Although fuse and circuit breaker conductors are rated for a maximum operating current, they should not be operated continuously at more than 80% of their rating. **See Figure 5-11.** For example, a 15 A rated circuit should not draw more than 12 A continuously (15 A × 0.80 = 12 A). The amount of current on a circuit is tested using the following procedure:

Note: Always wear proper PPE for the location and type of test. Verify the test instrument (clamp-on ammeter or DMM with clamp-on current probe accessory) has the minimum CAT III rating for the measuring location.

1. Set the meter to measure AC current. Test the meter on a known energized source before taking a measurement to verify that the meter is in proper working condition. To do this with a clamp-on ammeter, use the ammeter to test several circuits within the panel, even when only one circuit must be tested. *Note:* Unlike voltage, which remains relatively constant, current varies continuously as loads are switched on and off.

2. Perform a circuit current measurement. Open the meter jaws and connect them around the conductor coming from the circuit breaker or fuse of the circuit under test. Verify that the meter is connected around only one conductor. Record and compare the meter reading of the current measurement with the fuse or circuit breaker rating and conductor rating. *Note:* Although it may only

be required to take a measurement on one circuit, if potential problems such as overheated or discolored conductor insulation on other circuits are visible, additional measurements should be taken.

3. Perform a circuit current measurement over time. Since current typically varies as loads are switched on and off, or motors are loaded to a greater or lesser extent, measure current over time. If the meter includes a MIN MAX recording mode, press the MIN MAX button when the current is being measured. At any time during the measurement, or after the meter is removed (but not turned off), the MIN MAX button can be pressed to view the following:

- The minimum (MIN) mode displays the lowest current draw on the circuit during the recording time.
- The maximum (MAX) mode displays the maximum current draw on the circuit during the recording time.
- The average (AVE) mode (on some meter models) displays the average current reading during the recording time. *Note:* This reading is helpful because the MIN and MAX mode cannot indicate how long the current is at the minimum and maximum values.

4. Retest the test instrument on the known energized source to verify that the meter is still working properly.

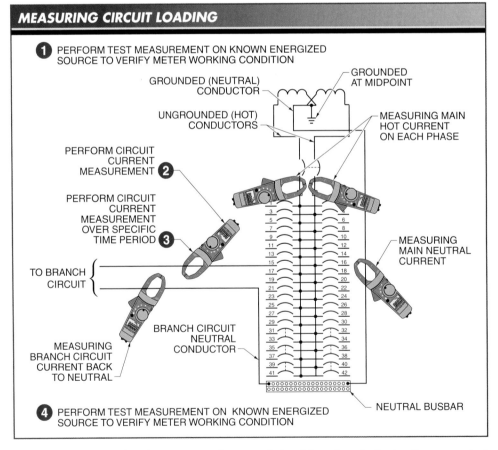

Figure 5-11. Measuring the current at fuses and circuit breakers indicates the amount of load on a circuit (current draw).

> **TECH FACT**
>
> Panelboards contain circuit breakers that, when the total current rating of each breaker (15 A, 20 A, etc.) is added, exceed the total listed rating of the panelboard. However, the current rating (200 A, etc.) of a panelboard is equal to the maximum continuous current that can be supplied through the main terminals and not the sum of the individual circuit breakers.

FUSES AND CIRCUIT BREAKERS

Fuses and circuit breakers are connected in series with conductors delivering power to electrical loads such as motors, receptacles, and lamps. A *fuse* is an overcurrent protection device with a fusible link that melts and opens a circuit when an overload condition or short circuit occurs.

A *circuit breaker* is an overcurrent protection device with a mechanical mechanism that may manually or automatically open a circuit when an overload condition or short circuit occurs. Fuses and circuit breakers are sized so that they are the weakest link in the circuit when the circuit is overloaded or there is a short circuit. Because of this, the testing of fuses and circuit breakers is a common starting point for troubleshooting circuits and loads that are not powered.

Testing Fuses

An ohmmeter or DMM with a continuity function can be used to test fuses. A good fuse has a measurement of 0 Ω when the ohmmeter leads are connected across the fuse. The ohmmeter indicates continuity by beeping when the continuity function is used. A bad fuse has a measurement of infinity (OL) resistance when the ohmmeter leads are connected across the fuse. The ohmmeter indicates no continuity by not beeping when the continuity function is used.

WARNING: An ohmmeter or continuity tester should be used to test fuses only if they have been removed from a circuit. Even if power in the circuit is off, an ohmmeter can give a false reading about the condition of a fuse that is connected to a circuit. False readings occur because the meter can read other parts of the circuit when it becomes part of the circuit by being connected in series with the circuit on an open fuse.

Using a DMM to test fuses that are not removed from an energized circuit is the best method to initially test a fuse since the fuse does not have to be removed from the circuit. A fuse that is suspected of being bad (open) can also be tested with an ohmmeter after it is removed from the circuit. **See Figure 5-12.** Fuses that are connected into a circuit are tested using the following procedure:

Note: Always wear proper PPE for the location and type of test. Verify the test instrument (DMM or ohmmeter) has the minimum CAT III rating for the measuring location.

1. Set the meter to measure AC voltage for an AC circuit or DC voltage for a DC circuit. Test the meter on a known energized source before taking a measurement to verify that the meter is in proper working condition.
2. Measure the voltage into the top of the fuse (line side) by connecting one test lead to the top of the fuse and the other test lead to another power line, neutral, or ground. Connect hot to neutral for 120 VAC, hot to hot for 240 VAC and 480 VAC, and positive (+) to negative (–) or hot to ground for DC.
3. Once a voltage measurement is taken from the top of the fuse, move one test lead to the bottom of the fuse. Move the test lead connected to the bottom of the fuse to the bottom of the other fuses in the panelboard, leaving the other test lead connected

Circuit Breaker Operation Media Clip

as is. *Note:* If the fuse is good, voltage going into the fuse is the same coming out of the fuse. If the fuse is bad (open), there is no voltage coming out of the fuse.

4. Retest the test instrument on the known energized source to verify that the meter is still working properly.

Note: Fuses open for a reason. Before replacing an open fuse, test the circuit for fault and ground conditions. Retest the circuit by measuring the circuit load using an ammeter after repairs are completed. When measuring the circuit load, turn on all loads that would normally be on at any one time in the circuit. This test is required if a replaced fuse continues to open.

Testing Circuit Breakers

Circuit breakers perform the same function as fuses and are connected into circuits in the same manner as fuses. An open circuit breaker can be identified by the movement of the circuit breaker switch lever away from the ON position. Some circuit breakers also include an indicator window that is red when the circuit breaker is tripped. The indicator window is either white or black when it is not tripped. When a circuit is overloaded, the circuit breaker trips open, but the red in the indicator window may not be visible. When the circuit has a short circuit (very high fault current), the circuit breaker opens and the red is visible in the indicator window.

Circuit breakers that are suspected of having a problem can be tested in the same manner as fuses using a DMM or voltage tester. **See Figure 5-13.** Circuit breakers that are connected into a circuit are tested using the following procedure:

Note: Always wear proper PPE for the location and type of test. Verify the test instrument (DMM or voltage tester) has the minimum CAT III rating for the measuring location.

1. Set the meter to measure AC voltage for an AC circuit or DC voltage for a DC circuit. Test the meter on a known energized source before taking a measurement to verify that the meter is in proper working condition.

2. Measure the voltage into the top of the circuit breaker (line side) by connecting one test lead to the top of the circuit breaker and the other test lead to another power line, neutral, or ground. Connect hot to neutral for 120 VAC, hot to hot for 240 VAC and 480 VAC, and (+) to (−) or hot to ground for DC.

3. Once there is a voltage measurement into the top of the circuit breaker, move one test lead to the bottom of the circuit breaker. Move the test lead connected to the bottom of the circuit

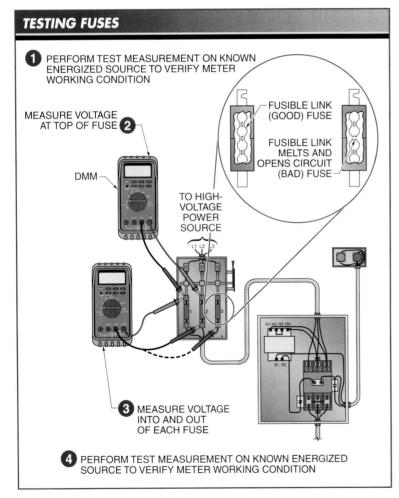

Figure 5-12. Fuses are tested with an ohmmeter or DMM with a continuity function.

breaker to the bottom of the other circuit breakers in the panelboard, leaving the other test lead connected as is. *Note:* If the circuit breaker is good, the voltage going into the circuit breaker is the same as the voltage coming out of the circuit breaker. If the circuit breaker is bad (open), there is no voltage coming out of the circuit breaker.

4. Retest the test instrument on the known energized source to verify that the meter is still working properly.

Harmonic distortion and other power quality problems can cause false tripping of circuit breakers. If the measured current over time does not indicate that the circuit is overloaded, but the circuit breaker trips, the voltage drop across the circuit breaker should be measured. If there is a high voltage drop (more than 100 mV), the circuit breaker is malfunctioning. If the voltage drop is low, there may be high harmonics or transient voltages on the line.

Note: Circuit breakers open for a reason. The circuit breaker should be reset only after the circuit is tested for fault and ground conditions. The circuit should be retested by measuring the circuit load using an ammeter after repairs are completed. When measuring the circuit load, all loads that would normally be on at any one time in the circuit should be turned on. This test is required if a reset circuit breaker continues to trip.

TECH FACT

All circuit breakers have a current rating, which should be listed on the circuit breaker switch lever and/or clearly marked. The conductor (wire) connected to the breaker should have an amperage rating equal to the breaker current rating. A conductor with a higher current rating than the circuit breaker can be used, but a conductor with a lower current rating than the circuit breaker must not be used.

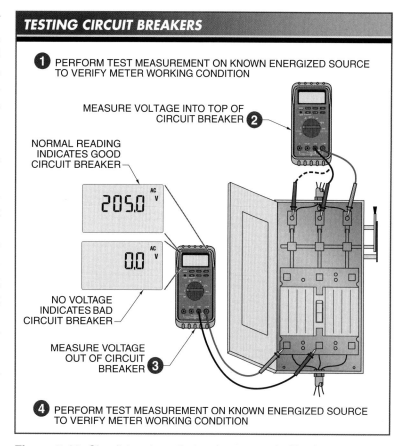

Figure 5-13. Circuit breakers that are suspected of having a problem can be tested using a DMM or voltage tester.

TEMPERATURE PROBLEMS

Heat is produced in electrical systems when current flows through a resistance. The higher the resistance of an electrical connection, conductor, etc., the greater the heat produced. *Ambient temperature* is the temperature of the air surrounding a piece of equipment. *Temperature rise* is the increase of equipment temperature over ambient temperature after the equipment is energized and loaded. A temperature rise above ambient temperature is expected on electrical equipment carrying current, but the higher the temperature rise, the greater the possibility of equipment failure.

Temperature rises of 50°F (28°C) must be investigated. Temperature rises of 100°F (56°C) or more require immediate action, such as shutting down the system and repairing the fault. Infrared (IR) thermometers and thermal imagers can be used to identify problems in building power distribution systems without making physical contact with the equipment to be tested. **See Figure 5-14.**

POWER QUALITY PROBLEMS

Electrical power distribution systems must deliver quality power to loads if the loads are to operate properly for their rated life and performance. Good power quality is power delivered to a load that is within the load specified voltage, is capable of delivering enough current under any operating condition, and includes minimal, not damaging changes such as voltage drops, voltage unbalance, voltage fluctuations, current unbalance, transients, and harmonic distortion. Poor power quality is power delivered to a load that includes excessive or damaging changes such as voltage drops, voltage unbalance, voltage fluctuations, current unbalance, transients, and harmonic distortion. **See Figure 5-15.**

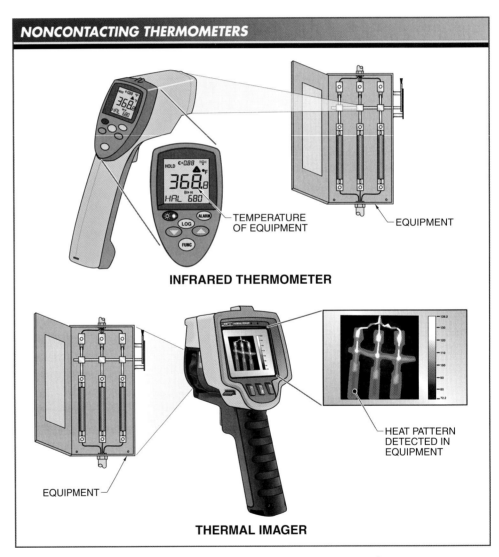

Figure 5-14. Infrared (IR) thermometers and thermal imagers can be used to identify problems in building power distribution systems without making physical contact with the equipment to be tested.

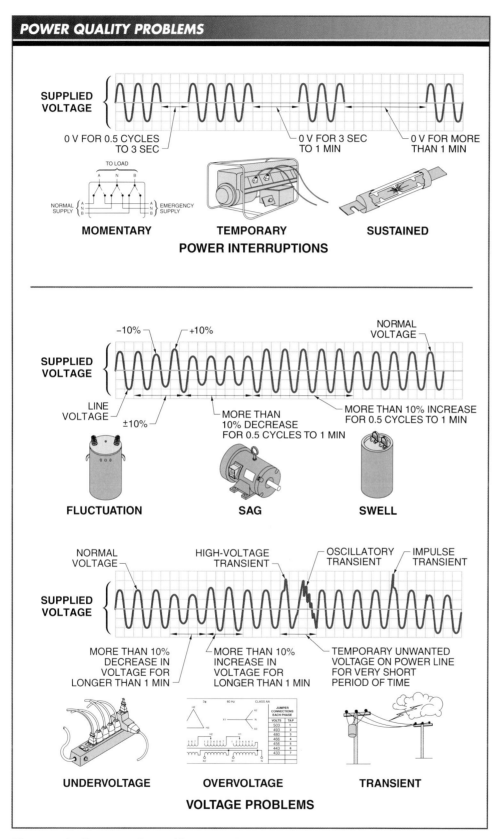

Figure 5-15. Power quality problems can damage electrical equipment and lead to unsafe operating conditions.

Proper PPE must always be worn when taking test measurements in an electrical enclosure.

Voltage Drop

Electrical and electronic equipment are rated for operation at a specific voltage. If the operating voltage is too high, the equipment can be damaged. If the operating voltage is too low, loads may unintentionally turn off. When loads unintentionally turn off, problems can be caused such as computers losing data and increased energy consumption through the restarting of electrical equipment. Power can be lost for reasons such as momentary utility power interruptions, voltage sags, the energizing of large current-drawing loads, and tripped fuses or circuit breakers. Such problems are common and easily corrected. However, loads may also be unintentionally turned off when circuit runs are too long, the conductors are too small, or the circuit is overloaded.

All circuit conductors have a voltage drop across them. The longer the conductor run and/or the higher the circuit current, the higher the voltage drop. At no time should a circuit have more than a 3% voltage drop from the start of the circuit (panel) to the farthest point. **See Figure 5-16.** A circuit percentage voltage drop is determined using the following procedure:

Note: Always wear proper PPE for the location and type of test. Verify the test instrument (DMM or voltage tester) has the minimum CAT III rating for the measuring location.

1. Set the meter to measure AC voltage for an AC circuit or DC voltage for a DC circuit. Test the meter on a known energized source before taking a measurement to verify that the meter is in proper working condition.
2. Measure the voltage at the start of the circuit (in circuit breaker panel) with all circuit loads turned off.
3. Measure circuit voltage at the farthest point away from the circuit breaker panel with circuit loads turned off (no load).
4. Measure circuit voltage at the farthest point away from the circuit breaker panel with all circuit loads turned on (full load).
5. Divide the voltage reading from the measurement taken with all loads on by the voltage reading from the measurement taken with no loads on to calculate the percentage voltage drop between no load and full load conditions at the furthest circuit point.
6. Divide the voltage reading from the measurement taken with all loads on (full load) by the voltage reading from the measurement taken at the start of the circuit to calculate the percentage voltage drop across the entire circuit.
7. If either answer in step 5 or 6 is greater than 3%, check the circuit for improper conductor size, too long of a circuit run, and too many loads on the circuit (even if the circuit breaker does not trip).
8. Retest the test instrument on the known energized source to verify that the meter is still working properly.

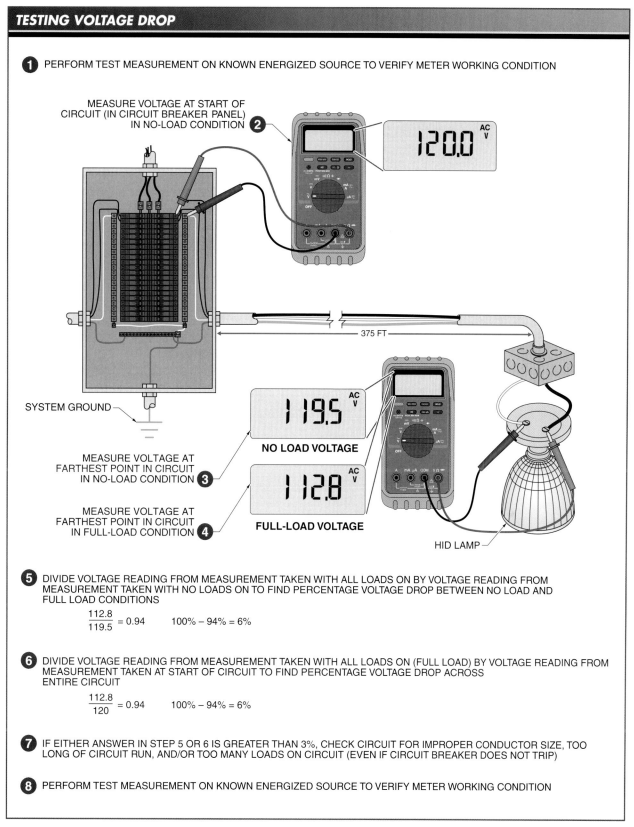

Figure 5-16. At no time should an electrical circuit have more than a 3% voltage drop from the start of the circuit (panel) to the farthest point.

Voltage Unbalance

Voltage unbalance, also known as voltage imbalance, is the unbalance that occurs when voltage at the terminals of an electric motor or other three-phase load are not equal. Voltage unbalance causes motor windings to overheat, resulting in thermal deterioration of the windings. When a 3φ motor fails due to voltage unbalance, one or two of the stator windings become blackened. **See Figure 5-17.**

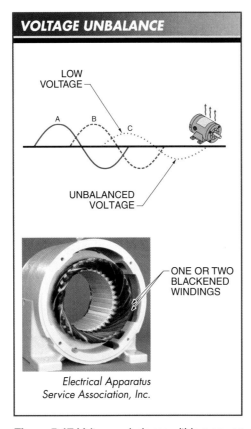

Figure 5-17. Voltage unbalance within a power distribution system can cause high current unbalance in loads such as electric motors.

TECH FACT

Electrical unbalances raise the amperage and temperature of electrical devices, such as three-phase motors. Small-horsepower three-phase motors are more sensitive to electrical unbalances than high-horsepower three-phase motors. Standard-efficiency three-phase motors are more sensitive to electrical unbalances than high-efficiency three-phase motors.

The problem with voltage unbalance within a power distribution system is that a small amount of voltage unbalance can cause a high current unbalance in loads such as electric motors. In general, voltage unbalance should not be more than 1%. Whenever there is a 2% or greater voltage unbalance, corrective action should be taken. This may include repositioning loads to balance the current draw on the three power lines if the problem is within the building. The problem is within the building if the unbalance deteriorates when loads are on and improves when loads are off. If the unbalance is at the main power entrance at all times, the problem is most likely with the utility system and the utility company should be notified. **See Figure 5-18.** A circuit voltage unbalance is tested using the following procedure:

Note: Always wear proper PPE for the location and type of test. Verify the test instrument (DMM or voltage tester) has the minimum CAT III rating for the measuring location.

1. Set the meter to measure AC voltage for an AC circuit or DC voltage for a DC circuit. Test the meter on a known energized source before taking a measurement to verify that the meter is in proper working condition.
2. Measure the voltage on each of the incoming power lines.
3. Add all the voltage measurements together.
4. Calculate the voltage average by taking the sum of the voltage measurements and dividing by the number of measurements taken.
5. Calculate the largest voltage deviation by subtracting the single lowest voltage measurement from the voltage average.
6. Calculate the voltage unbalance by dividing the largest voltage deviation by the voltage average and multiplying by 100.

7. Retest the test instrument on the known energized source to verify that the meter is still working properly.

Current Unbalance

Current unbalance, also known as current imbalance, is the unbalance that occurs when current on each of the three power lines of a three-phase power supply are not equal. Current unbalances from overloading one or two of the three-phase power lines can cause voltage unbalances. This can cause voltage unbalances on all loads connected within the building. A 2% voltage unbalance can cause an 8% or higher current unbalance.

TECH FACT

Unless otherwise marked, the power supply phase arrangement of the busbars in switchboards, panelboards, and motor control centers should be phase A, phase B, and phase C as viewed from front to back, top to bottom, or left to right.

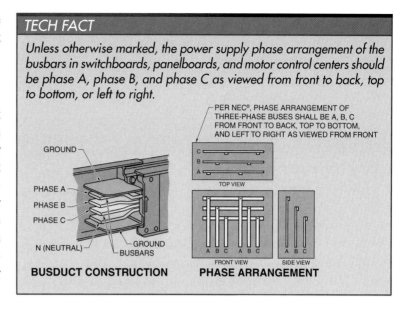

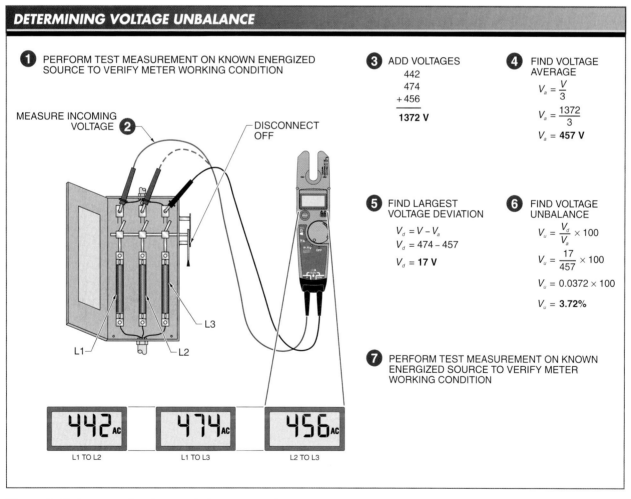

Figure 5-18. In general, voltage unbalance should not be more than 1%. Whenever there is a 2% or greater voltage unbalance, corrective action should be taken.

Current unbalances should not exceed 10%. Any time current unbalance exceeds 10%, the system should be tested for voltage unbalance. Likewise, any time a voltage unbalance is more than 1%, the system should be tested for a current unbalance. Current unbalance is determined in the same manner as voltage unbalance, except that current measurements are used. **See Figure 5-19.** A circuit percentage current unbalance is tested using the following procedure:

Note: Always wear proper PPE for the location and type of test. Verify the test instrument (clamp-on ammeter or DMM with current clamp-on attachment) has the minimum CAT III rating for the measuring location.

1. Set the meter to measure AC current for an AC circuit or DC current for a DC circuit. Test the meter on a known energized source before taking a measurement to verify that the meter is in proper working condition.
2. Measure current on each of the incoming power lines.
3. Add all current values together.
4. Calculate the current average by taking the sum of current measurements and dividing by the number of measurements taken.

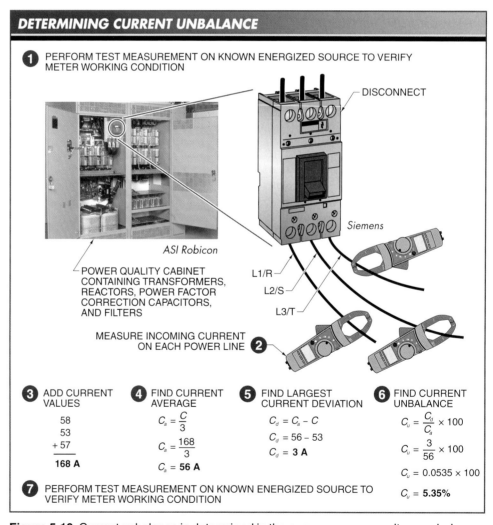

Figure 5-19. Current unbalance is determined in the same manner as voltage unbalance, except that current measurements are used.

5. Calculate the largest current deviation by subtracting the single lowest current measurement from the current average.
6. Calculate the current unbalance by dividing the largest current deviation by the current average and multiplying by 100.
7. Retest the test instrument on the known energized source to verify that the meter is still working properly.

Improper Phase Sequence

Three-phase circuits include three individual ungrounded (hot) power lines. The three power lines are referred to as phase A (L1/R), phase B (L2/S), and phase C (L3/T). Phases A, B, and C must be connected to switchboards and panelboards per NEC® requirements. Unless otherwise marked, the power supply phase arrangement of the busbars in switchboards, panelboards, and motor control centers should be phase A, phase B, and phase C as viewed from front to back, top to bottom, or left to right. The phase sequence of power lines can be verified using a phase sequence tester. **See Figure 5-20.** The phase sequence of power lines is tested using the following procedure:

Note: Always wear proper PPE for the location and type of test. Verify the test instrument (phase sequence tester) has the minimum CAT III rating for the measuring location.

1. Connect the three test leads from the phase sequence tester to the three power lines being tested. Connect phase A test lead to what should be power line phase A (L1/R), phase B test lead to B (L2/S), and phase C test lead to C (L3/T). *Note:* The test leads on a phase sequence tester are typically color coded and include alligator clips.
2. Verify that all three, three-phase indicator lights are ON. If one or more light(s) is not on, use a voltmeter to find why the phase is not powered (for example, a fuse may be blown). When all phase indicator lights are on, there is no open phase.
3. Check the phase sequence lights. *Note:* When the phases are in the correct order (test leads connected to the system A to A, B to B, and C to C) the phase sequence tester L1, L2, and L3 lights are ON.
4. Remove the phase sequence tester from the circuit.

Harmonic Distortion

Electrical loads are used to produce heat, light, sound, and motion. The two basic types of electrical loads are linear loads and nonlinear loads. A *linear load* is any load in which current increases proportionately as voltage increases and current decreases proportionately as voltage decreases. Linear loads have been used since the first electrical circuits were in service. In an AC circuit that includes a linear load, voltage and current are both a smooth undistorted sine wave (called a sinusoidal waveform) even when both voltage and current are out of phase. A *sinusoidal waveform* is a waveform that is not distorted during its rise from zero to peak value and back down to zero.

The Lincoln Electric Company
An AC arc welding machine is an example of a nonlinear load.

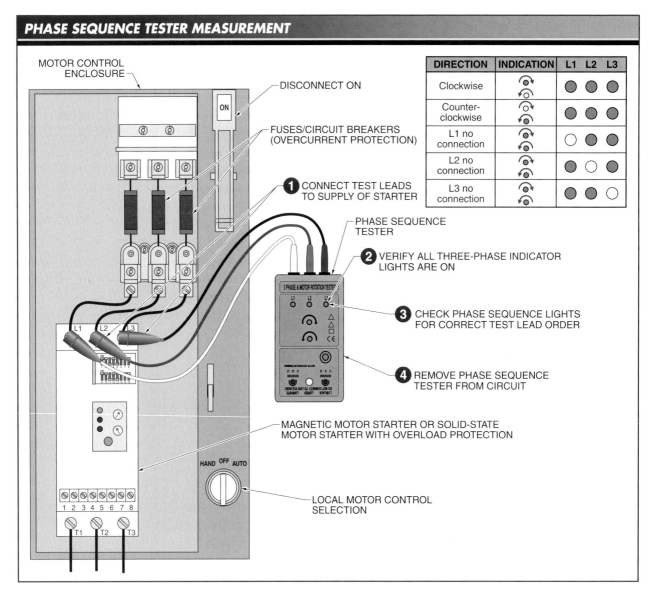

Figure 5-20. The phase sequence of power lines can be verified using a phase sequence tester.

Pure resistive, inductive, and capacitance loads, such as incandescent lamps, heating elements, motors, alarms (bells and horns), solenoids, and relay coils, are linear loads. Linear loads only cause problems on a power distribution system when shorted out, when oversized for the system, or when they are not operating properly. Linear loads do not cause problems such as harmonic distortion.

A *nonlinear load* is any load in which the current is not a pure proportional sine wave because current is drawn in short pulses rather than a smooth, even sine wave. Nonlinear loads became popular in the 1980s and have grown in number and percentage of total loads. Nonlinear loads include personal computers, copiers, printers, televisions, DVD players, electronic lighting ballasts, energy-efficient lamps, electric welding equipment, most medical equipment, electric motor drives, uninterruptible power supplies (UPSs), and programmable logic

Chapter 5—Electrical Distribution Systems 157

controllers (PLCs). **See Figure 5-21.** With nonlinear loads, the voltage is still a pure undistorted sine wave, but the current has been distorted. Nonlinear loads cause power quality problems because nonlinear loads produce harmonic distortion. Nonlinear loads cause electrical problems within commercial buildings because of their increased use.

TECH FACT
Power quality in buildings is usually best at the service entrance and gradually worsens as it is distributed to individual loads because of poor grounding, loads switching on and off, loose splices, undersized conductors, and loads on the lines that cause transients and harmonic distortion.

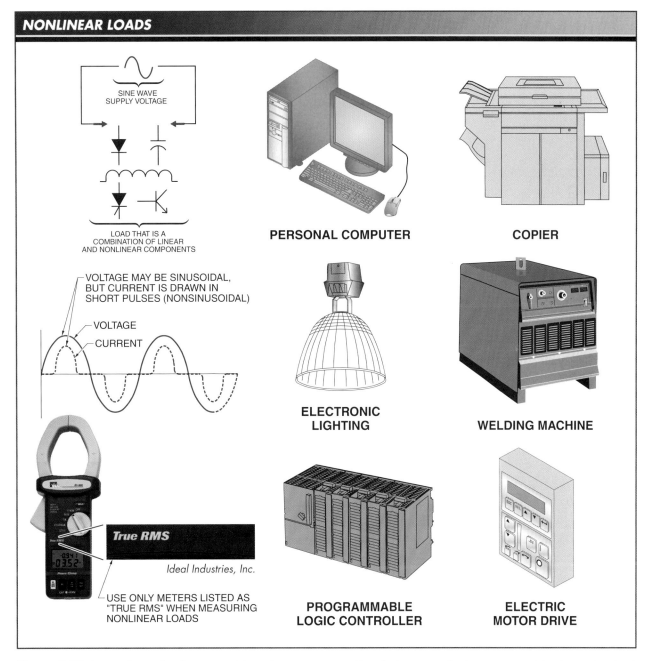

Figure 5-21. In nonlinear loads, current is not a pure proportional sine wave because current is drawn in short pulses.

Problems caused by harmonic distortion include overheated neutral conductors, transformer failure, voltage sags, and flat-topping of the voltage waveform (sine wave turning into a square wave). The problem with overheated neutral conductors is that they are a fire hazard because neutral conductors never have fuses or circuit breakers to open the line when the current exceeds the conductor amperage rating.

Voltage sags and flat-topping cause computers and other electronic equipment to continually reset due to the insufficient peak voltage needed to recharge their capacitors. The problem with transformer failure is the cost of replacement and downtime. Harmonic filters are used to reduce harmonic distortion, which is often identified by the noise that it creates in a panelboard.

Overheating of Neutral Conductors. Because current cancels out on the neutral conductor when 60 Hz linear loads are used, it is common practice in commercial building construction to run three hot conductors (phases A, B, and C) and one neutral conductor (all the same conductor size) with 120/208 V, three-phase, 4-wire, wye distribution systems. This is the most common wiring method used in commercial buildings because it is the best for distributing large amounts of 120 V, single-phase power required in these buildings. In this type of wiring method, the neutral conductor is the same size as the three hot conductors and never carries more current than any one hot conductor.

One of the major problems with high harmonic distortion on power lines is caused by the third harmonic frequency (180 Hz). This is because the third current harmonic frequency from loads does not cancel out on the neutral conductor as the fundamental frequency (60 Hz) does with linear loads. The third harmonic can cause current on the neutral conductors to be up to twice the amount of current on the hot/fused conductors. **See Figure 5-22.**

To prevent overheating of neutral conductors, an improvement to the common practice of running three hot conductors (phases A, B, and C) and one neutral conductor (all the same conductor size) is made by installing a neutral conductor twice the size of the hot conductors, which doubles its current-carrying capacity. To reduce harmonic distortion and overheating problems, however, the best type of wiring method is one in which each circuit has its own neutral conductor (no shared neutrals). **See Figure 5-23.** Circuits with harmonic distortion causing overloaded neutral problems are tested using the following procedure:

Note: Always wear proper PPE for the location and type of test. Verify the test instrument (clamp-on ammeter or DMM with clamp-on current probe accessory) has the minimum CAT III rating for the measuring location.

1. Set the meter to measure AC current. Test the meter on a known energized source before taking a measurement to verify that the meter is in proper working condition.
2. Measure the current on each hot conductor (conductor from circuit breaker) and each neutral conductor. *Note:* The current on each individual circuit should be the same on both the hot and neutral conductors. The current on a shared neutral conductor should be less than the current on any one of the three hot conductors. If the current on the shared neutral is higher than on any one hot conductor, there is a harmonic distortion problem. The greater the neutral current in comparison to the highest hot conductor measurement, the greater the harmonic distortion present on the circuit.
3. Retest the test instrument on the known energized source to verify that the meter is still working properly.

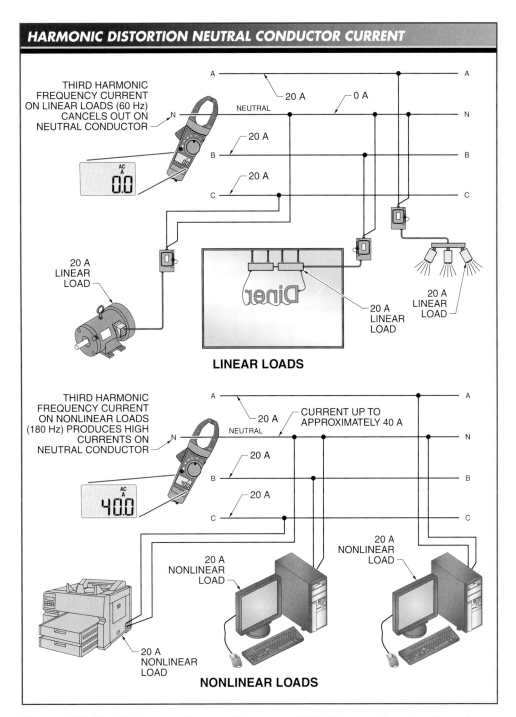

Figure 5-22. The third harmonic current frequency (180 Hz) on nonlinear loads produces high currents on the neutral conductor.

Harmonic Filters. To reduce harmonic distortion in commercial building power distribution systems and circuits, a harmonic filter can be connected to the system. A *harmonic filter* is a device used to reduce harmonic frequencies and total harmonic distortion in a power distribution system. Three-phase harmonic filters are installed between the transformer and the distribution panel and are used to reduce harmonic distortion produced by nonlinear loads connected to the system.

160 ELECTRICAL SYSTEMS for FACILITIES MAINTENANCE PERSONNEL

Harmonic filters should be installed as close as possible to the nonlinear loads, such as large motor drives. In most systems, a large system harmonic filter is installed in the main service panel. **See Figure 5-24.**

TECH FACT
Three-phase harmonic filters are also referred to as "trap filters" or "I-trap filters" with the "I" referring to current.

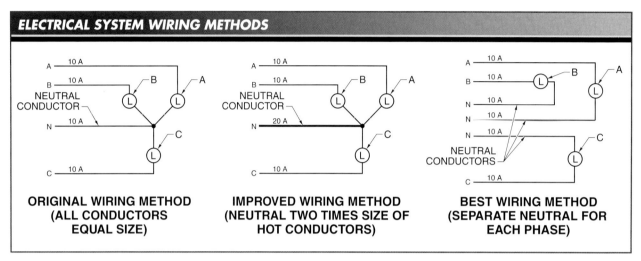

Figure 5-23. To reduce harmonic distortion and overheating problems, the best wiring method is one in which each circuit has its own neutral conductor (no shared neutrals).

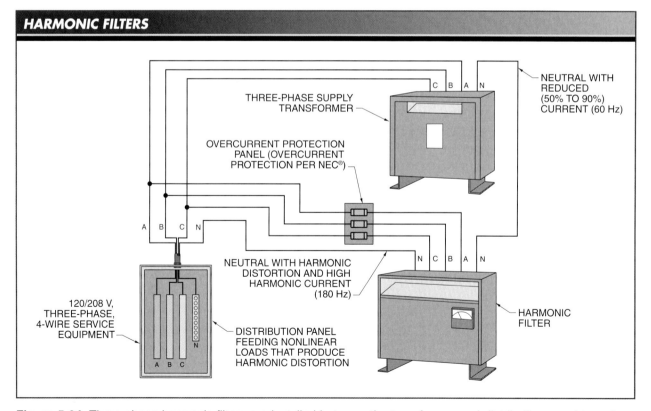

Figure 5-24. Three-phase harmonic filters are installed between the transformer and distribution panel to reduce harmonic frequencies and total harmonic distortion.

Electric Panel Noise from Harmonic Distortion. *Sound* is a series of pressure variations that produce an audible signal. An *audible signal* is any sound that can be heard. Sound is produced by a series of pressure vibrations originating from a vibrating object. Sound waves are produced and travel at different frequencies.

The 60 Hz (60 cycles per second) of a standard power frequency can be heard if the sound is amplified enough or the particular individual listening can hear 60 Hz. Perfect human hearing is in the range of 20 Hz to 20,000 Hz, but this range shortens with age and hearing damage from excessive loud noise exposure. Although the 60 Hz power frequency flowing through conductors is not heard, a noise problem can become noticeable when a 60 Hz power frequency vibrates loose mechanical parts such as the thin laminated iron strips that compose transformer coils. This noise is often the humming sound heard near large power transformers.

In addition to sounds produced by loose parts and connections within the power distribution system and components, sounds of high harmonic frequencies above 60 Hz can also cause problems because high frequencies are within the hearing range of most people. Electrical panels designed to carry 60 Hz current can become mechanically resonant to magnetic fields generated by high frequency harmonic distortion. When this occurs, the power panel starts to vibrate and emits a humming sound at the harmonic frequency.

> **TECH FACT**
> Nonlinear loads are becoming more common in commercial buildings. Always assume that a circuit contains nonlinear loads and has a separate neutral conductor for each phase back to the panel on three-phase systems feeding single-phase circuits (A-N, B-N, and C-N). Never install a shared neutral, as was the practice in the past.

To reduce noise produced by harmonic distortion, if the noise problem appears to be caused by a variable-speed motor drive (from an HVAC system), the motor drive carrier frequency can be changed by programming a new carrier frequency for the drive. *Carrier frequency* is the frequency that controls the number of times the solid-state switches in the inverter section of a drive switch on and off. Most variable-speed motor drives allow the carrier frequency to be changed to help with application-specific noise problems.

Changing the motor drive carrier frequency has advantages and disadvantages. Reducing noise is a major advantage when the noise is a problem within a building. A disadvantage is that higher carrier frequencies cause greater power losses (thermal losses) in an electric motor drive because of the high temperatures produced from faster switching and may require additional cooling at the drive such as more open space for airflow and/or derating of the drive.

TRANSFORMERS

Transformers are common pieces of electrical equipment installed throughout commercial buildings. Large power transformers are used to reduce the utility voltage into a building. Other power transformers are used to produce other required voltages such as 208 V, 240 V, or 480 V throughout a building. Control-circuit transformers are used to reduce the voltage to control circuits such as motor control circuits. Lighting transformers are used to increase the voltage for neon/fluorescent/HID lamps. Step-down transformers are used on electronic boards to lower the voltage before the voltage is rectified (changed from AC to DC) for electronic circuits. **See Figure 5-25.** Transformers installed in most commercial buildings are usually K-rated and tested for effects from overheating.

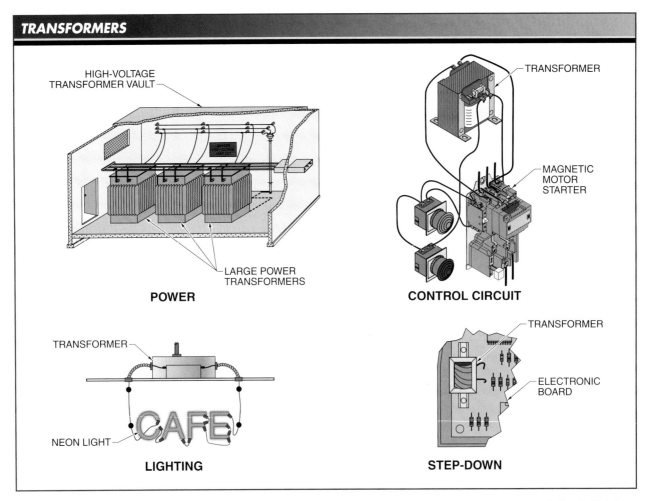

Figure 5-25. Common types of transformers used in commercial applications include power, control circuit, lighting, and step-down transformers.

K-Rated Transformers

Harmonics caused by nonlinear loads produce additional heat in transformers. In any system containing harmonic distortion, the K factor of the system is higher than 1. *K factor* is the measure of the extra heat caused by harmonic distortion in a transformer. K factor can be measured using a test instrument rated to measure K factor such as a power quality meter. Noncontact thermometers and thermal imagers can also be used to measure transformer overheating.

A *K rating* is a listed rating on a transformer that represents the ability of the transformer to operate properly with certain loads. A *K-rated transformer* is a transformer designed to withstand the extra heating effects caused by harmonic distortion. The higher the K rating of the transformer, the greater its heat dissipation capacity. K-rated transformers have a K rating listed on their nameplate, such as K-4, K-9, K-13, and K-20. If a transformer does not have a K rating listed, it is assumed to have a K rating of 1. A K-1 rated transformer has no additional harmonic distortion heat dissipation built into it. K-1 rated transformers are designed for use with linear loads. A K-1 rated transformer can be used with nonlinear loads, but must be derated to withstand the extra heat. **See Figure 5-26.**

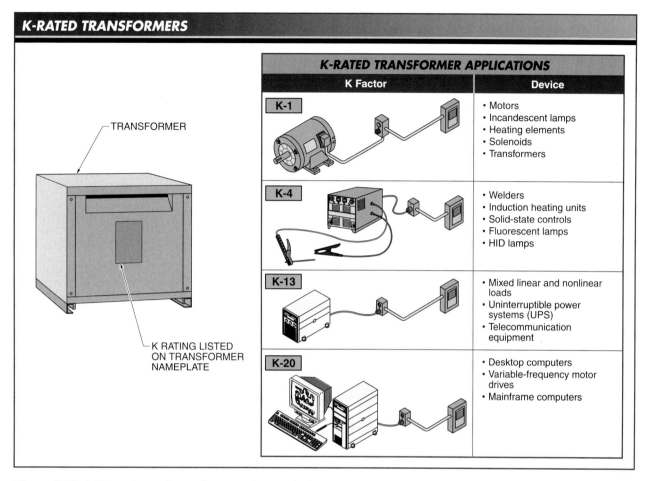

Figure 5-26. A K-rated transformer is a transformer designed to handle the extra heating effects caused by harmonic distortion.

Office buildings, schools, and other commercial buildings typically require derating of K-1 rated transformers to approximately 50% to 70% of their nominal kVA rating. When a meter that measures the K rating is not available, a clamp-on ammeter (or DMM with clamp-on current probe accessory) and thermometer can be used to approximate transformer overheating.

Transformer Test Measurements

All transformers are power supplies that are rated to continuously deliver a rated amount of power. Transformer power ratings are listed on the transformer nameplate as a kVA rating. It is possible for transformers to be overloaded for a short period of time. For this reason, it is required when taking test measurements on a transformer to take the measurements over incremental time periods. Transformer manufacturers list the length of time a transformer may be safely overloaded at a given peak level. For example, a transformer that is overloaded three times its rated current has a permissible overload time of about 6 min. **See Figure 5-27.**

When testing a transformer, both voltage and current measurements should be taken over incremental time periods. Low voltage readings on the secondary can cause equipment to drop out or reboot. Voltage on the secondary should not drop more than 10% below the transformer rated voltage output. High current readings on either the primary or secondary side of the transformer mean that the transformer is overloaded, which reduces voltage and increases transformer heating. Transformers are tested for overheating caused by harmonic distortion using the following procedure:

Note: Always wear proper PPE for the location and type of test. Verify the test instrument (clamp-on ammeter, DMM with clamp-on current probe accessory, or voltage tester) has the minimum CAT III (indoor) or CAT IV (outdoor) rating for the measuring location.

1. Set the meter to measure AC voltage and AC current. Test the meter on a known energized source before taking a measurement to verify that the meter is in proper working condition.

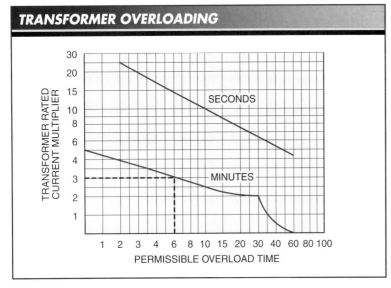

Figure 5-27. Because transformers can be overloaded for short periods, it is important when taking measurements to take them over incremental time periods.

2. Use the proper meter to measure the secondary winding voltage of the transformer over a predetermined time period. If there is an excessively low or excessively high voltage problem, also measure the transformer primary winding voltage over a predetermined time period. *Note:* If the voltage reading on the secondary winding is low and at the same time is correct on the primary winding, the transformer is overloaded. If the voltage reading is low on both the primary and secondary windings at the same time, the problem is upstream from the transformer.

WARNING: Individuals should never measure a transformer primary winding voltage over 480 VAC unless they have been trained in taking high-voltage measurements and are wearing the proper PPE. They must also use meters rated for the location/application and work with another qualified individual.

3. Determine the maximum amount of current the transformer secondary winding should draw. *Note:* The maximum amount of current the transformer secondary winding should draw is calculated by using the following formula:

$$I = \frac{P_A \times 1000}{V \times 1.732}$$

where

I = maximum current output on transformer primary (in A)

P_A = nameplate power rating of transformer (in kVA)

1000 = constant

V = voltage rating of transformer secondary output (in V)

1.732 = three-phase constant

Example: What is the maximum current of a 30 kVA transformer with a 480 V rated secondary winding at 108 A?

$$I = \frac{P_A \times 1000}{V \times 1.732}$$

$$I = \frac{30 \times 1000}{480 \times 1.732}$$

$$I = \frac{30,000}{831.36}$$

$$I = \mathbf{36.09\ A}$$

4. Calculate the transformer load. Use the proper test instrument to measure the current on the secondary winding of the transformer and compare it with the calculated value. Use the following formula to calculate the amount of load on a transformer:

$$T_L = \frac{I_M}{I_C} \times 100$$

where
T_L = transformer load (in %)
I_M = measured current (in A)
I_C = rated current (in A)
100 = conversion factor

Example: What is the load on a transformer with a measured current value of 25 A and a rated current value of 36.09 A?

$$T_L = \frac{I_M}{I_C} \times 100$$
$$T_L = \frac{25}{36.09} \times 100$$
$$T_L = 0.6927 \times 100$$
$$T_L = \mathbf{69.27\%}$$

5. Retest the meter on the known energized source to verify that the meter is still working properly.

If the current is higher on either side of the transformer than the rated value, the transformer is overloaded. If the transformer is overloaded, measurements should be taken over intermittent time periods (hours or days) to determine if the transformer is constantly overloaded or only overloaded at certain time periods.

Control-Circuit Transformer Testing. A *control-circuit transformer* is a transformer that is used to step down the voltage to the control circuit of a system or machine. After a transformer is installed in a circuit, it may operate without failure for a long time. One reason for this is that transformers have no moving parts. If a transformer does fail, it appears as either a short circuit or an open circuit in one of the coils. The two methods that can be used to determine if a control circuit transformer has failed are to measure the input and output voltages and to check the transformer resistance.

If a transformer is connected in a circuit, the transformer can be tested by measuring the input and output voltages. The transformer is good if the input and output voltages are reasonably close to the theoretical values. The current levels are tested if the voltage does not stay constant. Although the initial voltage may appear normal, it may not hold up when the transformer is fully loaded.

A DMM set to measure resistance can be used to check for open circuits in coils, short circuits between coils, or coils shorted to the core without power applied to the transformer. **See Figure 5-28.**

When checking for open circuits in coils, the resistance of each coil is checked with a DMM. The winding is open and the transformer is bad if any of the coils show an infinite resistance reading. Very low resistance readings do not indicate a short, just the resistance of the wire.

When checking for short circuits between primary and secondary coils, a DMM should show a low or 0 Ω resistance reading between the primary and secondary of shorted coils.

When checking for coils shorted to core, a resistance check is done from each transformer coil to the core of the transformer. All coils should show an infinite resistance reading. The transformer should not be used if a resistance is shown between any coil and the core.

Refer to Chapter 5 Quick Quiz® on CD-ROM.

TECH FACT
Most transformers can be built with additional features to help meet individual application requirements. Additional features add to the cost, but can solve specific problems. Additional features of a transformer include stainless steel enclosures, fungus proofing, sound reduction, double neutral conductors, increased energy efficiency ratings, increased temperature ratings, and electrostatic shielding.

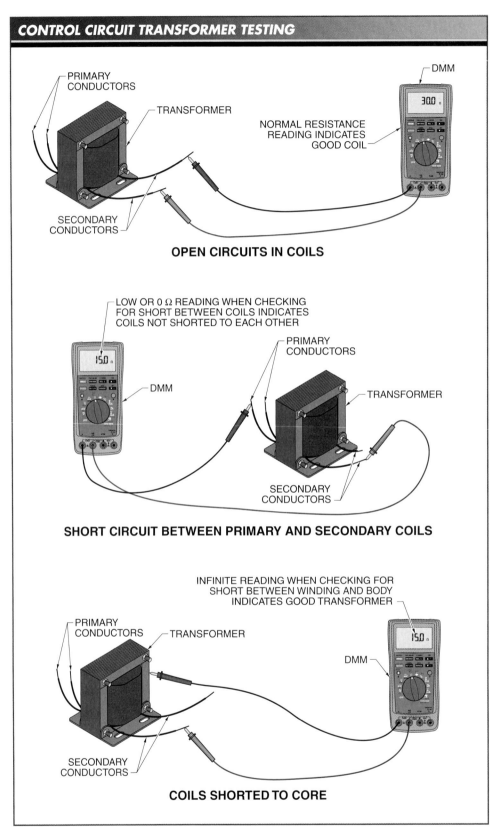

Figure 5-28. Control-circuit transformers are tested by checking for open circuits in the coils, short circuits between the primary and secondary coils, and coils shorted to the core.

Case Study—Checking Motor Voltage and Current Operating Condition

All motors are used to produce work. Within a building, some motors perform critical functions, such as operating water pumps in fire suppression systems. Other motors perform important functions, such as maintaining required HVAC system pressure and airflow. Motors that perform critical and important functions must be periodically tested to ensure they are working properly and to determine if any signs of future problems are present.

Voltage and current measurements can be used to determine if a motor is operating within its nameplate-listed specifications. Voltage or current unbalance measurements can be used to determine if any potential problems are present in a motor and can also indicate the operating condition of the motor.

A service call requires testing the operating condition of a sprinkler system pump motor to ensure it is still operating within electrical specifications. The motor is a dual voltage (230/460 V), 2 HP motor connected for 230 V.

The motor disconnect switch box is opened and the disconnect switch is placed in the on position so voltage and current measurements can be taken. While the motor is on, voltage measurements are taken from L1 to L2, L2 to L3, and L1 to L3. The voltage measurements taken and recorded are as follows:

L1 to L2 = 223.5 V
L2 to L3 = 224.8 V
L1 to L3 = 221.9 V

The measurements indicate a 223.4 V average [(223.5 + 224.8 + 221.9) ÷ 3 = 223.4], which indicates the voltage is a little low but within 3% of the motor nameplate rating of 230 V [223.4 ÷ (230 × 100) = 97.1%, 100% − 97.1% = 2.9%].

The measurements also indicate that the voltage unbalance is 0.67% [223.4 − 221.9 = 1.5 V, 1.5 ÷ (223.4 × 100)], which is less than the 2%

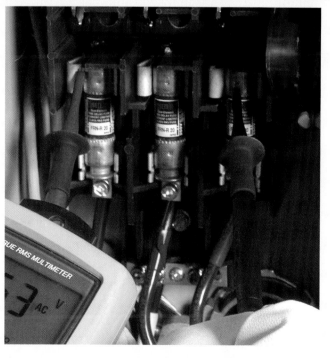

or more unbalance that may indicate a problem within the motor or the power distribution system.

However, even a 0.67% voltage unbalance could cause a higher current unbalance. Thus, the current draw of the motor is measured and compared to the motor nameplate-listed current of 230 V. These values are

Case Study (continued)

used to determine the motor current unbalance. The current measurements taken and recorded are as follows:

L1 = 6.22 A
L2 = 6.31 A
L3 = 6.14 A

The measurements indicate a 6.22 A average [(6.22 + 6.31 + 6.14) ÷ 3 = 6.22], which indicates that the current is less than the nameplate-rated current value of 6.8 A at 230 V. This indicates that the motor is operating at less than full rated power (HP), which is good.

The measurements also indicate that the current unbalance is 1.45% [6.31 − 6.22 = 0.09, 0.09 ÷ (6.22 × 100) = 1.45%], which is far less than the 10% or more unbalance that may indicate a problem within the motor or the power distribution system.

The motor electrical findings are recorded, and the motor is given additional mechanical checks to verify the motor alignment is good and the motor is clean and lubricated per its lubrication schedule.

Definitions

- *Ambient temperature* is the temperature of the air surrounding a piece of equipment.

- An *audible signal* is any sound that can be heard.

- *Carrier frequency* is the frequency that controls the number of times the solid-state switches in the inverter section of a drive switch on and off.

- A *circuit breaker* is an overcurrent protection device with a mechanical mechanism that may manually or automatically open a circuit when an overload condition or short circuit occurs.

- A *control-circuit transformer* is a transformer that is used to step down the voltage to the control circuit of a system or machine.

- *Current unbalance,* also known as current imbalance, is the unbalance that occurs when current on each of the three power lines of a three-phase power supply are not equal.

- An *equipment grounding conductor (EGC)* is an electrical conductor that provides a low-impedance ground path between electrical equipment and enclosures within a distribution system.

- A *fuse* is an overcurrent protection device with a fusible link that melts and opens the circuit when an overload condition or short circuit occurs.

- A *grounded conductor* is a conductor that has been intentionally grounded, and is typically known as the neutral conductor.

- A *grounding electrode conductor (GEC)* is a conductor that connects grounded parts of a power distribution system to an NEC® approved earth grounding system.

- A *ground* is an intentional or accidental electrical connection between a circuit or equipment and the earth or other conducting member.

- A *harmonic filter* is a device used to reduce harmonic frequencies and total harmonic distortion in a power distribution system.

- *K factor* is the measure of the extra heat caused by harmonic distortion in a transformer.

- A *K-rated transformer* is a transformer designed to withstand the extra heating effects caused by harmonic distortion.

- A *K rating* is a listed rating on a transformer that represents the ability of the transformer to operate properly with certain loads.

- A *linear load* is any load in which current increases proportionately as voltage increases and current decreases proportionately as voltage decreases.

- A *main bonding jumper (MBJ)* is a connection at the service equipment that connects the equipment grounding conductor, grounding electrode conductor, and grounded conductor (typically the neutral conductor).

- A *nonlinear load* is any load in which the current is not a pure proportional sine wave because current is drawn in short pulses rather than a smooth, even sine wave.

- A *receptacle,* also known as an outlet, is a device used to connect equipment with a cord and plug to an electrical system.

- A *separately derived system (SDS)* is a system that supplies a facility with electrical power derived, or taken, from a transformer, storage batteries, a photovoltaic system, or a generator.

- A *sinusoidal waveform* is a waveform that is not distorted during its rise from zero to peak value and back down to zero.

- *Sound* is a series of pressure variations that produce an audible signal.

- *Temperature rise* is the increase of equipment temperature over ambient temperature after the equipment is energized and loaded.

- *Voltage unbalance,* also known as voltage imbalance, is the unbalance that occurs when voltage at the terminals of an electric motor or other three-phase load are not equal.

Review Questions

1. Why must fuses and circuit breakers be connected in series between the loads and the main power lines feeding the panel?
2. What is the purpose of testing a 120/240 V, single-phase receptacle?
3. Why must the voltage between every slot of a three-phase receptacle be tested?
4. List the effects of poor power quality being delivered to a load.
5. What is the main advantage of using an infrared thermometer or thermal imager when troubleshooting building power distribution systems?
6. What is the main effect that voltage unbalance has on an electric motor?
7. List the procedural steps for testing fuses.
8. What are the differences between a linear load and a nonlinear load?
9. List the procedural steps for performing phase testing of power lines.
10. What are the advantages of using isolated-ground receptacles with sensitive electronic equipment?
11. Describe the best wiring method used to reduce harmonic distortion and overheating problems.
12. Why is an overheated neutral conductor considered a fire hazard?
13. What is the main disadvantage of increasing the carrier frequency of a motor drive?
14. Under what conditions can a K-1 rated transformer be used with a nonlinear load?
15. When checking coils in control circuit transformers with a DMM, what does an infinite reading indicate?

chapter 6
LIGHTING SYSTEMS

Lighting is one of the most important aspects of any building. Proper lighting can increase productivity, increase safety, and affect environmental conditions in a specific area. Poor lighting leads to unsafe conditions and reduced productivity.

Lighting requires energy and can account for the highest percentage of electrical usage, depending on a building's geographical location, intended use, and time of occupancy. Lamps used to produce light have varying efficiencies. Installation, troubleshooting, and maintenance procedures for lighting systems require complete knowledge and understanding of light sources, systems, and lighting control devices such as switches, sensors, and contactors.

OBJECTIVES

- Identify the different types of light sources.
- Test lighting system ballast voltage.
- Identify the different types of special-purpose lighting.
- Troubleshoot manual lighting switches.
- Identify the different types of lighting control devices.

LIGHT

Light is the portion of the electromagnetic spectrum that produces radiant energy. The electromagnetic spectrum ranges from cosmic rays with extremely short wavelengths to electric power frequencies with extremely long wavelengths. Light can be in the form of visible light or invisible light. *Visible light* is the portion of the electromagnetic spectrum to which the human eye responds. Visible light includes the section of the electromagnetic spectrum that ranges from violet to red light. *Invisible light* is the portion of the electromagnetic spectrum on either side of the visible light spectrum. Invisible light includes ultraviolet and infrared light. **See Figure 6-1.**

The color of light is determined by its wavelength. Visible light with the shortest wavelengths produces the color violet. Visible light with the longest wavelengths produces the color red. Wavelengths between violet and red produce the colors blue, green, yellow, and orange depending on the exact wavelength of the light. The combination of colored light produces white light when a light source, such as the sun, produces energy over the entire visible spectrum in approximately equal quantities. A nonwhite light is produced when a light source, such as a low-pressure sodium lamp, produces light that lies mostly in a narrow band of the visible spectrum, such as in the yellow-orange range. **See Figure 6-2.**

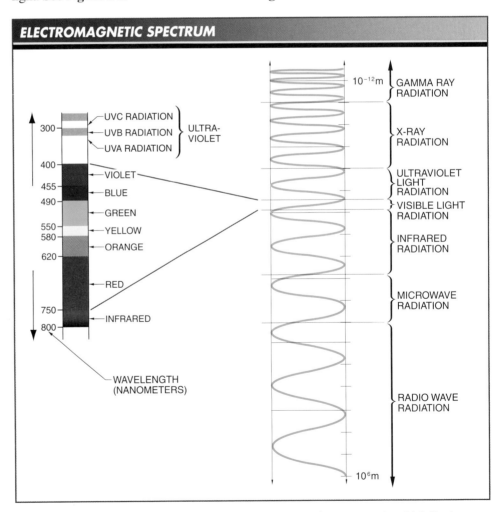

Figure 6-1. Visible light is the section of the electromagnetic spectrum to which the human eye responds and ranges from violet to red light.

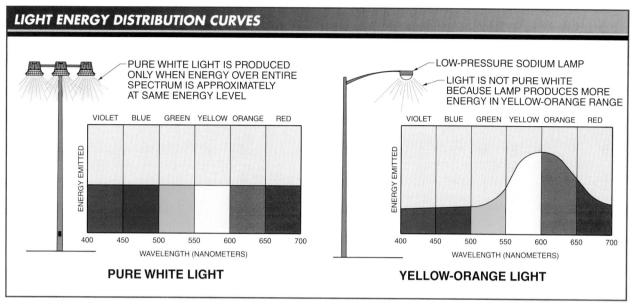

Figure 6-2. The color of light is determined by its wavelength.

TECH FACT

Color rendering is the appearance of a color when illuminated by a light source. Various lamps have different color rendering characteristics, which should be considered when selecting lamps for different applications.

A *lamp* is an output device that converts electrical energy into light. The amount of light a lamp produces is expressed in lumens. A *lumen (lm)* is the unit used to measure the total amount of light produced by a light source. For example, a standard 40 W incandescent lamp produces about 480 lm, and a standard 40 W fluorescent lamp produces about 3100 lm. Manufacturers rate lamps (light bulbs) in the total amount of light (lumens) produced by the lamp.

The light produced by a light source causes illumination. *Illumination* is the effect that occurs when light contacts a surface. The unit of measure of illumination is the footcandle. A *footcandle (fc)* is the amount of light produced by a lamp (lumens) divided by the area that is illuminated.

Light spreads as it travels farther from the light source. One lumen produces an illumination of 1 fc on a 1 sq ft area. If the amount of light falling on a surface 1′ away from the light source produces 1 fc, then the light source produces ¼ fc at a distance of 2′ and ⅑ fc at a distance of 3′. This is because the light produced at the source covers four times the area at a 2′ distance and nine times the area at a 3′ distance than it does at a 1′ distance.

The amount of light produced on a surface depends on the amount of lumens produced by the light source and the distance the surface is from the light source. The light level required on a surface varies widely. For example, an office in which regular office work is performed requires more light than an active warehouse area. The amount of light produced by a lamp is measured with a light meter using the following procedure:

1. Switch lamps on in the area where light is to be measured. *Note:* If natural light is to be measured, several measurements must be taken at different time periods over several days.
2. Turn the light meter ON.

3. Set the function switch to the measurement mode required, such as footcandle or lux.
4. Set the selector switch to expected light range, such as 40 fc or 400 fc.
5. Set the light meter to record light measurements with either MIN MAX or PEAK mode.
6. Take several light level measurements at various locations and at various angles.
7. Record each light level measurement.
8. Record any support information, such as time when measurements were taken, if shades were open or closed, if cloud cover present, etc.
9. Turn the light meter OFF.

After taking light measurements, the measurements should be compared to the recommended levels. Light measurements should be taken and recorded at different times because the amount of light can change over time, after lamps are changed, and because room layouts change. Recorded light measurements are also required before and after making any lamp changes, such as replacing incandescent lamps for light-emitting diode (LED) lamps or compact fluorescent lamps (CFLs) to conserve energy. Conservation of energy should not affect the amount of light output. **See Figure 6-3.**

LIGHT SOURCES

Lamps are the main source of light in most buildings. Outdoor commercial lighting is required for parking areas, sports complexes, airports, buildings, piers, and railroad yards. Outdoor commercial lighting systems may use low- or high-pressure sodium, metal-halide, or mercury-vapor lamps in voltages of 120 V, 208 V, 240 V, 277 V, or 480 V. Power ratings range from 35 W to over 1500 W. Indoor commercial lighting is required for offices, warehouses, assembly areas, and hospital/medical areas. Indoor commercial lighting systems may use incandescent, halogen, or fluorescent lamps in voltages of 120 V, 208 V, 220 V, 240 V, or 277 V. Tungsten-halogen lighting can be used in both indoor and outdoor locations.

Incandescent Lamps

An *incandescent lamp* is an electric lamp that produces light by means of the flow of current through a tungsten filament inside a gas-filled, sealed glass bulb. In the past, incandescent lamps were the standard lamps used in commercial and industrial applications. Today, incandescent lamps are being replaced by light-emitting diode (LED) lamps and compact fluorescent lamps (CFLs). Incandescent lamps have a low initial cost and are easily installed and serviced. However, they are not energy efficient because they produce only a few lumens per watt. Incandescent lamps are the most common lamp used in residential applications and are also used in various commercial applications. **See Figure 6-4.**

An incandescent lamp operates as current flows through the filament. A *filament* is a conductor with a resistance high enough to cause the conductor to heat. The filament glows white-hot and produces light. The filament is composed of tungsten, which limits the current to a safe operating level. Inrush current is higher than operating current because the filament has a low resistance when cold. The high inrush current, when the lamp is first switched on, is the major cause of most incandescent lamp failures. The air inside the bulb is removed before the bulb is sealed to prevent oxidation of the filament. The filament burns out quickly if oxygen is present. A mixture of nitrogen and argon gas is placed inside most incandescent bulbs to increase the life of the lamp.

RECOMMENDED LIGHT LEVELS

Interior Lighting		Exterior Lighting		Sports Lighting	
Area	fc	Area	fc	Area	fc
Assembly		**Airports**		**Baseball**	
Rough, easy seeing	30	Terminal Apron—		Outfield	100
Medium	100	Loading	2	Infield	150
Fine	500				
Auditorium		**Building Construction**		**Basketball**	
Exhibitions	30	General	10	College and professional	50
		Excavation	2	Recreational	10
Banks		**Buildings**		**Billiards**	
Lobby, general	50	Light surface	15	Recreational	30
Waiting areas	70	Dark surface	50	Tournament	50
Teller station	150				
Clothing Manufacturing		**Loading areas**	20	**Boxing Ring**	
Pattern making	50			Professional	200
Shops	100			Championship	500
Hospital/Medical		**Parking areas**		**Golf**	
Lobby	30	Industrial	2	Tee	5
Dental chair	1000	Shopping	5	Fairway	2
Operating table	2500			Green	5
				Miniature	10
Machine Shop		**Piers**		**Racing**	
Rough bench	50	Freight/Passenger	20	Auto, horse	20
Medium bench	100			Dog	30
Offices		**Railroad Yards**		**Ski Slope**	1
Regular office work	100	Switch points	2		
Accounting	150				
Detailed work	200				
Printing		**Service Station**		**Volleyball**	
Proofreading	150	Pump island area	25	Recreational	10
Color inspecting	200	Service areas	5	Tournament	20
Warehouses		**Streets**		**Tennis Courts**	
Inactive	5	Local	0.9	Recreational	15
Active	30	Expressway	1.4	Tournament	30

Figure 6-3. Recommended light levels vary based on interior, exterior, and sports lighting applications.

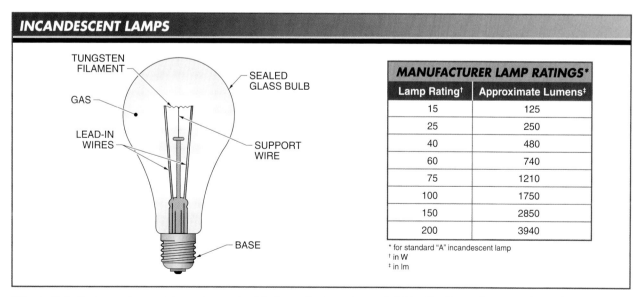

Figure 6-4. An incandescent lamp is an electric lamp that produces light by the flow of current through a tungsten filament inside a gas-filled, sealed glass bulb.

Incandescent Lamp Bases. The base of an incandescent lamp holds the lamp firmly in the lamp receptacle and connects electricity from the outside circuit to the filament. The ends of the filament are brought out to the base. The base simplifies the replacement of an incandescent lamp. Most incandescent lamp bases are threaded with a knuckle thread. A *knuckle thread* is a rounded thread normally rolled from sheet metal and used in various forms for electric bulbs and bottle caps.

Recessed incandescent fixtures are designed with an internal thermal protective device to sense excessive operating temperatures and open the circuit to the fixture.

The standard incandescent lamp base has right-hand threads that allow removal of the lamp by turning counterclockwise (CCW) and installation by turning clockwise (CW). The base of a lamp is used to hold the lamp in place and to connect the lamp to power. The size of the lamp and its current rating normally determine the type of base used. The two basic base configurations are bayonet and screw bases. Small lamps are fitted with bayonet bases. A *bayonet base* is a lamp base that has two pins located on opposite sides. The pins slide into slots located in the lamp receptacle. The advantage of a bayonet base is that it firmly holds the lamp in place.

Bayonet base lamps are used in applications that have high vibration, such as sewing machines and automobile lighting. Screw bases are used in low-vibration applications. The medium screw and mogul screw are the most common bases used for general lighting service lamps. The medium screw base is normally used for lamps up to 300 W. The mogul screw base is normally used for lamps over 300 W. The miniature and candelabra bases are normally used for low-wattage decorative lamps. **See Figure 6-5.**

Incandescent Lamp Bulb Shapes. Lamp bulb shape and size are determined by economics, aesthetics, and performance. For any wattage rating, the larger the lamp bulb, the greater the cost. Large lamp bulbs produce more light over time because a large lamp bulb produces less blackening of the glass. This is due to the large area over which the vaporized tungsten burning off of the filament can dissipate. Incandescent lamp bulbs are available in a variety of shapes and sizes. **See Figure 6-6.**

Incandescent lamp manufacturers identify different bulb shapes using letters. The most common bulb shape is the A bulb. The A bulb is used for most residential and commercial indoor lighting applications. The A bulb is normally available in sizes from 15 W to 200 W. Two common variations of the A bulb are the PS and the P bulbs. The PS bulb has a longer neck than the A bulb. The PS bulb is available in sizes from 150 W to 2000 W and is used for commercial light applications and certain industrial lighting applications. The P bulb has a shorter neck than the A bulb and is used in indoor lighting applications that require a smaller bulb than the A bulb. Other bulbs are used for decorative or special-purpose applications.

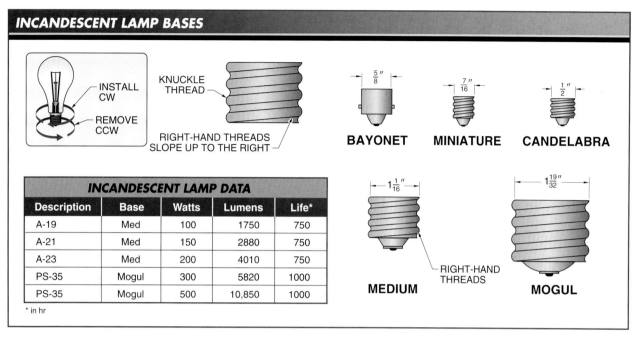

Figure 6-5. Incandescent lamp bases may be bayonet, miniature, candelabra, medium, or mogul.

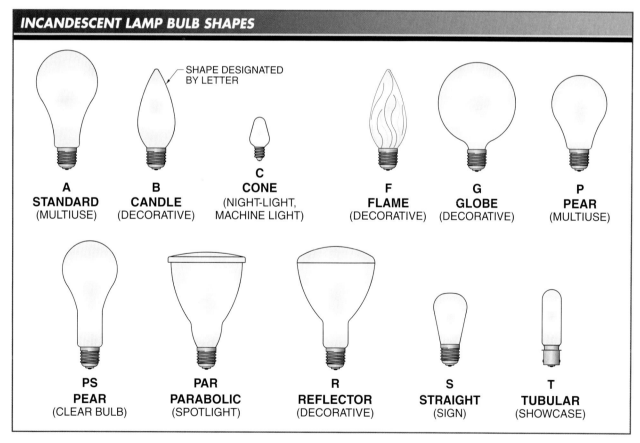

Figure 6-6. Incandescent lamp bulbs are available in a variety of shapes and sizes and are selected based on the application.

The PAR bulb is normally used as an outdoor spotlight. Floodlights normally use a reflector R bulb indoors or PAR bulb outdoors. The PAR bulb is preferred outdoors because its shape allows for a watertight seal. The R bulb is also used as a decorative light. Its high wattage is useful for applications such as showroom displays and advertising boards. The S bulb is a straight bulb that is normally used in sign applications. The T bulb is a tubular decorative light that is used in display cabinets. T bulbs containing bayonet bases are used in high-vibration applications such as automobile lighting.

Incandescent Lamp Sizes and Ratings. The size of an incandescent lamp is determined by the diameter of the lamp bulb. The diameter of a lamp bulb is expressed in eighths of an inch ($\frac{1}{8}''$). For example, an A-21 bulb is 21-eighths of an inch ($\frac{21}{8}''$) or $2\frac{5}{8}''$ in diameter at the maximum dimension of the bulb. A scale calibrated in eighths of an inch is used to measure the size of a lamp. The scale is placed at the largest diameter of the bulb. **See Figure 6-7.**

The power rating of incandescent lamps is provided in watts. A *watt (W)* is the unit of electrical power equal to the power produced by a current of 1 A across a potential difference of 1 V. Power is calculated using the following formula:

$P = E \times I$

where

P = power (in W)

E = voltage (in V)

I = current (in A)

Example: How many watts are produced by a 120 V bulb drawing 0.5 A of current?

$P = E \times I$

$P = 120 \times 0.5$

$P = \mathbf{60\ W}$

The higher the wattage rating, the greater the lumens produced. The higher the wattage rating of a lamp, the greater the lumens per watt output produced by the lamp. However, the amount of light produced by an incandescent lamp is not directly proportional to the wattage rating. **See Figure 6-8.**

Tungsten-Halogen Lamps

A *tungsten-halogen lamp* is an incandescent lamp filled with a halogen gas such as iodine or bromine. The gas combines with tungsten evaporated from the filament as the lamp burns. Tungsten-halogen lamps are used for display lighting, outdoor lighting, and in photocopiers because they produce a large amount of light instantly. A tungsten-halogen lamp lasts about twice as long as a standard incandescent lamp.

Tungsten-halogen lamp wattages range from 15 W to 1500 W. However, because they are not as widely used as the standard incandescent lamp, they are not available in as wide a selection of sizes and wattages. Because of the high amount of heat produced by a tungsten-halogen lamp, care must be taken when replacing a burned-out lamp. The lamp

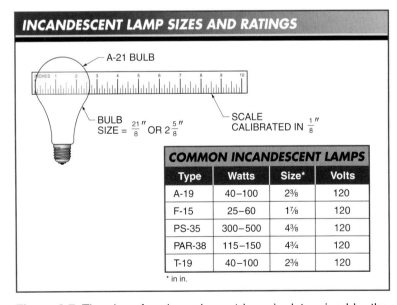

Figure 6-7. The size of an incandescent lamp is determined by the diameter of the lamp bulb.

bulb is composed of quartz to withstand the high temperature. High-wattage tungsten-halogen lamps are available in T or PAR shapes. The replacement cost of a tungsten-halogen lamp is about three times the cost of an incandescent lamp. **See Figure 6-9.**

Light-Emitting Diode (LED) Lamps

A *diode* is a semiconductor device that offers high opposition to current flow in one direction and low opposition to current flow in the opposite direction. A *light-emitting diode (LED)* is a diode that emits light when forward direct current is applied. Because LEDs require direct current, an AC to DC converter, that normally consists of diodes, must be included with the lamp base on lamps designed to be connected to AC power. LEDs last a long time, with certain types rated at over 100,000 hr. However, the AC to DC converter may not last as long as the LEDs, which reduces the rated lamp life.

LED lamps consist of many individual LEDs, each with a diffuser lens to help direct the light produced. LED lamps are categorized based on the number of individual LEDs that they contain and the direction in which they direct light. **See Figure 6-10.** Because LED lamps have a long service life and are available in various colors, they are used in special applications such as airport runway lighting and tower warning lights in addition to general lighting use. LED lamps are energy efficient and have a long service life, therefore, they are also often installed in difficult to reach locations within a building such as high passageways, hallways, ceilings, and walls. They are also installed as decorative lighting on building exteriors.

LED lamps are not designed to be repaired and should be replaced when they fail. However, premature failure normally results from poor power quality, such as an excessively high voltage or transient voltages. Adding a surge suppressor at the lighting panel helps reduce transient voltages entering the panel. LED lamps are available in a variety of sizes and shapes.

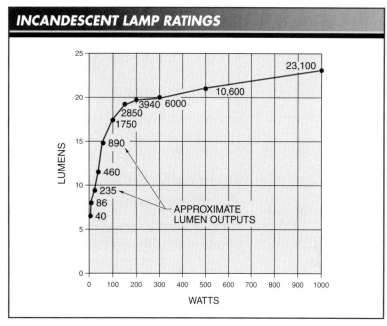

Figure 6-8. The higher the wattage rating of a lamp, the greater the lumens per watt output produced by the lamp.

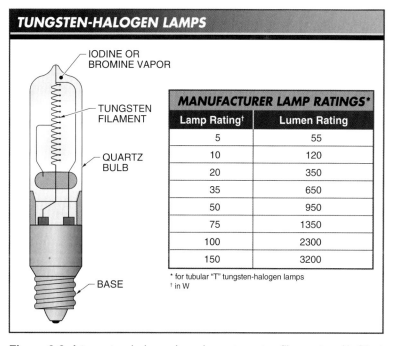

Figure 6-9. A tungsten-halogen lamp has a tungsten filament and is filled with a halogen gas that produces a large amount of light instantly.

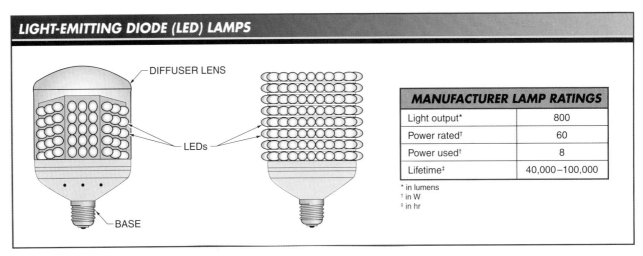

Figure 6-10. Light-emitting diode (LED) lamps are the most energy-efficient and longest lasting type of lamp.

Fluorescent Lamps

A *fluorescent lamp* is a low-pressure discharge lamp in which ionization of mercury vapor transforms ultraviolet energy generated by the discharge into light. The current in a fluorescent lamp flows through a gas such as mercury or argon that emits ultraviolet (UV) light. When the emitted UV light strikes the phosphor-coated outer glass tube, it glows. The type of gas and phosphor coating used determines the shade of colored light that the lamp produces. Fluorescent lamps are available in standard, U-shaped, or compact designs.

Standard fluorescent lamps are designed to deliver a peak light output under ambient temperature conditions of approximately 75°F.

Regardless of lamp size and shape, fluorescent lamp bulbs are available for cool (blue light), moderate (white light), and warm (yellow light) environments depending on their color temperature rating. *Color temperature* is a description of the coolness or warmth of a light source and is measured in Kelvin (K). High Kelvin ratings of 3600 K and above represent a cool light while low Kelvin ratings of 2700 K and below represent a warm light. **See Figure 6-11.** Cool light sources are recommended for spaces that require high brightness and provide a more natural appearance to most objects. Warm light sources are recommended for spaces that require low brightness and outdoor lighting.

Fluorescent lamps are also available with properties that are required for special applications. For example, shatter-resistant bulbs are required for use in foodservice and food processing areas to eliminate the possibility of broken-glass contamination.

A fluorescent lamp is not connected directly to power lines because high inrush current from power lines can destroy the lamp. A *ballast* is an autotransformer that delivers a specific voltage and limits the electric current supplied to a lamp.

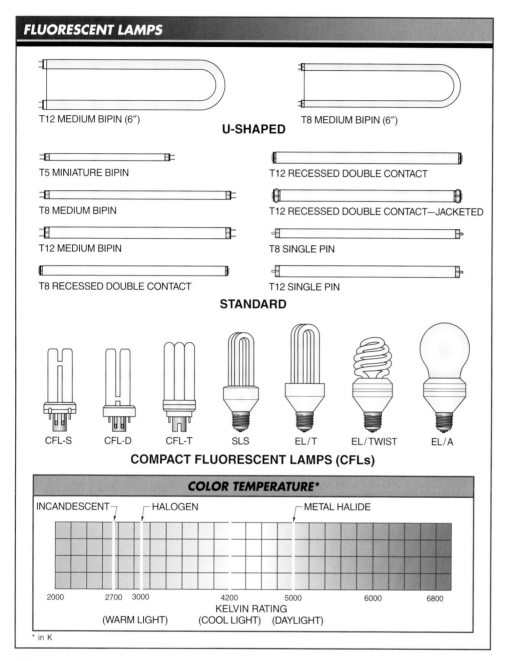

Figure 6-11. Fluorescent lamps are available in different sizes and are rated in color temperature, which is measured in Kelvin (K).

TECH FACT

Radio and sound interference caused by fluorescent lamps can be eliminated by separating radios and lamps by 10′ or more, providing a good ground for the radio or sound system, or connecting radio and lamp fixtures to separate branch circuits.

A supply voltage higher or lower than the rating of a ballast affects the service life of the lamp and ballast. The voltage supplied to a fluorescent lamp should be within ±7% of the lamp ballast rating. Voltages higher than the rated voltage cause more problems than voltages lower than the rated voltage. **See Figure 6-12.**

182 ELECTRICAL SYSTEMS for FACILITIES MAINTENANCE PERSONNEL

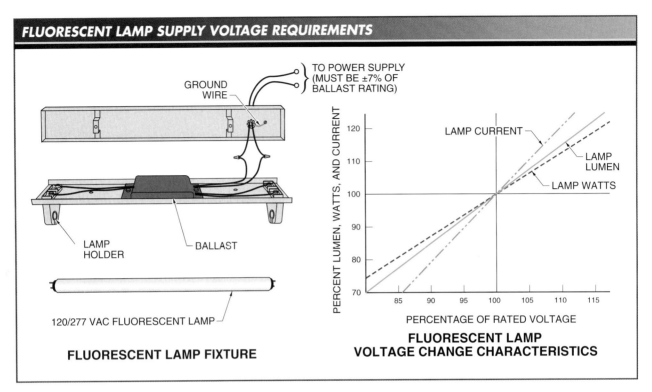

Figure 6-12. The voltage supplied to a fluorescent lamp should be within ±7% of the lamp ballast rating.

A *compact fluorescent lamp (CFL)* is a fluorescent lamp that has a smaller diameter than a conventional fluorescent lamp and has a folded bulb configuration. CFLs provide high light output in smaller sizes than conventional fluorescent lamps.

CFLs are available without a ballast or with an integrated ballast. CFLs without ballasts have standard fluorescent lamp bases and are connected to standard fluorescent lamp circuits. CFLs with integrated ballasts are available with standard incandescent lamp bases and are connected to standard incandescent lamp circuits.

Properly operating CFLs do not make any sound that can be heard. However, large fluorescent lamp fixtures can produce a humming sound when an electromechanical ballast is used. The humming sound is normally not a problem because of the distance and insulating properties of the lamp location. However, the humming sound can become a nuisance if amplified. If a humming sound can be heard, the equipment must be inspected for improper mounting and loose parts. Mounting fluorescent fixtures near air ducts and any other open area that can carry sound should be avoided, and the manufacturer sound rating should be verified to ensure that the sound rating matches the application. The lower the sound rating number, the lower the noise level.

WARNING: A ballast produces heat during operation and must be open and not covered with insulation in order to dissipate the heat.

A uniform blackening throughout the length of a fluorescent lamp is normal as the lamp ages. The blackening starts gradually and is not normally detectable until the ends of the lamp blacken heavily. Heavy end blackening soon after a new lamp is installed indicates a problem. The problem may be a defective lamp but is normally caused by excessive or low line voltage outside the ±7% range. **See Figure 6-13.**

Chapter 6—Lighting Systems   **183**

FLUORESCENT LAMP DISCOLORATION

- BLACKENING OF ENDS IS NORMAL AND MAY OCCUR AT END OF LAMP LIFE
- HEAVY SPOTTING MAY DEVELOP GRADUALLY DURING NORMAL LAMP LIFE
- END BANDS MAY DEVELOP GRADUALLY DURING NORMAL LAMP LIFE
- LIGHT SPOTTING MAY OCCUR IN NEW LAMPS AT ANY POINT ON BULB

Figure 6-13. A uniform blackening or spotting throughout the length of a fluorescent lamp is normal as the lamp ages.

TECH FACT

Cold weather negatively affects fluorescent lamp operation. An outdoor-rated lamp should be used in areas that are regularly exposed to temperatures below 32°F. Fluorescent lamps do not start well in cold conditions because the electrons released by the heated cathode are thermally released. The colder the lamp, the longer the time required to heat the cathode and release electrons into the mercury vapor. Even after the lamp is turned on, it delivers less light in cold conditions as low temperatures affect the mercury-vapor pressure inside the bulb. The colder the lamp, the less light output.

Troubleshooting Fluorescent Lamps

Fluorescent lamps may turn on before they are properly seated in the lamp holder. The lamp may fall out, and lamp life is normally reduced when a lamp is not properly seated. Fluorescent lamps are seated properly by aligning the base mark on the lamp with the center of the lamp holder. Both ends of the lamp must be checked for correct seating. **See Figure 6-14.**

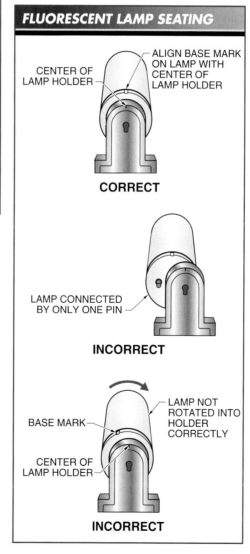

Figure 6-14. Fluorescent lamps must be seated correctly by aligning the pins and rotating the lamp into the correct position.

A fluorescent lamp cannot be operated directly from a standard electrical circuit. All fluorescent lamps require a means of starting. The method used to start the lamp determines the circuit name. The methods of fluorescent lamp starting include preheat, instant start, and rapid start.

Preheat Circuits. A *preheat circuit* is a fluorescent lamp starting circuit that heats the cathode before an arc is created. A preheat circuit uses a separate starting switch that is connected in series with the ballast. The electrodes in a preheat circuit require a few seconds to reach the proper temperature to start the lamp after the starting switch is pressed. The opening of the starting switch breaks the path of current flow through the starting switch, leaving the gas in the lamp as the only path for current to travel. A high-voltage surge is produced by the collapsing magnetic field as the starting switch is no longer connected to power. This high-voltage surge starts the flow of current through the gas in the lamp. Once started, the ballast limits the flow of current through the lamp.

In a preheat circuit, a starting switch is used to preheat the lamp filament. As the lamp ages, the starting switch has to be pressed to start and restart the lamp several times before the lamp stays on. Ballast life is reduced when a hard-to-start lamp is not replaced. The lamp should be replaced any time a hard-to-start lamp is allowed to remain in operation for more than one week. In two-lamp circuits, premature ballast failure occurs when only one lamp is operating. The inoperative lamp should be replaced immediately. The ballast may require replacement on any circuit in which only one lamp is allowed to operate for longer than a few days. **See Figure 6-15.**

A preheat circuit is tested by applying the following procedure:

1. Wear proper PPE for the measurement to be taken and area where work is to be performed.
2. Verify that the clamp-on ammeter is rated for the current level and type to be tested.
3. Verify that the clamp-on ammeter is in proper operating condition by testing it on a known, energized current source.
4. Turn power on by placing the line switch in the ON position.
5. Connect the clamp-on ammeter around one of the incoming power lines.
6. Press the starting switch and release after 2 sec. Record the starting current as soon as the switch is released. *Note:* The starting current is the current value measured just after the starting switch is released. The starting current should be about 20% to 80% higher than the operating current when the circuit is working properly. The operating current is the current value measured after the lamp is energized. No current reading indicates an open circuit.

The same procedure may be used when an automatic starting switch is used. The starting current is read immediately after the line switch is closed. There is a problem with the starter or lamp when the lamp does not ignite properly and continues to try to start. A lamp does not start when it is at the end of its normal life. The starter continues to try to start the lamp if the lamp is not replaced. This damages the starter. An old lamp that cycles during starting must be replaced. The starter requires replacement when a new lamp continues to cycle.

Instant-Start Circuits. An *instant-start circuit* is a fluorescent lamp starting circuit that provides sufficient voltage to strike an arc instantly. The instant-start circuit was developed to eliminate the starting switch and overcome the starting delay of preheat circuits. An arc strikes without preheating the cathodes when a high enough voltage is applied across a fluorescent lamp.

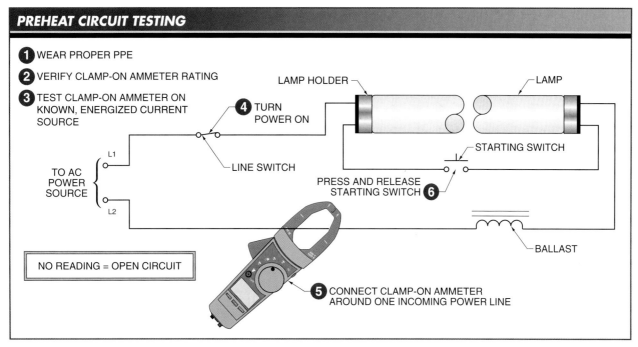

Figure 6-15. A preheat circuit is a fluorescent lamp starting circuit that heats the cathode before an arc is created.

The high initial voltage requires a large autotransformer as an integral part of the ballast. The autotransformer delivers an instant voltage of 270 V to 600 V, depending upon bulb size and voltage rating. Instant-start lamps require only one pin at each base because no preheating of the electrode is required.

A safety circuit is used with an instant-start lamp to prevent electrical shock due to the high starting voltage. When the lamp is removed, the base pin acts as a switch, which interrupts the circuit to the ballast. The lamp is replaced by pushing the lamp into the spring lamp holder at the high voltage end of the fixture and then inserting it into the rigid lamp holder at the low-voltage end of the fixture. Both lamp ends must be in place before current can flow through the ballast winding. The circuit does not operate when the lamp is not in place.

In an instant-start circuit, light is produced instantly. Short lamp life is normally due to low supply voltage, improper lamp and lamp holder contact, or improper wiring of the ballast. In two-lamp circuits, any lamp that burns out should be replaced immediately. Both lamps should be replaced when the good lamp shows signs of blackening. Ballast problems may exist when there is a problem shortly after replacing a lamp in any circuit in which one lamp was allowed to operate alone for more than a few weeks.

WARNING: Care must be taken to prevent electrical shock when measuring high voltages. The DMM must be set to the high-voltage range. **See Figure 6-16.**

An instant-start circuit is tested by applying the following procedure:

1. Wear proper PPE for the measurement to be taken and area where work is to be performed.
2. Verify that the DMM is rated for the voltage level and type to be tested.
3. Verify that the DMM is in proper operating condition by testing it on a known, energized voltage source.
4. Turn the power to the circuit off by placing the line switch in the OFF position.
5. Remove the lamp.

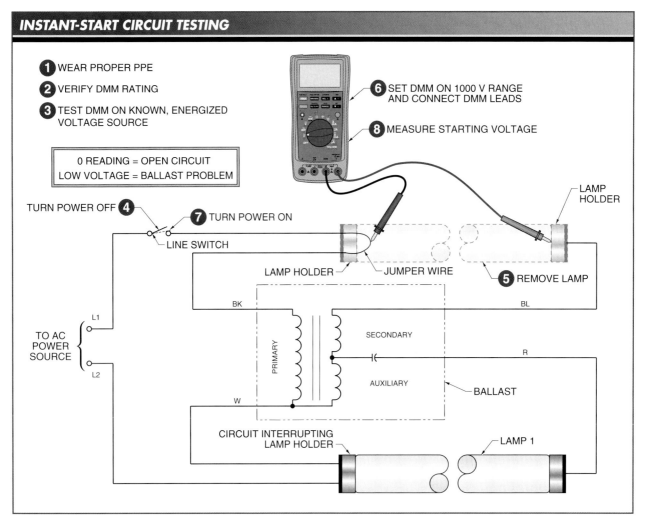

Figure 6-16. An instant-start circuit is a fluorescent lamp starting circuit that provides sufficient voltage to strike an arc instantly.

6. Set the DMM on the 1000 V range. Connect the DMM leads to each end of the lamp holder. *Note:* The DMM must complete the primary circuit at the lamp holder having two conductors. The two conductors may be shorted using a jumper wire. **WARNING:** Ensure that power is OFF before shorting the pins.
7. Turn the power to the circuit on.
8. Measure the starting voltage. *Note:* The ballast must provide an open circuit voltage of about three times the operating voltage to produce the arc required to light the lamp. There is a problem with the circuit when there is not a high starting voltage present.

A zero voltage reading indicates an open circuit. A low voltage normally indicates a ballast problem.

Rapid-Start Circuits. A *rapid-start circuit* is a fluorescent lamp starting circuit that brings the lamp to full brightness in about 2 sec. A rapid-start circuit is the most common circuit used in fluorescent lighting. A rapid-start circuit uses lamps that have short, low-voltage electrodes that are automatically preheated by the lamp ballast. The rapid start ballast preheats the cathodes by means of a heater winding. The heater winding continues to provide current to the lamp after ignition.

In a rapid-start circuit, light is produced in less than 1 sec. Long start times normally indicate a lamp problem. There is normally a ballast problem when start time is still long after replacing a lamp. In two-lamp circuits, any lamp that burns out should be replaced immediately. Both lamps should be replaced when the good lamp shows signs of blackening. **See Figure 6-17.**

A rapid-start circuit is tested by applying the following procedure:
1. Wear proper PPE for the measurement to be taken and area where work is to be performed.
2. Verify that the DMM is rated for the voltage level and type to be tested.
3. Verify that the DMM is in proper operating condition by testing it on a known, energized voltage source.
4. Turn the power to the circuit off by placing the line switch in the OFF position.
5. Measure the filament or starting voltage. Remove the lamp and connect a DMM across the opposing lamp holders to measure starting voltage. Connect the DMM across the end of each holder (normally red-red, blue-blue, and yellow-yellow) to measure the filament voltage.

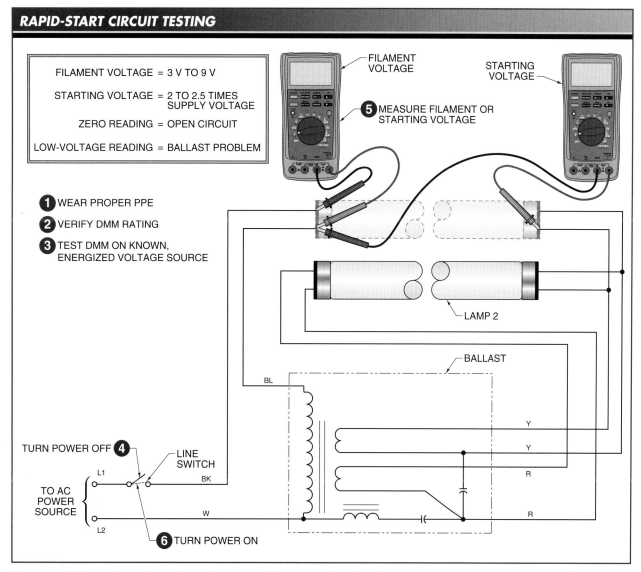

Figure 6-17. A rapid-start circuit is a fluorescent lamp starting circuit that has separate windings to provide continuous heating voltage on the lamp cathodes.

6. Turn power to the circuit on. *Note:* Filament voltage should equal 3 V to 9 V, depending upon the lamp. Starting voltage is given by the manufacturer and is normally about 2 to 2.5 times the supply voltage. A zero voltage reading indicates an open circuit. A low-voltage reading normally indicates a ballast problem.

To minimize maintenance equipment setup time and disruption to business activities, all lamps in high bays or otherwise difficult to reach locations should be replaced at the same time based on a planned scheduled maintenance program.

High-Intensity Discharge Lamps

A *high-intensity discharge (HID) lamp* is a lamp that produces light from an arc tube. An *arc tube* is the light-producing element of an HID lamp. An arc tube contains metallic and gaseous vapors and electrodes. An arc is produced in the tube between the electrodes. The arc tube is enclosed in a bulb that may contain phosphor or diffusing coating that improves color rendering, increases light output, and reduces surface brightness.

HID lamps include low-pressure sodium, mercury-vapor, metal-halide, and high-pressure sodium lamps. HID lamps are electric discharge lamps. An *electric discharge lamp* is a lamp that produces light by means of an arc discharged between two electrodes. High vapor pressure is used to convert a large percentage of the energy produced into visible light. Arc tube pressure for most HID lamps normally ranges from one to eight atmospheres (15 psia to 120 psia). HID lamps provide an efficient, long-lasting source of light and are used for street, parking lot, exterior building, and warehouse lighting applications. *Color rendering* is the appearance of a color when illuminated by a light source. HID lamps do not have as high a quality of color rendering as incandescent and fluorescent lamps.

HID lamps are available with a variety of lamp shapes and base types. Because HID lamps produce more light per watt, they are energy and cost effective when used in areas where the quality of the light is not as important as the cost of the lighting. The exact type of lamp used depends on the required color rendering, initial cost, operating cost, and expected operating life of the lamp. **See Figure 6-18.**

Another factor considered when selecting a lamp for an application is the start and restart time of the lamp. HID lamps require several minutes to warm up before full light output is obtained. Any short interruption in the power supply can extinguish the arc. An HID lamp cannot restart immediately after being turned off. The lamp must cool enough to reduce the vapor pressure in the arc tube to a point where the arc can restrike.

TECH FACT

HID lamps require ballasts to limit the current in the lamps to the correct operating level and provide the proper starting voltage to strike and maintain the arc. Each HID ballast is designed for a specific lamp, bulb size, voltage range, and line frequency. Different lamps are not interchangeable without changing the ballast because lamp wattage is controlled by the ballast and not the lamp.

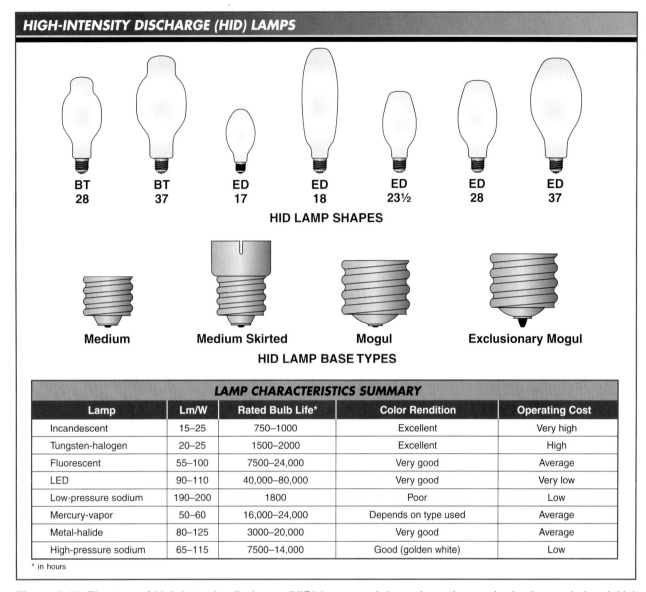

Figure 6-18. The type of high-intensity discharge (HID) lamp used depends on the required color rendering, initial cost, operating cost, and expected operating life of the lamp.

Ballast voltage at the input of an HID lamp fixture must be measured to verify that it is at the proper level. Low light output and reduced lamp service life result when voltage is not within a specified limit. For constant wattage-rated ballasts, the line voltage must be within ±10% of the ballast or lamp fixture nameplate rating. For reactor or high-reactance ballasts, line voltage should be within ±5% of the ballast or lamp fixture nameplate rating. Ballast output voltage should be measured when the incoming voltage is correct, but the lamp is not operating properly. **See Figure 6-19.** Ballast voltage is measured using the following procedure:

Note: It is recommended to work with one or more partners when performing any type of electrical measurement procedure.
1. Wear proper PPE for the measurement to be taken and area where work is to be performed.
2. Verify that the DMM is rated for the voltage level and type to be tested.

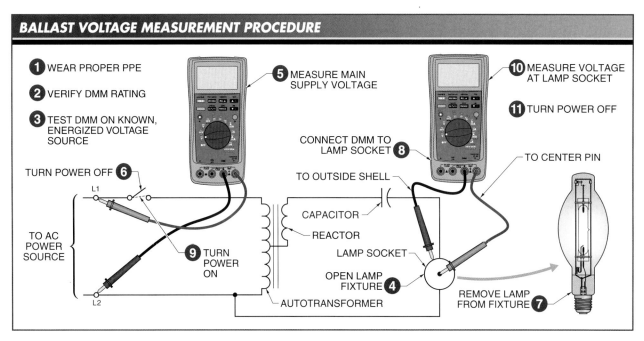

Figure 6-19. Ballast voltage should be measured to ensure that it is at the proper level.

3. Verify that the DMM is in proper operating condition by testing it on a known, energized voltage source.
4. Open the lamp fixture so that voltage may be measured directly at the lamp input.
5. Measure the main supply voltage between the two incoming power lines. *Note:* Voltage measurement readings should be within ±10% for constant-rated ballast and ±5% for reactor or high-reactance ballasts.
6. Turn the power to the fixture off.
7. After the lamp cools, remove from fixture.
8. Connect the DMM to the lamp socket. Verify that the DMM is set to a voltage setting higher than maximum expected voltage. See recommended ballast output voltage limits for minimum expected voltage measurements. **See Appendix.**
9. Turn the power on.
10. Measure the voltage at the lamp socket.
11. Turn the power off. Replace fixture if voltage is not within specified limits.

Low-Pressure Sodium Lamps. A *low-pressure sodium lamp* is an HID lamp that operates at a low vapor pressure and uses sodium as the vapor. A low-pressure sodium lamp has a U-shaped arc tube. The arc tube has both electrodes located at the same end. The arc tube is placed inside a glass bulb and contains a mixture of neon, argon, and sodium metal. On startup, an arc is discharged through the neon, argon, and sodium metal. As the sodium metal heats and vaporizes, the amber color of sodium is produced. **See Figure 6-20.**

A low-pressure sodium lamp is named for its use of sodium inside an arc tube. A low-pressure sodium lamp has the highest efficiency rating of any lamp. Certain types of low-pressure sodium lamps deliver up to 200 lm/W of power, which is 10 times the output of an incandescent lamp. Low-pressure sodium lamps must be operated on a ballast that is designed to meet the lamp starting and running requirements. Low-pressure sodium lamps do not have a starting electrode. The ballast must provide an open-circuit voltage of approximately three to seven times the lamp rated voltage to start and sustain the arc.

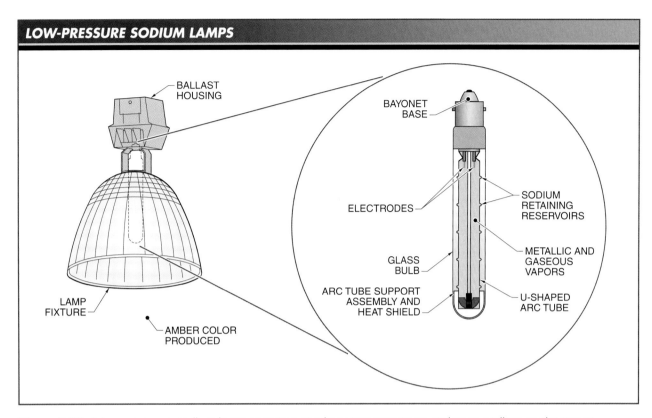

Figure 6-20. A low-pressure sodium lamp operates at a low vapor pressure and uses sodium as the vapor.

Low-pressure sodium lamp fixtures are available in a wide variety of lamp, ballast, and reflector configurations. Lamp fixtures include high profile, low profile, and twin optical fixtures. The variety of fixture configurations provides a wide range of light distribution patterns. The pattern chosen is based on the lighting requirements of the application.

Mercury-Vapor Lamps. A *mercury-vapor lamp* is an HID lamp that produces light by means of an electrical discharge through mercury vapor. Mercury-vapor lamps are used for general lighting applications. Phosphor coating is added to the inside of the bulb to improve color-rendering characteristics. **See Figure 6-21.** A mercury-vapor lamp contains a starting electrode and two main electrodes. An electrical field is set up between the starting electrode and one main electrode when power is first applied to the lamp. The electrical field causes current to flow and an arc to strike. Current flows between the two main electrodes as the heat vaporizes the mercury.

Metal-Halide Lamps. A *metal-halide lamp* is an HID lamp that produces light by means of an electrical discharge through mercury vapor and metal halide in the arc tube. Metal halide (normally sodium and scandium iodide) is added to the mercury in small amounts. A metal-halide lamp produces more lumens per watt than a mercury-vapor lamp.

The light produced by a metal-halide lamp has less color distortion than a mercury-vapor lamp. A metal-halide lamp is an efficient source of white light but has a shorter life than the other HID lamps. A metal-halide ballast uses the same basic circuit as the constant-wattage autotransformer mercury-vapor ballast. The ballast is modified to provide the high starting voltage required by metal-halide lamps. **See Figure 6-22.**

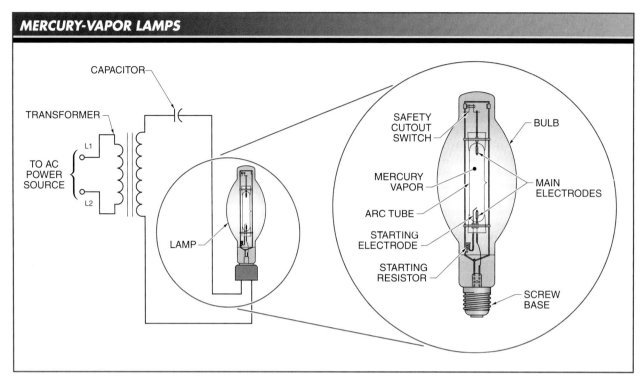

Figure 6-21. A mercury-vapor lamp produces light by an electrical discharge through mercury vapor and is used for general lighting applications.

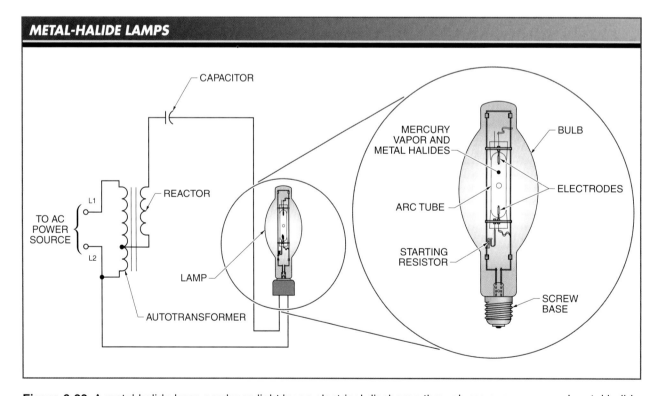

Figure 6-22. A metal-halide lamp produces light by an electrical discharge through mercury vapor and metal halide in the arc tube.

High-Pressure Sodium Lamps. A *high-pressure sodium lamp* is an HID lamp that produces light when current flows through sodium vapor under high pressure and high temperature. A high-pressure sodium lamp is a more efficient lamp than a mercury-vapor or metal-halide lamp. The light produced from a high-pressure sodium lamp is golden-white in color. A high-pressure sodium lamp is constructed with a bulb and an arc tube. The arc tube is composed of ceramic to withstand high temperatures. The bulb is composed of weather-resistant glass to prevent heat loss and protect the arc tube. High-pressure sodium lamps have a ballast with an ignitor to deliver a voltage pulse high enough to start and maintain the arc.

An *ignitor* is a component inside a lamp ballast that produces a high starting voltage. A high starting voltage pulse must be delivered every cycle and must be 4000 V to 6000 V for 1000 W lamps and 2500 V to 4000 V for smaller lamps. **See Figure 6-23.**

A high-pressure sodium ballast is similar to a mercury-vapor reactor ballast. The main difference is the added starter. The reactor ballast is used when the input voltage meets the lamp requirements. A transformer or autotransformer is added to the ballast circuit when the incoming voltage does not meet the lamp requirements.

WARNING: Mercury-vapor and metal-halide lamps can cause serious skin and eye damage in the form of burns when the lamp bulb is cracked or shattered because they can release ultraviolet radiation (UV). Always replace any bulb that is damaged or cracked.

SPECIAL-PURPOSE LIGHTING

In addition to general-purpose lighting, commercial buildings normally include special-purpose lighting. Special-purpose lighting includes exit lighting systems and emergency exit lighting systems. Exit lighting systems and emergency exit lighting systems are required in commercial buildings.

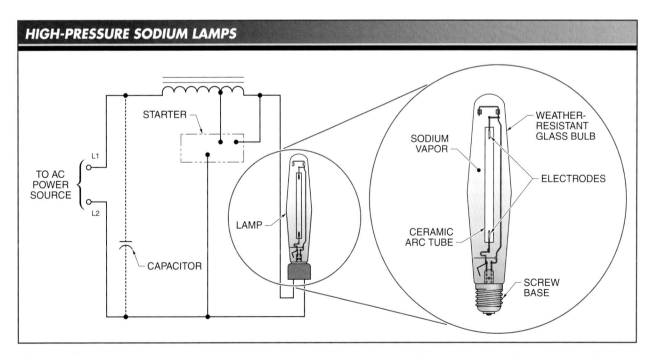

Figure 6-23. A high-pressure sodium lamp produces light when current flows through sodium vapor under high pressure and high temperature.

Exit Lighting Systems

An *exit lighting system* is illuminated signage that indicates the safest and quickest path out of a building. Local and state municipal codes mandate the amount and placement of exit lighting. Exit lighting must be constantly illuminated. Because exit lighting is constantly illuminated, it requires constant power. To avoid frequent bulb replacement and increase energy efficiency, LED lamps should be used in place of incandescent lamps. LED lamp retrofit kits are available for replacement of incandescent bulb exit signs. **See Figure 6-24.**

Note: A comparison of exit lamp specifications can be used to determine the type of lamp that operates best in certain commercial applications. **See Appendix.**

TECH FACT

The U.S. Occupational Safety and Health Administration (OSHA) includes standards for exit lighting that states that the line-of-sight to an exit sign must be clear at all times and that it must be illuminated to a surface value of at least 5 fc by a reliable, distinctively colored light source.

Emergency Exit Lighting Systems

An *emergency exit lighting system* is a lighting system that is activated when a power failure occurs and provides adequate lighting for egress from a building. Local and state municipalities require that emergency exit lighting comply with building codes in commercial buildings and be tested and documented on a regular basis. Testing procedures can be performed on emergency exit lighting systems without removing or opening the emergency lighting system. However, additional tests can be performed to ensure that each section of the system is operating properly. **See Figure 6-25.** Emergency exit lighting systems are tested by using the following procedure:

1. Wear proper PPE for the measurement to be taken and area where work is to be performed.
2. Verify that the DMM is rated for the voltage level and type to be tested.
3. Verify that the DMM is in proper operating condition by testing it on a known, energized voltage source.

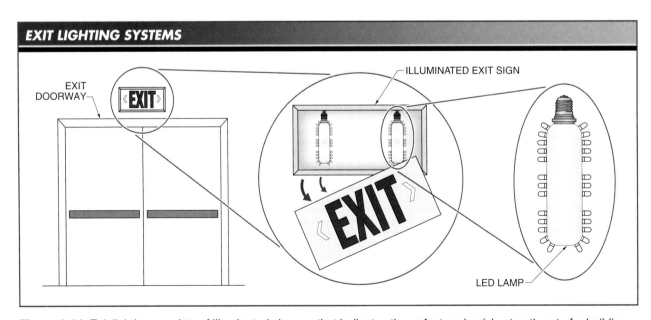

Figure 6-24. Exit lighting consists of illuminated signage that indicates the safest and quickest path out of a building.

4. Open the lamp fixture so that voltage can be measured directly at the power supply input, battery, and lamps.
5. Measure the voltage between the two incoming power lines. *Note:* Line voltage must be within ±10% of the unit nameplate rating.
6. Set the DMM to measure DC voltage and measure the voltage at the battery. *Note:* A 6 VDC rated battery should measure about 6.8 VDC to 7 VDC and a 12 VDC battery should measure about 13.6 VDC to 14 VDC.
7. Turn AC power off. *Note:* The emergency exit light should automatically turn on (or the TEST button can also be used).
8. Measure the voltage at the lamps. *Note:* For a properly operating system, lamps must remain on with voltage at the lamps at a minimum of 10% of lamp voltage rating after 10 min.
9. Retest battery output voltage. *Note:* Battery output voltage must be equal to or higher than the battery rated voltage.
10. Verify that the DMM is in proper operating condition by testing it on a known, energized voltage source.

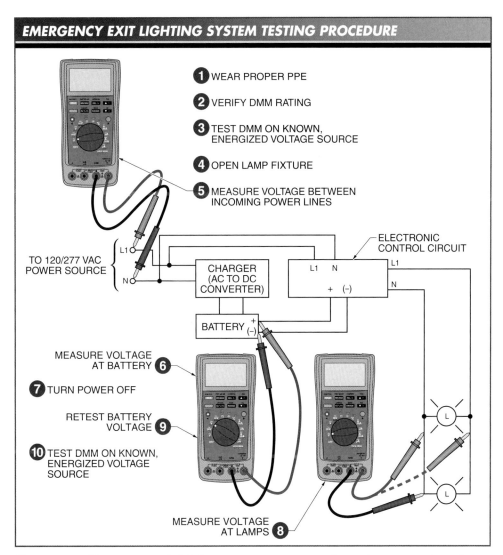

Figure 6-25. Emergency exit lighting system tests are performed to ensure that all sections of the system are operating properly.

LIGHTING CONTROL DEVICES

Regardless of the type of lighting installed in a building and the surrounding area, lighting systems require control devices. Lighting control devices can be as basic as manually operated ON/OFF switches or can include timers, motion sensors, light level sensors, and contactors. In order to conserve energy and improve security, modern building lighting systems can be fully automated. Fully automated lighting systems include both manual and automated switch controls. Troubleshooting manually operated lighting switches requires identification of unknown voltages.

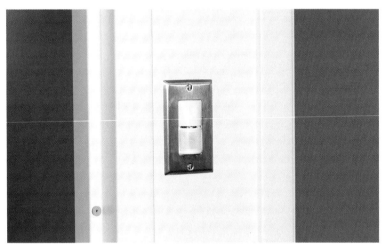

Manual lighting switches can be replaced with motion sensor switches that automatically turn lights on and off in building spaces to help conserve energy.

Manual Lighting Switches

Manual lighting switches are used to manually start, stop, or redirect the flow of current to lamps. Manually operated switches normally have maximum voltage and current limits listed on the switch along with the type of conductor material that must be used with the circuit in which the switch is installed. The most common types of manually operated light switches are two-way, three-way, and four-way switches. The type of switch used depends on the number of switch locations required to control the lamp(s). Manual switches are connected into the positive (hot) leg of a circuit. Two-way, three-way, and four-way switches are also specified according to the number of poles and throws they have.

A *pole* is the number of completely isolated circuits that a switch can switch. A *single-pole (SP) switch* is an electrical control device used to turn lights or appliances on and off from a single location. A single-pole (SP) switch can carry current through only one circuit at a time. A *double-pole (DP) switch* is a control device that consists of two switches in one and is used for controlling two separate loads.

A *throw* is the number of closed switch positions per pole. A single-throw (ST) switch can control only one circuit, while a double-throw (DT) switch can control two circuits. With a double-throw (DT) switch, the two circuits are mechanically connected to open or close simultaneously and are electrically insulated from each other. **See Figure 6-26.**

Two-Way Switches. A *two-way switch* is a single-pole, single-throw (SPST) switch. A two-way switch can be placed in one of two different positions. The switch allows current to flow through it in the ON position and does not allow current to flow through it in the OFF position. Two-way switches have ON and OFF position markings on them because they have distinct ON and OFF positions. Two-way switches are normally used to control a lamp from one location. **See Figure 6-27.**

> **TECH FACT**
> Only switches that are approved by Underwriters Laboratories (UL) should be used as lighting control devices. A switch that is UL approved is stamped with a UL listing. All lighting switches also have their ampere and voltage rating stamped on them and often include their horsepower rating.

MANUAL SWITCHES				Terms
SPST NO	SPST NC	SPDT		SPST: SINGLE-POLE, SINGLE-THROW
Two-Way	Two-Way	Three-Way		
				SPDT: SINGLE-POLE, DOUBLE-THROW
				DPST: DOUBLE-POLE, SINGLE-THROW
DPST, 2NO	DPST, 2NC	DPDT		DPDT: DOUBLE-POLE, DOUBLE-THROW
Four-Way	Four-Way	Four-Way		
				NO: NORMALLY OPEN
				NC: NORMALLY CLOSED

Figure 6-26. Manual switches are two-way, three-way, or four-way switches and are specified according to the number of poles and throws they have.

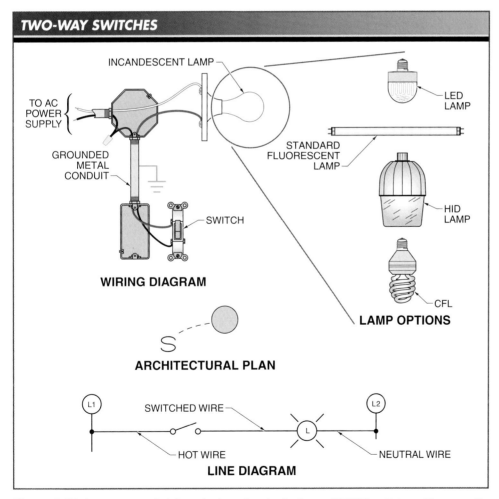

Figure 6-27. A two-way switch is a single-pole, single-throw (SPST) switch and is normally used to control a lamp from one location.

Three-Way Switches. A *three-way switch* is a single-pole, double-throw (SPDT) switch. A three-way switch has one common terminal and two traveler terminals. The common terminal is dark in color to distinguish it from the two light-colored traveler terminals. The switch connects the common terminal to the top traveler terminal in one position and connects the common terminal to the bottom traveler terminal in the other position. A three-way switch does not have a designated ON and OFF position because the common terminal is always connected to one of the traveler terminals. Two separate three-way switches are used to control a lamp from two different locations. **See Figure 6-28.**

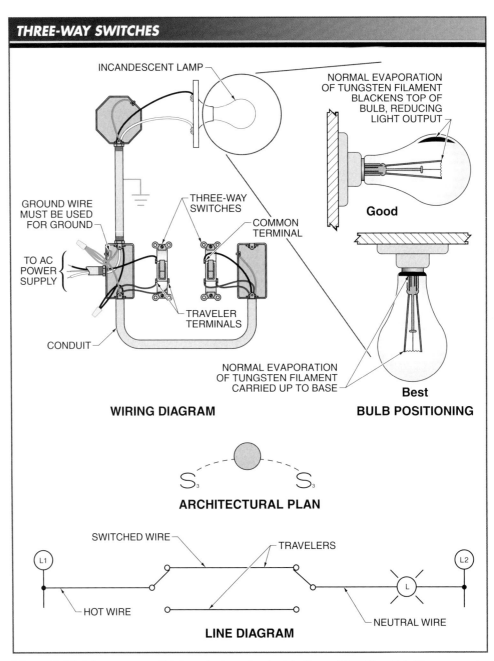

Figure 6-28. A three-way switch is a single-pole, double-throw (SPDT) switch. Two separate three-way switches are used to control a lamp from two different locations.

Four-Way Switches. A *four-way switch* is a control device that is used in combination with two three-way switches to allow control of a load from three locations. Four-way switches do not have ON and OFF position markings on them. When controlling a lamp from three locations, the four-way switch is placed between the two three-way switches. **See Figure 6-29.**

Troubleshooting Manually Operated Lighting Switches

Manually operated lighting switches can be tested before they are installed in a circuit by using the continuity or resistance function on a DMM. A closed switch indicates little or no resistance (continuity/beeping), and an open switch indicates infinite (OL) resistance (no continuity/no beeping). **See Figure 6-30.**

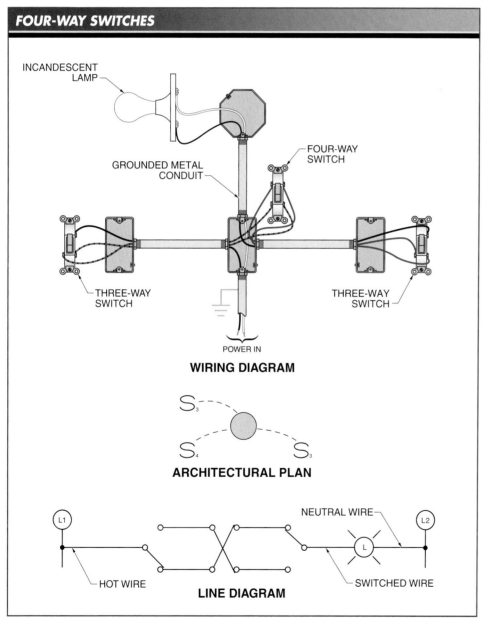

Figure 6-29. When controlling a lamp from three locations, a four-way switch is placed between two three-way switches.

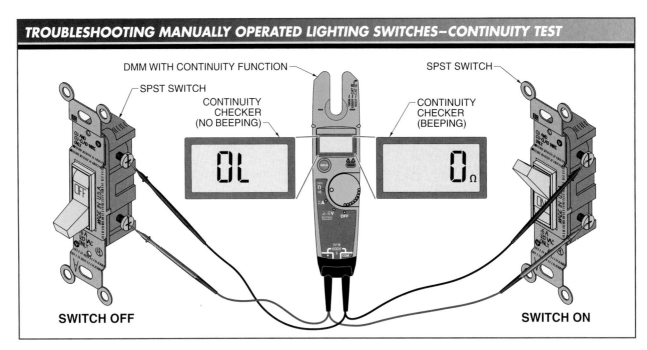

Figure 6-30. Manually operated lighting switches can be tested before they are installed in a circuit by using the continuity or resistance function on a DMM.

Switches should be tested for continuity prior to connection in a circuit. This is because certain types of switches are designed with the position of the common terminal in different locations on the switch, and the common terminal is identified by using a different color screw. Also, certain types of four-way switches are designed to use the two top and two bottom terminals for connecting the travelers in and out, while other switches are designed to use the two left and two right terminals.

Switches are connected to hot conductors and control the flow of current to lamps. The conductor between the power source (normally a circuit breaker in the panel) and the switch is referred to as the "hot" conductor, while the conductor between the switch and the lamp is referred to as the "switched" conductor. **See Figure 6-31.** Circuits that contain switches can be tested for proper connections using the following procedure:

1. Wear proper PPE for the measurement to be taken and area where work is to be performed.
2. Verify that the DMM is rated for the voltage level and type to be tested.
3. Verify that the DMM is in proper operating condition by testing it on a known, energized voltage source.
4. Test for proper grounding by connecting a DMM set to measure AC voltage between the metal box (or the green/bare ground conductor in a plastic box) and the hot conductor (ungrounded conductor) going into the switch. *Note:* A ground test must be performed at the switch box, lamp box, and any other electrical box in the circuit. If the circuit is not properly grounded, any open grounds must be corrected and the circuit must be retested.
5. Test for correct circuit voltage by connecting a DMM between the hot conductor and the neutral conductor. *Note:* System voltage must be present regardless of the switch position. The voltage must be within +5% to –10% of the lamp voltage rating.

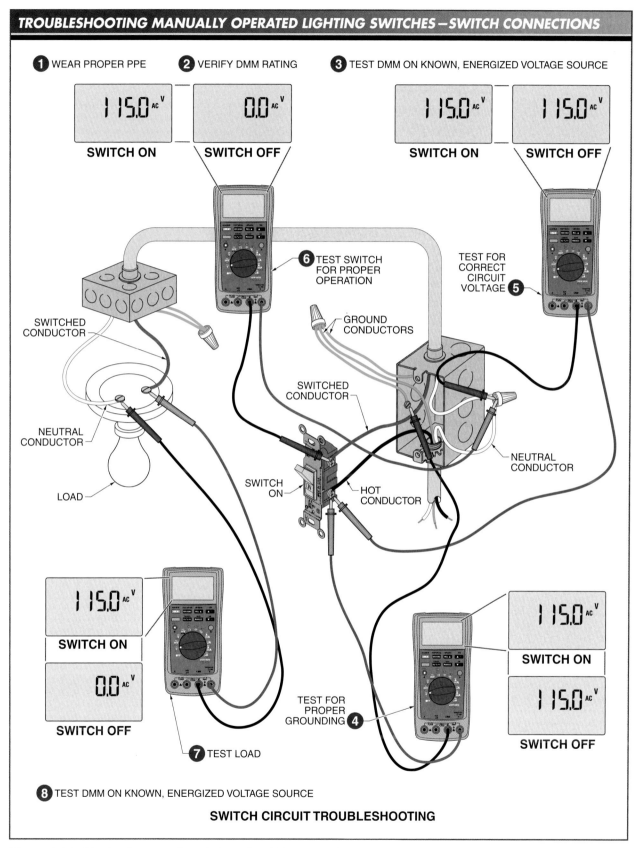

Figure 6-31. Circuits that contain switches should be tested for proper connections by using a DMM.

6. Test switch for proper operation. Connect a DMM between the neutral conductor and switched conductor. *Note:* There must be no voltage when the switch is in the OFF position. Voltage must be present when switch is in the ON position.
7. Test the load. Connect the DMM between load terminals. *Note:* There must be no voltage when the switch is in the OFF position. Voltage must be present when switch is in the ON position.
8. Verify that DMM is in proper operating condition by testing it on a known, energized voltage source.

The same procedure performed on a two-way switching circuit can also be used to test three-way and four-way switching circuits. When testing three-way or four-way switching circuits, the switch is tested to ensure that power is coming into the switch and going out of the switch as the switch position is changed.

Although voltage must be within +5% to −10% of the lamp rating, even a small voltage variation from the lamp rated voltage can affect lamp operation. For example, operating a 120 V rated lamp on 125 V power (+4%) produces approximately 15% more lumens, uses 7% more power, and shortens lamp life by 40%. Operating the same lamp on 115 V power (−4%) produces approximately 15% less lumens, uses 7% less power, and increases lamp life by 70%. Operating a lamp at higher than rated voltage is more destructive than operating it at lower than rated voltage and should be taken into consideration as a problem when lamps have a shortened life. **See Figure 6-32.**

Several switches can be used to control different lamps located together for convenience. When testing switches, lamps, and grounds, voltage measurements can be taken at several different locations. Normally, testing at the switch is easier than testing at a load because it does not require additional equipment such as ladders or lifts that are needed when performing tests on an overhead lamp.

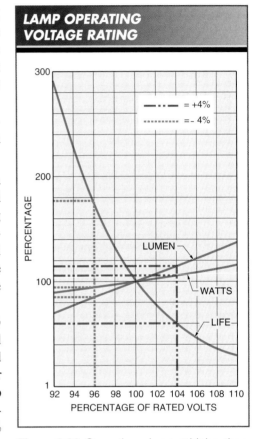

Figure 6-32. Operating a lamp at higher than rated voltage is more destructive than operating it at lower than rated voltage and should be taken into consideration as a problem when lamps have a shortened life.

TECH FACT
Use dimmer switches to reduce the amount of light when full brightness is not required and to conserve energy and bulb life.

When testing a lighting circuit, system grounding must be verified. Verifying that the system is grounded helps prevent electrical shocks and allows for easier troubleshooting because there is not always a neutral conductor in the switch box. System voltage must be measured between the hot conductor entering and exiting the switch, the neutral conductor, and/or the ground. **See Figure 6-33.**

Motion Sensors

A *motion sensor* is a sensing device that detects the movement of a temperature variance and automatically switches when the movement is detected. Temperature variance is normally caused by the movement of an individual but can also occur from other temperature changes such as during a fire. While most motion sensors are infrared-sensing devices, other motion sensors use microwave or ultrasonic waves to detect a temperature variance.

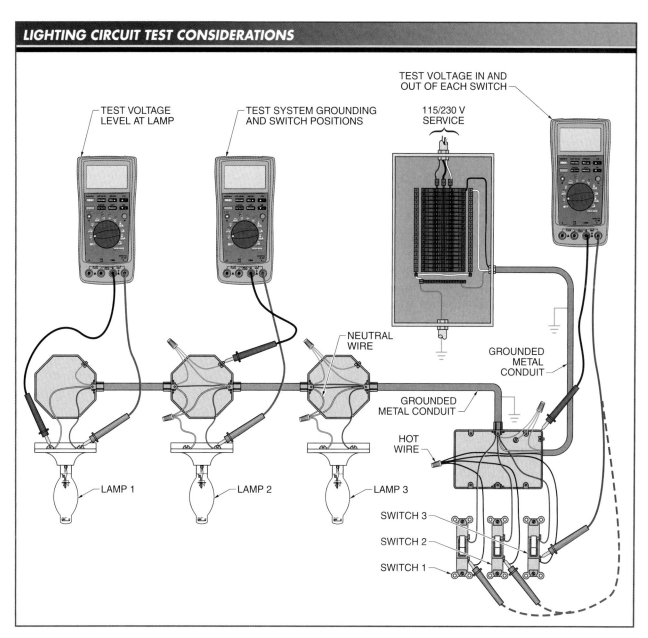

Figure 6-33. When testing switches, lamps, and grounds, voltage measurements can be taken at several different locations.

Motion sensors are used in commercial areas such as parking lots, walkways, hallways, stairwells, restrooms, building entrances and exits, and for general room security. Motion sensors are used to conserve energy since lamps connected to motion sensors are only on as required.

Commercial motion sensors are connected in the same manner as a two-way switch and normally include settings for variables such as timer adjustment, motion sensitivity, and airflow compensation. **See Figure 6-34.** The timer allows the switch to remain in the ON position for a set period of time after no additional motion is detected. The sensitivity setting allows for setting the amount of movement needed to activate the sensor, and airflow settings compensate for airflow from open doorways or HVAC equipment.

Light Level Sensors

A *light level sensor* is a control device that measures the amount of ambient light and activates a control switch at a preset light level. Light level sensors are normally used for automatic control of switching on and off lamps but are also used to control electrically operated window shades and other loads. They can also be used to turn off or dim lamps when the ambient light level is high and turn on or brighten lamps when the ambient light level is low.

As with motion sensors, light level sensors include adjustments for setting light-activation levels and delay to prevent nuisance activation during changes caused by cloud patterns, automobile headlights, etc. They can be used as stand-alone devices or with additional timers that automatically turn lamps on and off after a preset light level is reached and a set time has passed. When controlling

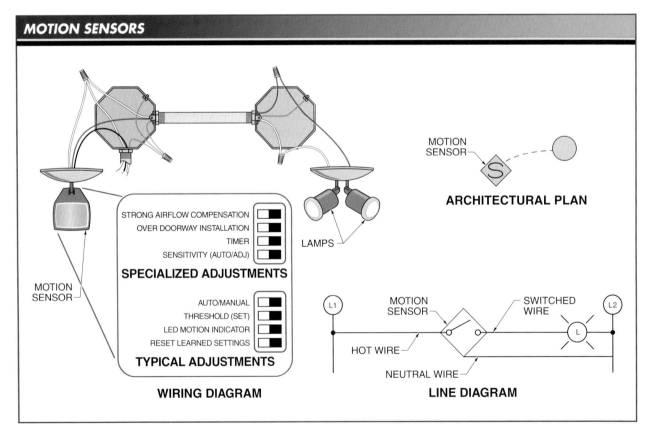

Figure 6-34. Motion sensors are connected in the same manner as a two-way switch and normally include settings for variables such as timer adjustment, motion sensitivity, and airflow compensation.

high wattage or multiple lamps, light level sensors are used to control lighting contactors that control lamps.

Contactors

A *contactor* is a control device that uses a small control current to energize or deenergize the load connected to it. Contactors are either heating contactors or lighting contactors, depending on whether they control heating elements or lamps. Contactors include a coil and one or more control devices. Control devices can be manually operated switches, motion sensors, or light level sensors and are used to control contactor coils. Commercial lighting contactors are used to switch high-power lamps or multiple lamps by using a low-voltage control switch. **See Figure 6-35.**

Refer to Chapter 6 Quick Quiz® on CD-ROM.

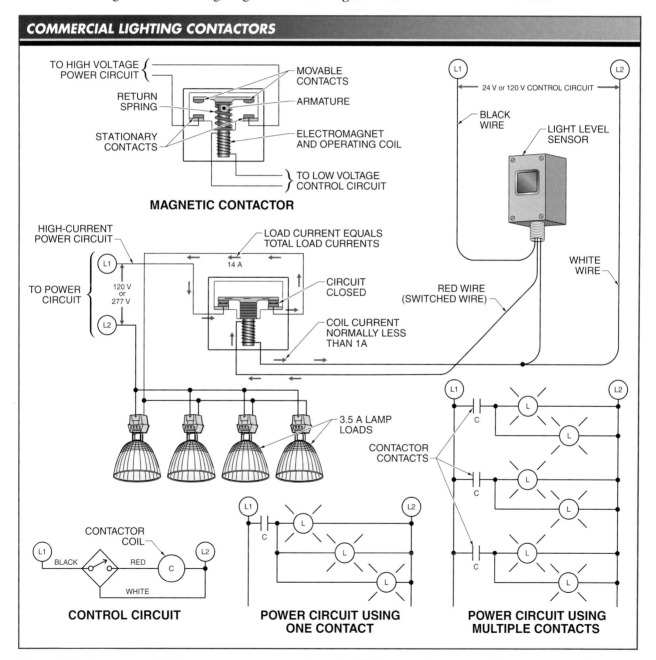

Figure 6-35. Commercial lighting contactors are used to switch high-power lamps or multiple lamps by using a low-voltage control switch.

Case Study—Troubleshooting Lighting Systems

In a commercial building, the maintenance staff have been changing the fluorescent lamps in several office areas. Two problems have been reported. The first problem is that the office staff has been complaining that some of the new bulbs produce less light than the old bulbs, making it harder to work under the reduced light. The second problem is that the bulbs seem to be burning out too quickly. In addition, many bulbs are blackening at their ends, rather than along the entire bulb surface as is normal.

A check of the old bulbs indicates that they have a color temperature that is a Kelvin rating of 5000 K, which is classified as natural light. The new bulbs have a color temperature that is a Kelvin rating of 3000 K, which is classified as warm light. This means that the replacement bulbs give off a more yellowish light. To solve this problem, it is decided that the office areas will use 5000 K rated bulbs and the hallways will use 3000 K rated bulbs.

The second problem that is addressed is the short bulb life and heavy blackening at the bulb ends. The warranty from the bulb manufacturer is checked and it is determined that the bulb warranty period is 2 years. Although most of the bulbs were failing at a little over 2 years, they should have been lasting longer.

Fluorescent lamp bulb life can be shortened by the lamps being turned on and off frequently, operated in low ambient temperatures, or supplied with low voltage. Since the lamps were normally turned on and off only once a day and were located indoors where low temperatures are not a problem, these issues can be eliminated as causes of the problem. Since short bulb life and heavy blackening of the bulb ends are both signs of low supply voltage, the voltage at the lighting circuits must be measured.

Wearing proper PPE, voltage measurements are taken by the maintenance staff at the switch controlling the lights. A DMM set to measure AC voltage is used to take the measurements.

When the lights are off, the meter reads the voltage of the circuit. For fluorescent lamp circuits, the voltage should not be more than 7% less than the rating of the bulbs. A supply voltage below 7% of the bulb rating shortens bulb life and can cause the bulb to flicker. Since the bulbs are rated at 277 V, the supply voltage should not be less than 257 V.

The measured voltage with the switch in the OFF position is 264 V. Although the measured voltage is low, it is still within the acceptable range. However, building voltage can fluctuate. For this reason, the meter is connected across the switch and the MIN/MAX record function is set on the meter.

The measurements over time show that the minimum measured voltage was 258 V and the maximum was 266 V. Thus, it is determined that there is a low voltage problem that is causing reduced lamp life. The transformer feeding the circuit must be tested. The transformer taps may need to be adjusted to increase the voltage in the circuit.

Definitions

- An *arc tube* is the light-producing element of an HID lamp.
- A *ballast* is an autotransformer that delivers a specific voltage and limits the electric current supplied to a lamp.
- A *bayonet base* is a lamp base that has two pins located on opposite sides.
- *Color rendering* is the appearance of a color when illuminated by a light source.
- *Color temperature* is a description of the coolness or warmth of a light source and is measured in Kelvin (K).
- A *compact fluorescent lamp (CFL)* is a fluorescent lamp that has a smaller diameter than a conventional fluorescent lamp and has a folded bulb configuration.
- A *contactor* is a control device that uses a small control current to energize or deenergize the load connected to it.
- A *diode* is a semiconductor device that offers high opposition to current flow in one direction and low opposition to current flow in the opposite direction.
- A *double-pole (DP) switch* is a control device that consists of two switches in one and is used for controlling two separate loads.
- An *electric discharge lamp* is a lamp that produces light by means of an arc discharged between two electrodes.
- An *emergency exit lighting system* is a lighting system that is activated when a power failure occurs and provides adequate lighting for egress from a building.
- An *exit lighting system* is illuminated signage that indicates the safest and quickest path out of a building.
- A *filament* is a conductor with a resistance high enough to cause the conductor to heat.
- A *fluorescent lamp* is a low-pressure discharge lamp in which ionization of mercury vapor transforms ultraviolet energy generated by the discharge into light.
- A *footcandle (fc)* is the amount of light produced by a lamp (lumens) divided by the area that is illuminated.
- A *four-way switch* is a control device that is used in combination with two three-way switches to allow control of a load from three locations.
- A *high-intensity discharge (HID) lamp* is a lamp that produces light from an arc tube.
- A *high-pressure sodium lamp* is an HID lamp that produces light when current flows through sodium vapor under high pressure and high temperature.
- An *ignitor* is a component inside a lamp ballast that produces a high starting voltage.
- *Illumination* is the effect that occurs when light contacts a surface.
- An *incandescent lamp* is an electric lamp that produces light by means of the flow of current through a tungsten filament inside a gas-filled, sealed glass bulb.
- An *instant-start circuit* is a fluorescent lamp starting circuit that provides sufficient voltage to strike an arc instantly.
- *Invisible light* is the portion of the electromagnetic spectrum on either side of the visible light spectrum.
- A *knuckle thread* is a rounded thread normally rolled from sheet metal and used in various forms for electric bulbs and bottle caps.
- A *lamp* is an output device that converts electrical energy into light.

- *Light* is the portion of the electromagnetic spectrum that produces radiant energy.
- A *light-emitting diode (LED)* is a diode that emits light when forward direct current is applied.
- A *light level sensor* is a control device that measures the amount of ambient light and activates a control switch at a preset light level.
- A *low-pressure sodium lamp* is an HID lamp that operates at a low vapor pressure and uses sodium as the vapor.
- A *lumen (lm)* is the unit used to measure the total amount of light produced by a light source.
- A *mercury-vapor lamp* is an HID lamp that produces light by means of an electrical discharge through mercury vapor.
- A *metal-halide lamp* is an HID lamp that produces light by means of an electrical discharge through mercury vapor and metal halide in the arc tube.
- A *motion sensor* is a sensing device that detects the movement of a temperature variance and automatically switches when the movement is detected.
- A *pole* is the number of completely isolated circuits that a switch can switch.
- A *preheat circuit* is a fluorescent lamp starting circuit that heats the cathode before an arc is created.
- A *rapid-start circuit* is a fluorescent lamp starting circuit that brings the lamp to full brightness in about 2 sec.
- A *single-pole (SP) switch* is an electrical control device used to turn lights or appliances on and off from a single location.
- A *three-way switch* is a single-pole, double-throw (SPDT) switch.
- A *throw* is the number of closed switch positions per pole.
- A *tungsten-halogen lamp* is an incandescent lamp filled with a halogen gas such as iodine or bromine.
- A *two-way switch* is a single-pole, single-throw (SPST) switch.
- *Visible light* is the portion of the electromagnetic spectrum to which the human eye responds.
- A *watt (W)* is the unit of electrical power equal to the power produced by a current of 1 A across a potential difference of 1 V.

Review Questions

1. Define light and discuss its two forms.
2. What are the common types of lamps used in commercial buildings?
3. Describe a tungsten-halogen lamp and its operation, and list several common uses.
4. Define a ballast as it relates to fluorescent lamps.
5. What are the steps for measuring ballast voltage?
6. Define and list several requirements of an exit lighting system.
7. What are the steps for testing an emergency exit lighting system?
8. What is the procedure for testing circuits that contain switches for proper operation?
9. Describe a motion sensor and its operation, and list several common uses.
10. What is a contactor?

chapter 7
HVAC SYSTEMS

Heating, ventilating, and air conditioning (HVAC) systems are designed to ensure the air in a building is clean, safe, and at a comfortable temperature. Although different systems are designed for producing heated air and water, cooled air and water, and clean and properly circulated air, each system has many of the same components and devices. Understanding how to maintain, test, and troubleshoot system components and devices ensures that HVAC systems are maintained in proper operating condition. The main electrical components and devices that perform work in commercial HVAC systems are motors, heating elements, solenoid-operated directional control valves, and dampers.

OBJECTIVES

- Explain the procedure for troubleshooting split-phase and capacitor motors.
- Calculate the percentage of voltage unbalance in a three-phase motor.
- Calculate the percentage of current unbalance of a three-phase motor.
- Test motor windings using a megohmmeter.
- Test contactor coils.
- Measure the current flowing in an electric heating element.

ELECTRIC MOTORS IN HVAC SYSTEMS

HVAC systems include electric motors for circulating air, water, and coolant. In heating and cooling systems, electric motors are used to rotate pumps and compressors that move air and liquids through a system regardless of whether the system is a gas, oil, electric, boiler, geothermal, solar, or hybrid system. Troubleshooting electric motors requires testing for various malfunctions, environmental or mechanical problems, voltage unbalance, and current unbalance.

The three main types of electric motors are DC, single-phase AC, and three-phase AC motors. DC motors produce higher torque than similarly sized AC motors. However, DC motors are normally not installed in HVAC systems because they have high maintenance requirements because of brush wear and because motor space is normally not a concern in most HVAC systems.

> **TECH FACT**
>
> *Electric motors produce torque by means of magnetic force that is created within the motor. The strength of the magnetic force determines the amount of torque that the motor can produce. The torque that a motor produces is used to perform work.*

Although AC motors have certain electrical and mechanical problems that must be periodically corrected, they are the most common motor used in HVAC systems. In commercial HVAC systems, single-phase and three-phase AC motors are used to drive equipment that circulates air and liquids.

Single-phase motors are used when a motor with a fractional horsepower rating is required or if there is only single-phase power available. Three-phase motors are used when a motor rated higher than 1 HP is required or if a motor is controlled by a motor drive for speed control or energy savings.

Although single-phase motors can be used in HVAC systems, the most common AC motors used in commercial HVAC systems are three-phase motors. Three-phase motors are more energy efficient than single-phase motors of the same horsepower rating. Three-phase motors have three, six, nine, or twelve leads inside the motor connection box. Motors that have three leads are single-voltage motors and must have the supply voltage match the motor nameplate rated voltage. Motors that have six leads are dual-voltage motors that can be connected in either a delta configuration (low voltage) or a wye configuration (high voltage). Motors that have nine and twelve leads are dual-voltage motors that can also be wired for either low-voltage or high-voltage operation. Motors with three, six, or nine leads are normally installed in commercial HVAC systems. **See Figure 7-1.**

WARNING: HVAC systems contain capacitors that can remain charged and cause an electrical shock even after power has been turned off. To help reduce the risk of electrical shock, proper PPE must be worn. In addition, for systems that operate on 208 V to 240 V, a period of at least 5 min should pass before working on electrical components after the power has been turned off. For systems that operate on 380 V to 460 V, a period of at least 8 min should pass, and for systems operating on 525 V to 600 V, a period of at least 12 min should pass before working on the electrical components after the power has been turned off.

Electric Motor Malfunctions

Electric motors are designed and sized to operate in HVAC systems for a number of years with minimal malfunctions or failures. For a motor to operate without malfunctions or failures, the motor electrical operating conditions must be within the original equipment manufacturer (OEM) operating specifications. The motor operating specifications are listed on the motor nameplate. Motors are rated to operate at a specified voltage and current to deliver full horsepower without producing excessively high temperatures. In addition to the voltage at the motor being within the acceptable range of the nameplate rating, high transient voltages must be avoided because they can cause deterioration of motor insulation and electrical malfunctions.

Voltage Malfunctions. Electric motors have an operating voltage range in which they can perform satisfactorily. The OEM specifies the operating voltage range in the electrical specifications provided with the HVAC system or motor. It is standard practice to use the OEM specifications because the listed values are based on data from actual motor use and operating conditions. If OEM specifications are not available, the voltage range is normally +5% to –10% of the nameplate rated voltage. **See Figure 7-2.**

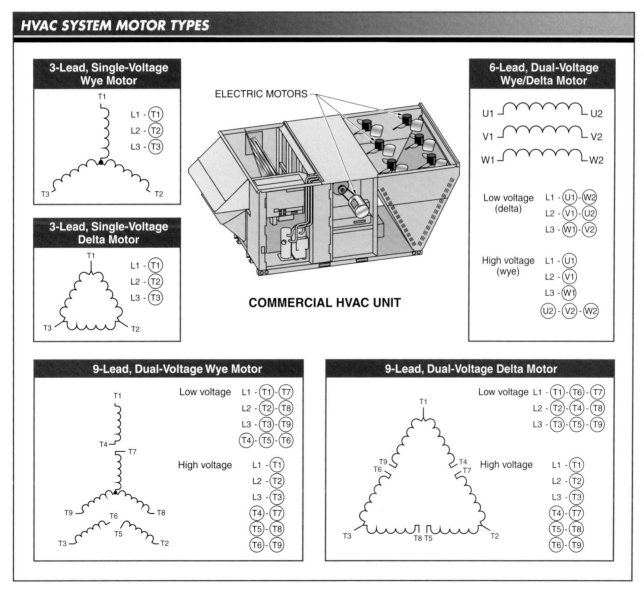

Figure 7-1. The most common motors used in commercial HVAC systems are three-phase motors with three, six, or nine leads.

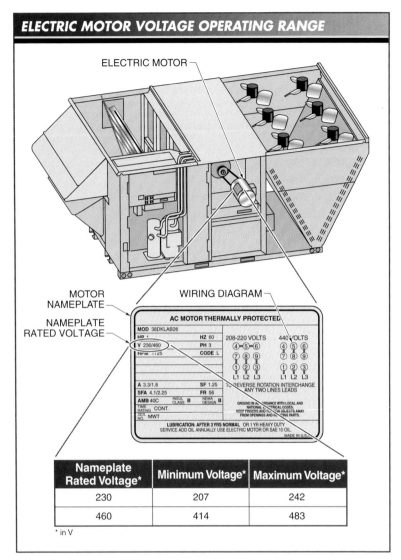

Figure 7-2. If original equipment manufacturer (OEM) specifications are not available, the motor operating voltage range is normally +5% to −10% of the nameplate rated voltage.

When measuring voltage, it is best to measure the voltage over time. A test instrument MIN MAX operating function can be used. Voltage measurements taken over time should not vary by more than 3%. A high voltage fluctuation is an indication that the system is overloaded, the conductors connecting the system are too small, or the conductor run is too long.

For three-phase motors, the voltage between each of the power lines connected to the motor (T1 to T2, T2 to T3, and T1 to T3) should be measured. The voltage measured between any two lines should not vary more than 3%.

Excessive Heat Due to Motor Acceleration Time. When full power is applied to a motor, the motor accelerates to full speed. A motor is connected to full power when a motor starter (across-the-line starter) is used to start the motor. When a motor starter is used to start a motor, the motor must accelerate to its rated speed within a limited time period. The longer a motor accelerates, the higher the temperature rise in the motor. The larger the load a motor must operate, the longer the acceleration time. The maximum recommended acceleration time depends on the motor frame size. Large motor frames dissipate heat faster than small motor frames.

When a motor drive is used to control a motor, acceleration and deceleration time can be programmed to best match the requirements of the application. The programmed acceleration and deceleration time must not overheat the motor. Motor drives automatically control the voltage applied to a motor to keep it from overheating at all speeds. However, in HVAC systems, the motor acceleration time should be as short as possible. Normally, the OEM default settings for acceleration and deceleration times are acceptable. **See Figure 7-3.**

Overcycling. *Overcycling* is the process of turning a motor on and off repeatedly. Overcycling occurs when a motor is at its operating temperature and continues to cycle on and off. Starting current is several times the running full-load current (FLC) of the motor. Regardless if a motor is started using a motor starter or motor drive, most motors are not designed to start more than 10 times per hour because it increases the temperature of the motor, which destroys the motor wire insulation.

Overcycling in HVAC units occurs when the controlling temperature switch (thermostat) differential is set too low. The *differential* is the difference between the temperature at which a switch turns on a unit and the temperature at which it turns off the unit. For example, a 1°F differential keeps the temperature in a room within 1°F, but requires the unit to continuously cycle on and off. Thermostats have a typical default differential setting of approximately 6°F (4.5°C).

When a motor application requires a motor to be cycled often, the following guidelines should be applied:
- Install a motor that has a 122°F (50°C) ambient temperature rating rather than a standard 104°F (40°C) rating.
- Install a motor with a service factor of 1.25 or 1.35 rather than a service factor of 1.00 or 1.15.
- Provide additional cooling by forcing air over the motor.
- Install a motor drive to control the motor so the motor speed can be controlled instead of cycling the motor fully on and off.

TECH FACT

The National Electrical Manufacturers Association (NEMA) standard MG1 sets the basic requirements for information to be marked on electric motor nameplates.

Motor Insulation Failure. An *ohmmeter* is a test instrument that measures resistance. A *megohmmeter* is a high-resistance ohmmeter used to measure insulation deterioration on conductors by measuring high resistance values using high-voltage test conditions. A megohmmeter can detect motor insulation deterioration before a motor fails. Typical test voltages range from 50 V to 5000 V. A megohmmeter is used to measure the condition of motor wiring by detecting insulation failure caused by excessive moisture, dirt, heat, cold, corrosive vapors or solids, vibration, and aging. **See Figure 7-4.**

MOTOR ACCELERATION TIME

MAXIMUM ACCELERATION TIME – MAGNETIC MOTOR STARTER

Frame Number	Maximum Acceleration Time*
48 and 56	8
143–286	10
324–326	12
364–506	15

MAXIMUM ACCELERATION TIME – MOTOR DRIVE

Parameter	MIN MAX Setting*	Factory Default Setting*
Acceleration Time	0.1–600	10.0
Deceleration Time	0.1–600	10.0

* in sec   acceptable range

Figure 7-3. In HVAC systems, motor acceleration time should be as short as possible.

A megohmmeter is used to measure the resistance of different motor windings or the resistance from a motor winding to ground. Several test measurements should be taken and recorded over time to provide a complete analysis of the insulation condition. The minimum acceptable insulation resistance depends on the motor voltage rating. **See Figure 7-5.**

Note: A motor with good insulation may have readings 10 times to 100 times the minimum acceptable resistance. Service the motor if the resistance reading is less than the minimum value.

CAUTION: A megohmmeter uses high voltage for testing (up to 5000 V). Avoid touching the meter leads to the motor frame. Always follow the OEM recommended service and safety procedures. After performing insulation test measurements, connect the motor windings to ground through a 5 kΩ, 5 W resistor. The motor winding must be connected to ground for 10 times the motor testing time in order to discharge the energy stored in the wiring.

214 ELECTRICAL SYSTEMS for FACILITIES MAINTENANCE PERSONNEL

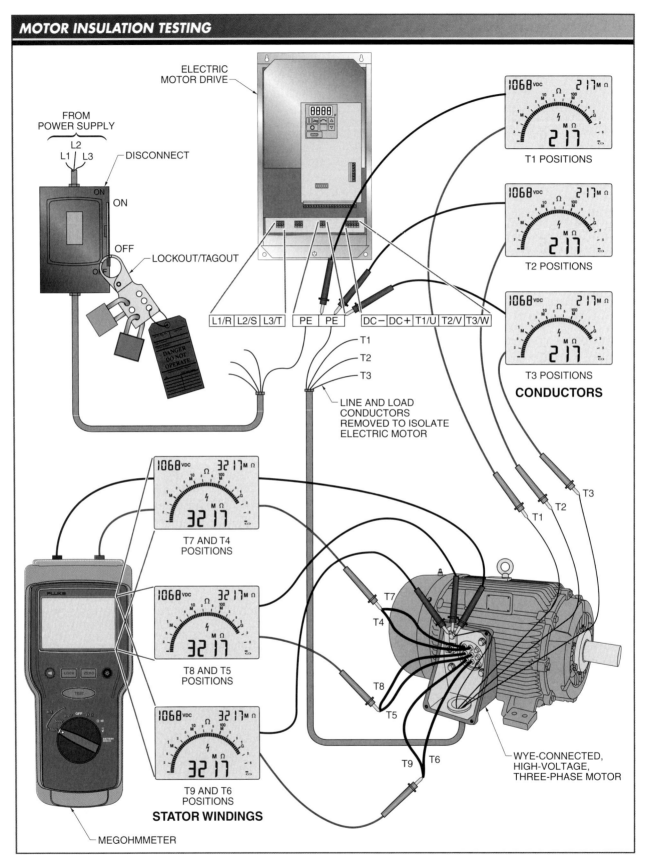

Figure 7-4. A megohmmeter is used to perform tests on motor insulation.

| MOTOR INSULATION RESISTANCE ||
Motor Nameplate Rated Voltage*	Minimum Acceptable Resistance†
<208	100,000
208–240	200,000
241–600	300,000
601–1000	1,000,000

* in V
† in Ω

Figure 7-5. The minimum acceptable insulation resistance of an electric motor depends on the motor voltage rating.

Motor Environmental and Mechanical Problems

In order for a motor to operate properly over time, the motor must be operated in an environment for which it is designed and within its mechanical limits. Motors used in HVAC systems have been selected and installed by the OEM to operate properly for an indefinite period of time. However, problems can occur when the HVAC system is subjected to environmental problems such as poor ventilation, high moisture, or poor air quality, or mechanical problems such as over tightened belts and equipment vibration. Problems can also occur when the system is not operated as designed.

Heat. Excessive heat is a major cause of motor failure and is a symptom of other motor problems. Heat destroys the insulation on motor wiring. When insulation is destroyed, the motor windings become short-circuited, and the motor is no longer functional.

Electric motors produce heat as they convert electrical energy to mechanical energy, which must be removed to prevent destruction of motor wiring insulation. Motors are designed with air passages that permit a free flow of air over and through the motor because airflow removes the heat from the motor. Any restrictions to airflow through a motor cause the motor to operate at a higher temperature than it was designed for.

Airflow through a motor may be restricted by the accumulation of dirt, dust, lint, grass, insect or rodent matter, or rust. If a motor is coated with oil from leaking seals or from overlubrication, airflow is more heavily restricted.

As the speed of a motor is decreased, the airflow over the motor windings is also decreased. In HVAC systems that include a variable-speed drive, decreased airflow is not a problem because the OEM designed the system to operate with the motor running at different speeds. However, when a magnetic motor starter is replaced with a variable-speed drive, motor overheating may result because the drive allows motor speed to be reduced, which reduces the airflow over the motor. The reduced airflow was not originally designed into the system with the magnetic motor starter.

The life of the insulation is shortened when the heat in a motor increases beyond the temperature rating of the insulation. The higher the temperature, the sooner the insulation fails. The temperature rating of motor insulation is listed as the insulation class on the nameplate of the motor. **See Figure 7-6.** The temperature rating of the insulation class is listed in Fahrenheit (°F) and/or Celsius (°C). Heat buildup in a motor can be caused by the following conditions:
- incorrect motor type or size for the application
- improper cooling caused by dirt buildup
- excessive load from improper use
- excessive friction from misalignment or vibration
- electrical problems such as voltage unbalance, phase loss, or voltage surge

Overloads. An *overload* is the condition that occurs when the load connected to a motor exceeds the full-load torque rating of the motor. An electric motor attempts to drive any load connected to it when the power is on. The larger the load, the more power required to move the load. Every motor has a limit to the load it can drive.

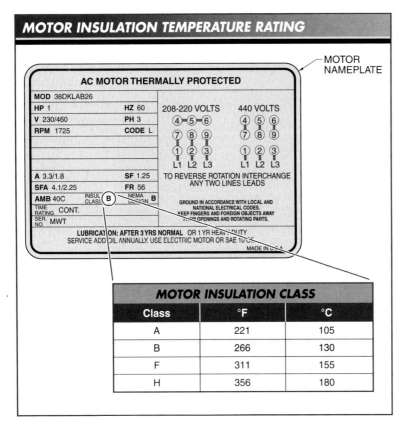

Figure 7-6. The temperature rating (maximum temperature) of motor insulation is listed as the insulation class on the nameplate of the motor.

overloaded if it draws more than nameplate rated current. A larger motor may be used or the load on the motor decreased if overloads are causing motor malfunctions.

Altitude. Temperature is normally not a problem for motors that operate at altitudes of 3300′ or less. A motor is derated when it operates at altitudes above 3300′. *Derating* is the reduction of the total rated operating output of a device to allow for safe operation under abnormal environmental conditions. For example, motors in standard HVAC systems that are installed at altitudes higher than 3300′ require derating to prevent motor overheating. **See Figure 7-8.**

Overloads should not harm a properly protected motor. Any overload present longer than the built-in time delay of the protection device is detected and power is removed from the motor. Properly sized heaters in magnetic motor starters and properly programmed overload parameter settings in motor drives ensure that a motor is removed from an overload before any damage to the motor occurs.

A motor that fails due to overloading can be identified by an even blackening of all motor windings. The blackening of motor windings is caused by the slow destruction of the motor over a long period of time. **See Figure 7-7.**

Current readings can be taken at the motor to determine if an overload problem is present. A motor is operating at its maximum recommended output if it draws nameplate rated current. A motor is

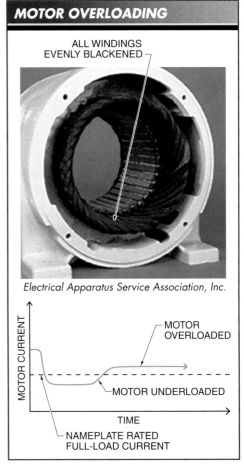

Figure 7-7. A motor that fails due to overloading can be identified by an even blackening of all motor windings.

Altitude Range*	Service Factor			
	1.0	1.15	1.25	1.35
3300–9000	93%	100%	100%	100%
9001–9900	91%	98%	100%	100%
9901–13,200	86%	92%	98%	100%
13,201–16,500	79%	85%	91%	94%
Over 16,500	Consult manufacturer			

* in ft

Figure 7-8. Motors in standard HVAC systems that are installed at altitudes higher than 3300′ require derating to prevent motor overheating.

Mounting and Positioning. Motors that are not properly mounted are more likely to experience mechanical failure than motors that are properly mounted. Motors must be mounted on a flat, stable base in order to reduce vibration and misalignment problems. In HVAC systems, the OEM mounts and adjusts the motor for best operation. However, over time, belts and motor mounts may loosen and may require adjustment. Adjustments are required whenever motor belts, couplings, or the motor is replaced.

An *adjustable motor base* is a mounting base that allows a motor to be easily moved a short distance. An adjustable motor base allows for easy motor installation, maintenance, mounting, alignment, belt tensioning, and replacement.

Belt tension is checked by placing a straightedge from the drive pulley to the driven pulley and measuring the amount of deflection at the midpoint. Belt tension can also be checked using a tension tester. Belt deflection should equal $\frac{1}{64}''$ per inch of span. For example, if the span between the center of a drive pulley and the center of a driven pulley is 16″, the belt deflection is $\frac{1}{4}''$ (16″ × $\frac{1}{64}''$ = $\frac{1}{4}''$). Belt tension is normally adjusted by moving the drive component (motor) away from or closer to the driven component. **See Figure 7-9.**

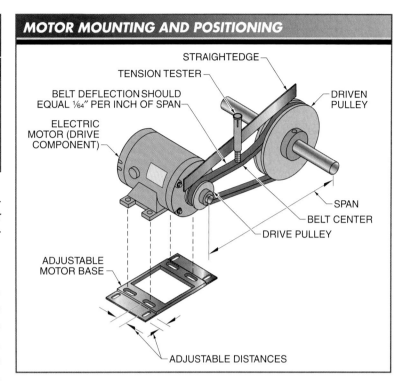

Figure 7-9. Motor mounting and positioning (belt tension adjustment) is required whenever motor belts, couplings, or the motor is replaced.

Electric motors produce vibration as they rotate. This vibration can loosen mechanical and electrical connections. Loose mechanical connections generally cause noise and are easily detected. Loose electrical connections do not cause noise but do cause a voltage drop to the motor. Loose electrical connections also cause excessive heat. Mechanical and electrical connections should be inspected when troubleshooting a motor.

Misalignment of a motor and its driven load is a common cause of motor failure. If a motor and driven load are misaligned, premature failure of the motor bearings, load, or both can occur. Equipment shafts must be properly aligned on new installations and checked during periodic maintenance inspections. Misalignment is normally corrected by placing shims under the feet of the motor. Shims are used to align the motor.

> **TECH FACT**
>
> After being turned off, electric motors remain warm for a period of time. Pests, such as insects and rodents, may be attracted to a warm motor. A special screened zip-seal jacket that allows airflow can be placed around the motor to help protect it against damage from pests.

Single-Phase Motors

Single-phase motors are designed to be connected to a single-phase AC power source. HVAC single-phase motors are normally designed to be connected to 120 VAC or 240 VAC. Single-phase motors include a rotor (rotating part) and a stator (stationary part) and require a method for starting the rotor in the correct direction. Single-phase motors are designated by their starting methods and are classified as shaded-pole, split-phase, capacitor-start, capacitor-run, and capacitor start-and-run motors.

Shaded-Pole Motors. A *shaded-pole motor* is a single-phase AC motor that uses a shaded stator pole for starting. Shaded-pole motors produce the least amount of starting torque and are not normally used in HVAC systems except for small air-circulating fans. Shaded-pole motors are small and inexpensive and are normally replaced when they fail.

Split-Phase Motors. A *split-phase motor* is an AC motor that can run on one or more phases. Split-phase motors produce higher starting torque than shaded-pole motors and are commonly installed in HVAC systems that use single-phase motors. Split-phase motors have a main (running) winding and an auxiliary (starting) winding. The two windings are placed in the stator slots and positioned 90° apart.

The running winding is composed of large gauge conductors and has a greater number of turns than the starting winding. When the motor is first energized, the inductive reactance of the running winding is higher and the resistance is lower than the starting winding. *Inductive reactance* is the opposition to the flow of alternating current in a circuit due to inductance.

The starting winding is composed of small gauge conductors and has a fewer number of turns than the running winding. When the motor is first energized, the inductive reactance of the starting winding is lower and the resistance is higher than the running winding.

When power is applied to a split-phase motor, both the running and starting windings are energized. The running winding current lags the starting winding current because of its different inductive reactance, which produces a phase difference between the starting and running windings. A 90° phase difference is required to produce maximum starting torque, but the phase difference is normally less than 90°. A rotating magnetic field is produced because the two windings are out of phase. The rotating magnetic field starts the rotor rotating. With the running and starting windings out of phase, the current changes in magnitude and direction, and the magnetic field moves around the stator. The movement of the magnetic field forces the rotor to rotate with the rotating magnetic field.

To minimize energy loss and prevent heat buildup in the starting winding once the motor is started, a centrifugal switch is used to remove power from the starting winding when the motor reaches a set speed. A *centrifugal switch* is a switch that opens to disconnect the starting winding when the rotor reaches a preset speed and closes to reconnect the starting winding when the rotor speed drops below the preset value. In most motors, the centrifugal switch is located inside the motor enclosure on the motor shaft. **See Figure 7-10.**

As the motor shaft accelerates, the switch is activated by centrifugal force. *Centrifugal force* is the force that moves rotating bodies away from the center of rotation. In certain motors, the centrifugal switch is located outside the motor enclosure for ease of repair.

Centrifugal Switch Media Clip

The centrifugal switch is connected in series with the starting winding and automatically deenergizes the starting winding at a predetermined speed. The predetermined speed is normally about 60% to 80% of the running speed. After power from the starting winding is removed, the motor continues to operate on the running winding only.

In addition to a centrifugal switch, certain split-phase motors contain a thermal switch. A *thermal switch* is a switch that operates its contacts when a preset temperature is reached. A thermal switch is located inside a motor and is used to protect the motor windings from burnout. **See Figure 7-11.** A thermal switch is activated by high temperatures and automatically removes power from both the starting and running windings. A higher-than-normal temperature can be caused by improper ventilation, excessive motor load, high ambient temperature, or a mechanical problem that prevents motor shaft rotation.

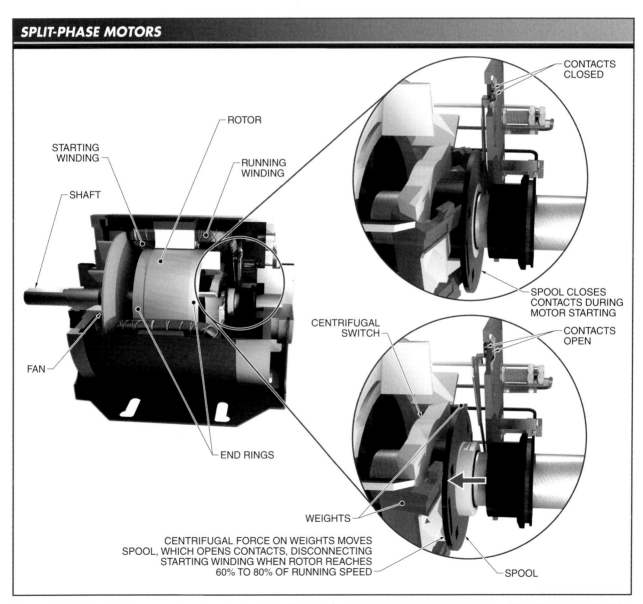

Figure 7-10. A split-phase motor uses an auxiliary (starting) winding for starting the motor, which is deenergized by a centrifugal switch after the motor reaches a preset speed.

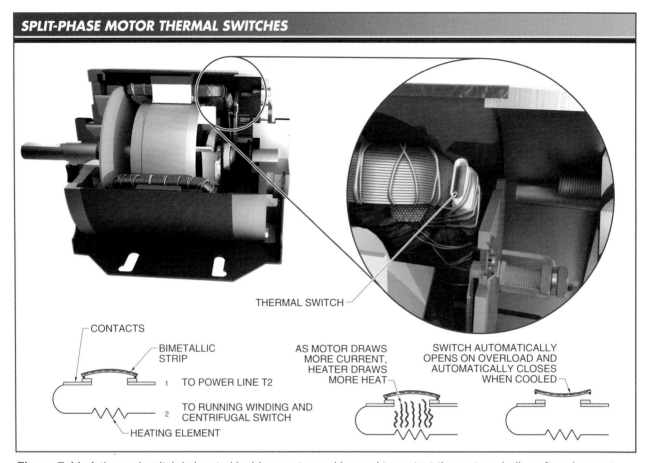

Figure 7-11. A thermal switch is located inside a motor and is used to protect the motor windings from burnout.

During normal operation, a thermal switch is in the closed position. As a higher-than-normal current passes through the motor windings, the switch begins to heat. At a preset temperature, the switch opens, which automatically removes the motor windings from power. When the motor windings are removed from power, they begin to cool. The thermal switch closes as it cools, and the motor is automatically restarted. The thermal switch constantly cycles the motor on and off if the problem that caused the original overheating is not corrected.

Troubleshooting Split-Phase Motors. Most problems with split-phase motors are caused by a malfunction with the centrifugal switch or thermal switch. A split-phase motor can be repaired if the problem is with the centrifugal switch or thermal switch but is normally replaced because the motor is normally a fractional horsepower motor. The motor is replaced if the windings have been burned or if the motor has been in service for more than five years. **See Figure 7-12.** Troubleshooting a split-phase motor is performed using the following procedure:

1. Shut off power to the motor and visually inspect the motor. Replace the motor if the windings are burned, the shaft is jammed, or there is any damage.
2. If the motor is controlled by a manual thermal switch, reset the thermal switch and turn the motor on.
3. Wear proper PPE and use a DMM to test for voltage at the motor terminals if the motor does not start. *Note:* Voltage measurements must be within 10% of the motor listed voltage. If the voltage measurement is not correct, troubleshooting should

be done on the circuit leading to the motor. If the voltage is correct, power to the motor should be shut off so the motor can be tested.

4. Turn off and lock out power to the safety switch or combination starter.
5. With power off, connect a DMM set to measure resistance to the same motor terminals from which the incoming power leads were disconnected to read the resistance of the starting and running windings. *Note:* The combined resistance of the windings should be less than the resistance of either winding alone because the windings are connected in parallel. A short circuit is present if the DMM displays a reading of 0 Ω (or MΩ). An open circuit is present if the DMM reads infinity (∞). If either condition is present, the motor will need to be replaced because the motor size is too small for a repair to be cost effective.
6. Visually inspect the centrifugal switch for signs of burning or broken springs. Service or replace the switch if any visual problems are present. Test the switch with a DMM set to measure resistance if no visual signs of problems are present. Manually operate the centrifugal switch. *Note:* The endbell on the switch side may have to be removed. The resistance measurement on the DMM decreases if the motor is in proper operating condition. If the resistance does not change, the motor must be subjected to additional testing to determine the problem.

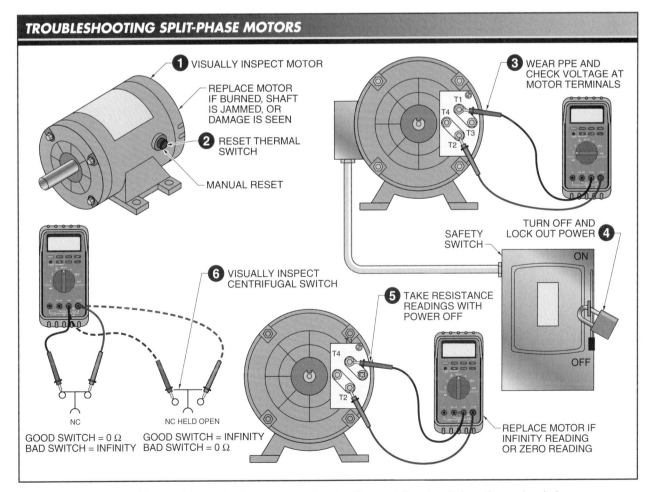

Figure 7-12. Most problems with single-phase motors involve the centrifugal switch or thermal switch.

WARNING: Certain types of split-phase motors can include a thermal switch that automatically shuts off the motor when it overheats. Thermal switches may be manual reset or automatic reset switches. Caution should be taken with any motor that has an automatic reset switch because the motor could automatically restart at any time.

A thermal switch removes the starting and running motor windings from the circuit at a preset temperature. When the windings cool, the contacts close. When a thermal switch is not operating properly, a motor either does not operate or operates without thermal protection. **See Figure 7-13.** A thermal switch is tested using the following procedure:

1. Remove the motor from power and allow approximately 15 min to 30 min for the motor to cool.
2. Remove the endbell of the motor from the side that includes the thermal switch.
3. Remove one lead running from the thermal switch to the motor windings.
4. In a motor containing a two-terminal thermal switch, test for continuity (low resistance) across the switch contacts using a DMM set on resistance or continuity. *Note:* The contacts are open and the switch is defective if a high-resistance measurement is obtained.

TECH FACT

Split-phase motors are a common type of single-phase motor used in light commercial buildings for applications that do not require very high starting torque.

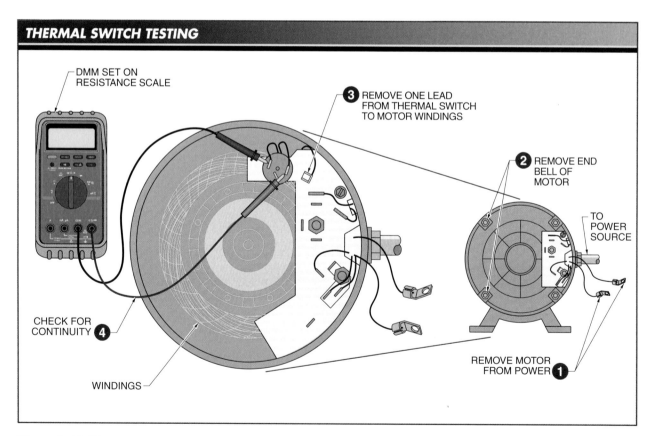

Figure 7-13. Thermal switches are tested when a faulty motor either does not operate or operates without thermal protection.

In a motor containing a three-terminal thermal switch, test for continuity (low resistance) across the switch contacts using a DMM set on resistance or continuity. The contacts are open and the switch is defective if a high-resistance measurement is obtained. Test for continuity across the heating element. The heating element is open and defective if continuity is not obtained.

In a motor containing a four-terminal thermal switch, test for continuity across the switch contacts and the heating element using a DMM set on resistance. The contacts or the heating element is open and defective if continuity is not obtained. The component should be replaced.

Capacitor Motors. A *capacitor motor* is a single-phase motor with a capacitor connected in series with the stator windings to produce phase displacement in the starting winding. The capacitor is added to provide a higher starting torque at lower starting current than is delivered by a split-phase motor. Capacitor motor applications include refrigerators, air conditioners, air compressors, and certain types of power tools.

In a capacitor motor, the capacitor causes current in the starting winding to lead the applied voltage by about 40°. Since the starting and running windings are about 90° out of phase, the motor operating characteristics are improved, a higher starting torque is produced, and the motor has a better power factor with a lower current draw. Three types of capacitor motors are capacitor-start, capacitor-run, and capacitor start-and-run motors. Capacitor-start motors are used in HVAC systems to provide the high starting torque required to move air and liquids. Capacitor-run and capacitor start-and-run motors produce a high running torque but are not normally used in HVAC systems because high starting torque is required for most HVAC system motors.

Capacitor-Start Motors. A *capacitor-start motor* is a motor that has the capacitor connected in series with the starting winding. The capacitor is connected only during starting. The capacitor and starting winding are disconnected from the power by a centrifugal switch at a set speed. **See Figure 7-14.**

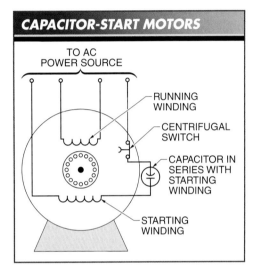

Figure 7-14. A capacitor-start motor has a capacitor connected in series with the starting winding.

A capacitor-start motor develops considerably more locked rotor torque per ampere than a split-phase motor because of the capacitor in the circuit. *Locked rotor torque* is the torque a motor produces when the rotor is stationary and full power is applied to the motor.

Troubleshooting Capacitor and Capacitor-Start Motors. Troubleshooting capacitor and capacitor-start motors is performed similarly to troubleshooting split-phase motors. The only additional device to be considered is the capacitor. Capacitors have a limited life and are often the source of a problem with capacitor motors. Capacitors may have a short or be open or may just deteriorate to the point that they must be replaced.

When a capacitor has a short, the winding in the motor can burn out. When a capacitor opens or deteriorates, the motor has poor starting torque. Poor starting torque can prevent the motor from starting, which normally trips the thermal overloads.

Capacitors are composed of two conducting surfaces separated by dielectric material. *Dielectric material* is a medium in which an electric field is maintained with little or no outside energy supply. Dielectric material is used to insulate the conducting surfaces of a capacitor. Capacitor dielectric material is either oil or electrolyte. An *oil capacitor* is a capacitor that uses oil or oil-impregnated paper as a dielectric. Oil capacitors are sealed in a metal container.

Fluke Corporation
The proper PPE must always be worn when using test instruments to perform measurements on electric motors.

Most motors use electrolytic capacitors. An *electrolytic capacitor* is a capacitor formed by winding together two sheets of aluminum foil separated by pieces of thin paper impregnated with an electrolyte. An *electrolyte* is a conducting medium in which the current flow occurs by ion migration. The electrolyte is used as the dielectric material. The aluminum foil and electrolyte are encased in a cardboard or aluminum cover. A vent hole is provided to prevent a possible explosion in the event the capacitor is shorted or overheated. AC capacitors are used with capacitor motors. Capacitors that are to be connected to AC power sources have no polarity. **See Figure 7-15.** A capacitor motor is tested using the following procedure:

1. Wear the proper PPE for the measurement to be taken and area where work is to be performed.
2. Verify that the DMM is rated for the voltage level and type to be tested.
3. Verify that the DMM is in proper operating condition by testing it on a known, energized voltage source.
4. Turn off and lock out power.
5. Use a DMM set to measure voltage to check for voltage at the motor terminals to verify that power is OFF.
6. Remove the capacitor cover and capacitor from the outside frame of the motor.
7. Visually inspect the capacitor for signs of damage such as leakage, cracks, or bulges. Replace the capacitor if damage is present.

   WARNING: A good capacitor stores an electrical charge and may be energized when power is removed. Before taking a capacitance measurement or touching the capacitor terminals, discharge the capacitor.
8. Discharge the capacitor. To safely discharge a capacitor, place a 20 kΩ, 5 W resistor across the capacitor terminals for 5 sec. Use a DMM set to measure voltage to confirm the capacitor is fully discharged. *Note:* On very large capacitors, the process may have to be repeated as the capacitor can build up a charge again after discharging.
9. After discharging the capacitor, connect a DMM set to measure resistance to the capacitor terminals and test the capacitor. The following DMM measurements indicate if a capacitor is good, short-circuited, or open:

- Good—The measurement changes from zero resistance to infinity. When the measurement reaches the halfway point, remove one of the leads and wait 30 sec. When the lead is reconnected, the measurement should change back to the halfway point and continue to infinity, which indicates that the capacitor can hold a charge.
- Short-circuited—The measurement remains at zero and does not change. The capacitor is defective and must be replaced.
- Open—The measurement remains at infinity and does not change. The capacitor is defective and must be replaced.
10. Verify that the DMM is in proper operating condition by testing it on a known resistance.

Certain DMMs include a capacitor test measurement mode, which can be used to directly measure the capacitance of a capacitor. The measured value should be within 15% of the capacitor listed rating (located on the capacitor).

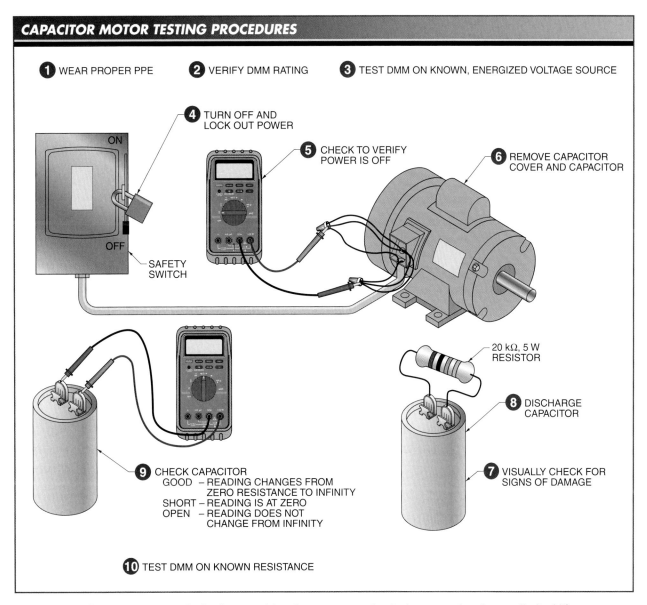

Figure 7-15. Capacitors are periodically tested for short or open circuits because they have a limited life.

Motor enclosures are chosen to fit the operating environment and to help avoid motor failure and system downtime.

Capacitor Failure. Single-phase motor capacitors fail because of excessive temperature, voltage, or duty cycles, internal corrosion, or an open fuse. A substitute capacitor is placed in the circuit to test the motor if a capacitor is suspected to be defective. The original capacitor is defective if the motor operation improves. Capacitors normally operate in ambient temperatures up to 176°F (80°C). The life of a capacitor is shortened at high temperatures. Operating at low temperatures does not harm a capacitor, however the capacitance of a capacitor decreases at temperatures below 32°F (0°C).

Excessive voltage causes arcing in the capacitor that leads to permanent damage. Excessive voltage is applied by connecting a motor to a supply voltage higher than the capacitor voltage rating. Excessive voltage can also be applied due to a faulty centrifugal switch. The voltage rating of a capacitor is normally listed on the capacitor. The capacitor voltage rating must be equal to or greater than the supply voltage applied to the motor.

WARNING: Never replace a capacitor with one that has a lower voltage rating than the original. Never use a capacitor with a voltage rating higher than 10% of the original. Using a capacitor with too high or too low of a voltage rating increases the amperage and wattage draw of the motor. An increase in current or power can burn out motor windings.

Faulty centrifugal switch contacts chatter before opening. Voltage on the capacitor is several times higher than the supply voltage when centrifugal switch contacts chatter. The high voltage is induced by the collapsing magnetic field of the starting winding.

A capacitor is damaged by normal voltage if the motor is subjected to excessive starting and stopping (excessive duty cycle) or if the load applied to the motor requires a long starting time. A starting capacitor is normally in a circuit for less than 3 sec. An acceleration time more than 3 sec causes excessive heat buildup in the capacitor, which reduces capacitor life. A capacitor with an open seal allows moisture absorption. Moisture absorption leads to corrosion that destroys the film inside the capacitor.

Certain capacitors have an internal fuse that opens when excessive voltage or current is applied to the capacitor. The internal fuse can also open because of improper servicing of the motor. A capacitor with an open internal fuse displays a resistance reading of infinity (∞) when checked with a DMM set to measure resistance.

Three-Phase Motors

Unlike single-phase motors, three-phase motors do not require a starting winding or centrifugal switch to start the motor rotating. A rotating magnetic field is set up automatically in the stator when the motor is connected to three-phase power. The coils in the stator are connected to form three separate windings (phases). Each phase contains one-third of the total amount of individual coils in the motor. The three windings (phases) are phase A, phase B, and phase C.

Each phase is positioned 120° from the other phases in the motor. A rotating magnetic field is produced in the stator when each phase reaches its peak value, 120° apart from the other phases. **See Figure 7-16.**

Interchanging any two of the three-phase power lines to a motor reverses the direction of rotation of a three-phase motor. The industrial standard is to interchange T1 and T3, which is implemented for safety reasons. When first connecting a motor, the direction of rotation is not normally known until the motor is started. It is common practice to temporarily connect the motor to determine the direction of rotation before making permanent connections. Motor leads of temporary connections are not completely taped. By always interchanging T1 and T3, T2 may be permanently connected to L2, creating an insulated barrier between T1 and T3.

Single-Voltage Three-Phase Motors. A *single-voltage three-phase motor* is a motor that operates at only one voltage level in a three-phase system. Motor windings must be connected to the proper voltage to develop a rotating magnetic field in a single-voltage three-phase motor. The proper voltage level is predetermined by the motor OEM and listed on the motor nameplate.

Single-voltage three-phase motors are less expensive than dual-voltage three-phase motors but are limited to locations having the same voltage as the motor. Typical single-voltage three-phase motor ratings are 230 V, 460 V, and 575 V. Other single-voltage three-phase motor ratings are 200 V, 208 V, 220 V, and 280 V.

Wye-Connected Three-Phase Motors. A *wye-connected three-phase motor* is a motor that has one lead end of each of the three phases (phases A, B, and C) internally connected to the other two phases. The remaining end of each phase (phase A, B, and C) is externally routed and connected to the incoming power source (L1, L2, and L3). **See Figure 7-17.**

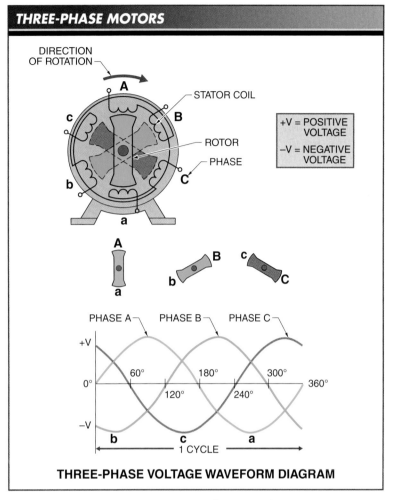

Figure 7-16. The coils in the stator of a three-phase motor are connected to form three separate windings (phases).

The leads, which are brought out externally, are labeled terminals one, two, and three (T1, T2, and T3). When connecting a wye-connected three-phase motor to three-phase power lines, the power lines and motor terminals are connected L1 to T1, L2 to T2, and L3 to T3.

Delta-Connected Three-Phase Motors. A *delta-connected three-phase motor* is a motor in which each phase is connected end to end to form a completely closed-loop circuit. At each point where the phases are connected, leads are externally routed to form T1, T2, and T3. **See Figure 7-18.**

228 ELECTRICAL SYSTEMS for FACILITIES MAINTENANCE PERSONNEL

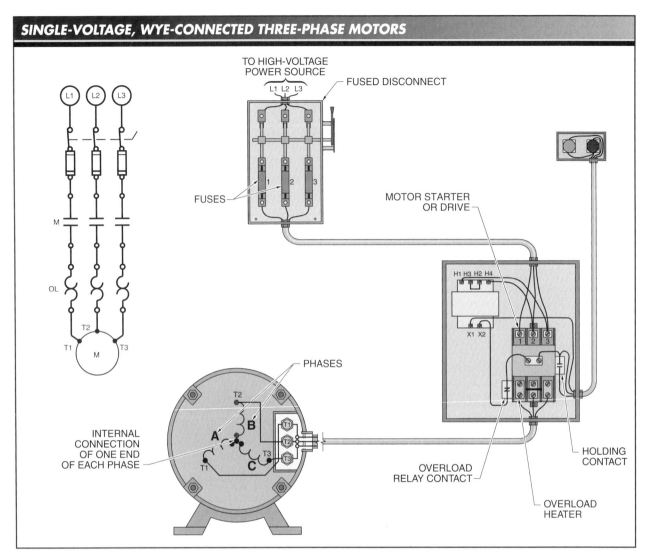

Figure 7-17. In a single-voltage, wye-connected three-phase motor, one end of each phase is internally connected to the other phases.

T1, T2, and T3 are connected to the three power lines. L1 is connected to T1, L2 to T2, and L3 to T3. The three-phase line supplying power to the motor must have the same voltage and frequency rating as the motor.

Dual-Voltage Three-Phase Motors. A *dual-voltage three-phase motor* is a three-phase motor that operates at more than one voltage level. Most three-phase motors are designed so that they can be connected to either of two different voltages. Motors that have connectivity for two different voltages can be used with two different power line voltages.

A typical dual-voltage three-phase motor rating is 230/460 V. Other dual-voltage, three-phase motor ratings are 240/480 V and 208/460 V to 230/460 V. The dual-voltage rating is listed on the nameplate of the motor. If both voltages are available, the higher voltage is normally preferred because the motor uses the same amount of power while providing the same horsepower output for either high or low voltage. However, as the voltage is doubled, the current draw on the power lines is reduced by 50%. With the reduced current, the conductor size is reduced and the material cost is decreased.

Dual-Voltage, Wye-Connected Three-Phase Motors. In a dual-voltage, wye-connected three-phase motor, each phase coil (A, B, and C) is divided into two equal sections. By dividing the phase coils equally, nine terminal leads are available. The terminal leads are marked terminals T1 to T9 and may be connected for high or low voltage.

To connect a dual-voltage, wye-connected three-phase motor for high voltage, L1 is connected to T1, L2 to T2, and L3 to T3. Wire nuts are used to connect T4 to T7, T5 to T8, and T6 to T9. By making these connections, the individual coils in each phase are connected in series. The applied voltage is divided equally among the coils because the coils are connected in series. **See Figure 7-19.**

TECH FACT

Electric motors are rated with 5 min, 15 min, 30 min, 60 min, and continuous duty cycle ratings. A motor with a 5 min rating is designed to operate at its rated horsepower for periods no more than 5 min.

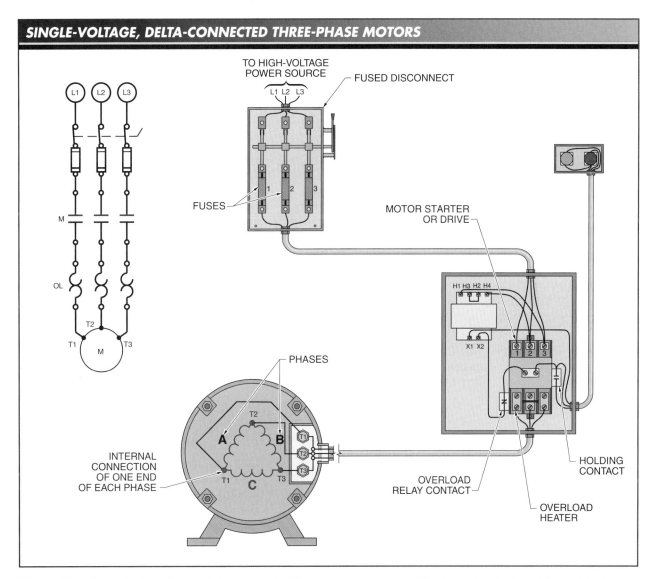

Figure 7-18. In a single-voltage, delta-connected three-phase motor, each phase is wired end to end to form a completely closed loop.

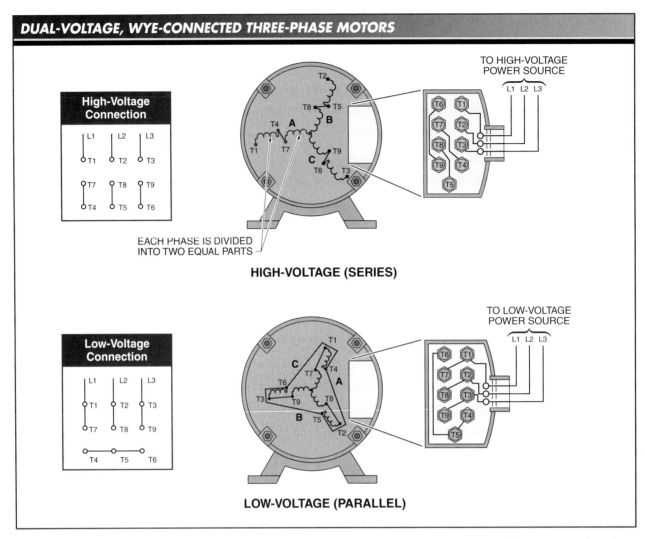

Figure 7-19. In a dual-voltage, wye-connected three-phase motor, each phase coil is divided into two equal sections and can be connected for high or low voltage.

To connect a dual-voltage, wye-connected three-phase motor for low voltage, L1 is connected to T1 and T7, L2 to T2 and T8, and L3 to T3 and T9. A wire nut is used to connect T4, T5, and T6 together. By making these connections, the individual coils in each phase are connected in parallel. The applied voltage is present across each set of coils because the coils are connected in parallel.

Dual-Voltage, Delta-Connected Three-Phase Motors. In a dual-voltage, delta-connected three-phase motor, each phase coil (A, B, and C) is divided into two equal sections. By dividing the phase coils equally, nine terminal leads are available. Motor leads are marked terminals one through nine (T1 to T9). The nine terminal leads can be connected for high or low voltage.

To connect a dual-voltage, delta-connected three-phase motor for high voltage, L1 is connected to T1, L2 to T2, and L3 to T3. Wire nuts are used to connect T4 to T7, T5 to T8, and T6 to T9. By making these connections, the individual coils in each phase are connected in series. Since the coils are connected in series, the applied voltage divides equally among the coils. **See Figure 7-20.**

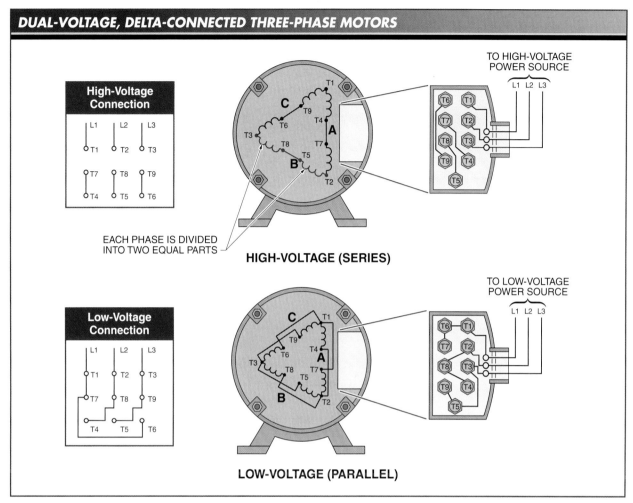

Figure 7-20. In a dual-voltage, delta-connected three-phase motor, each phase coil is divided into two equal sections and can be connected for high or low voltage.

To connect a dual-voltage, delta-connected three-phase motor for low voltage, L1 is connected to T1, T7, and T6, L2 to T2, T4, and T8, and L3 to T3, T5, and T9. By making these connections, the individual coils in each phase are connected in parallel. Since the coils are connected in parallel, the applied voltage is present across each set of coils.

Troubleshooting Three-Phase Motors

When troubleshooting a three-phase motor, it is best to measure both the voltage into the motor and the current draw of the motor. Measuring the voltage into the motor ensures the motor is powered with the proper voltage. Measuring the current draw indicates the amount of work the motor is performing. Both measurements must be taken over time and are compared to the motor nameplate ratings. **See Figure 7-21.**

TECH FACT

Three-phase motors are simple in construction, are capable of rugged performance, and require very little maintenance. Often, a three-phase motor can be used in an application for 10 or more years without a failure. Due to its long-lasting work life, when a three-phase motor experiences problems, it is often simply replaced.

232 ELECTRICAL SYSTEMS for FACILITIES MAINTENANCE PERSONNEL

TROUBLESHOOTING THREE-PHASE MOTORS

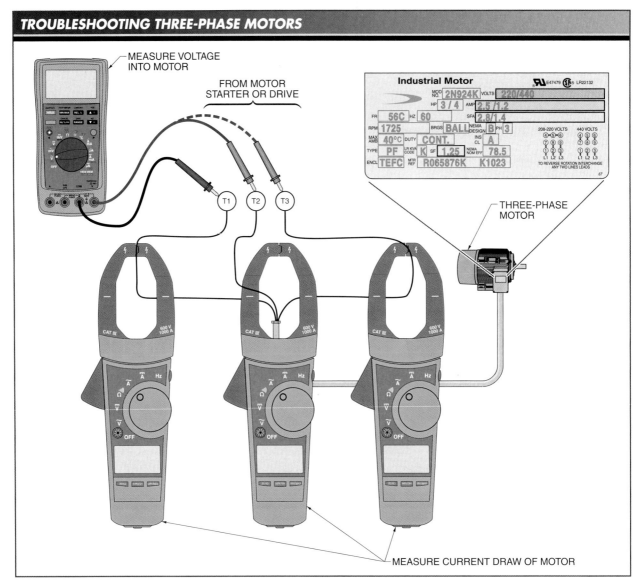

Figure 7-21. When troubleshooting a three-phase motor, it is best to measure both the voltage into the motor and the current draw of the motor.

Voltage Unbalance. *Voltage unbalance (imbalance)* is the unbalance that occurs when voltage at the terminals of an electric motor or other three-phase load are not equal. Voltage unbalance results in a current unbalance. Line (L1, L2, and L3) voltages must be tested for voltage unbalance periodically and when performing service. In general, voltage unbalance should not be more than 1%. Whenever there is a voltage unbalance of 2% or greater, the building power system should be tested for excessive loads connected to one of the phases (L1, L2, or L3). A power quality meter can be used to measure true power, apparent power, reactive power, power factor, and displacement power factor of a circuit. The load or motor rating is adjusted by reducing the load on the motor or by oversizing the motor if the voltage unbalance cannot be corrected. The local power utility should be notified if the voltage unbalance source is outside the facility.

The primary source of voltage unbalances of less than 2% is excessive loads on one phase of a three-phase distribution system. High voltage unbalances are normally the result of a blown fuse in one phase of a three-phase capacitor bank. **See Figure 7-22.** Voltage unbalance is calculated using the following procedure:

1. Measure the voltage between each incoming power line. Take measurements from T1 to T2, T1 to T3, and T2 to T3.
2. Calculate the total voltage. Total voltage is calculated by adding the voltages together.
3. Calculate the average voltage. Average voltage is calculated using the following formula:

$$E_a = \frac{E_T}{n}$$

where

E_a = average voltage (in V)

E_T = total voltage (in V)

n = number of lines measured

4. Calculate the largest voltage deviation. Largest voltage deviation is calculated using the following formula:

$$E_d = E_h - E_a$$

where

E_d = voltage deviation (in V)

E_h = highest voltage measurement (in V)

E_a = average voltage (in V)

5. Calculate the voltage unbalance. Voltage unbalance is calculated using the following formula:

$$E_u = \frac{E_d}{E_a} \times 100$$

where

E_u = voltage unbalance (in %)

E_d = voltage deviation (in A)

E_a = voltage average (in A)

100 = constant

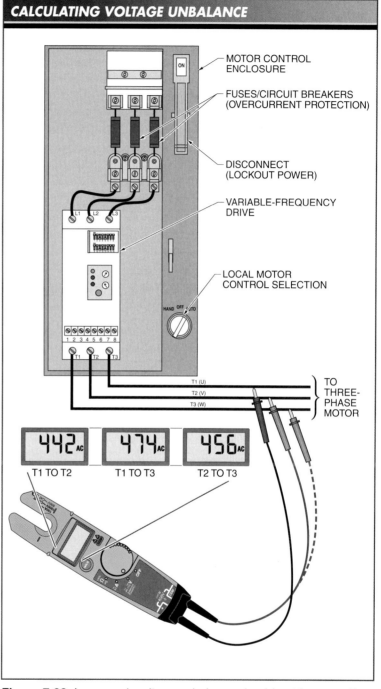

Figure 7-22. In general, voltage unbalance should not be more than 1%. Voltage unbalances that are 2% or higher must be eliminated.

TECH FACT

Power quality meters measure, display, and record voltage (V), current (A), true power (W), apparent power (VA), power factor (PF), phase unbalance, harmonics, and transients. Most power quality meters can also measure resistance and capacitance and check diodes.

Example: What is the voltage unbalance on a motor control enclosure that has voltage measurements on the power lines between the motor disconnect and the motor of 442 V (T1 to T2), 474 V (T1 to T3), and 456 V (T2 to T3)?

1. Measure the voltage between each incoming power line.
2. Calculate the total voltage.

$$E_T = 442 + 474 + 456$$
$$E_T = 1372 \text{ V}$$

3. Calculate the average voltage.

$$E_a = \frac{E_T}{n}$$
$$E_a = \frac{1372}{3}$$
$$E_a = 457 \text{ V}$$

4. Calculate the largest voltage deviation.

$$E_d = E_h - E_a$$
$$E_d = 474 - 457$$
$$E_d = 17 \text{ V}$$

5. Calculate the voltage unbalance.

$$E_u = \frac{E_d}{E_a} \times 100$$
$$E_u = \frac{17}{457} \times 100$$
$$E_u = 0.0372 \times 100$$
$$E_u = \mathbf{3.72\%}$$

The voltage unbalance is greater than 2%. For this reason, the power system should be tested for excessive loads connected to a power line. When a three-phase motor fails due to voltage unbalance, one or two of the stator windings become blackened. The darkest winding is the winding with the largest voltage unbalance. **See Figure 7-23.**

Current Unbalance. *Current unbalance (imbalance)* is the unbalance that occurs when current on each of the three power lines of a three-phase power supply are not equal. The problem with small voltage unbalances is that small voltage unbalances cause high current unbalances. High current unbalances cause excessive heat, resulting in motor insulation breakdown. Normally, voltage unbalances cause current unbalances at a rate of about 8:1. For example, a 2% voltage unbalance creates a 16% current unbalance.

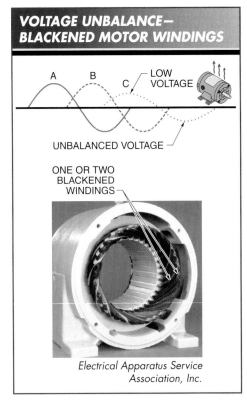

Figure 7-23. When examining blackened motor stator windings, the darkest winding is the winding with the largest voltage unbalance.

Current unbalance must not exceed 10%. When current unbalance exceeds 10%, the circuit must be tested for voltage unbalance. Likewise, any time there is a voltage unbalance of more than 1%, the circuit must be tested for current unbalance. Current unbalance is calculated in the same manner as voltage unbalance, except with current measurements. **See Figure 7-24.** Current unbalance is calculated using the following procedure:

1. Measure the current between each incoming power line. Take measurements from T1 to T2, T1 to T3, and T2 to T3.

2. Calculate the total current. Total current is calculated by adding the current values together.
3. Calculate the average current. Average current is calculated using the following formula:

$$I_a = \frac{I_T}{n}$$

where

I_a = average current (in A)
I_T = total current (in A)
n = number of lines measured

4. Calculate the largest current deviation. Largest current deviation is calculated using the following formula:

$$I_d = I_h - I_a$$

where

I_d = current deviation (in A)
I_h = highest current measurement (in A)
I_a = average current (in A)

5. Calculate the current unbalance. Current unbalance is calculated using the following formula:

$$I_u = \frac{I_d}{I_a} \times 100$$

where

I_u = current unbalance (in %)
I_d = current deviation (in A)
I_a = average current (in A)
100 = constant

Example: What is the current unbalance on a motor control enclosure that has current measurements on the power lines between the motor disconnect and the motor of 21 A (T1 to T2), 27 A (T1 to T3), and 24 A (T2 to T3)?

1. Measure the current between each incoming power line (T1 to T2, T1 to T3, and T2 to T3) entering the circuit disconnect.
2. Calculate the total current.
$I_T = 21 + 27 + 24$
$I_T = 72$ A

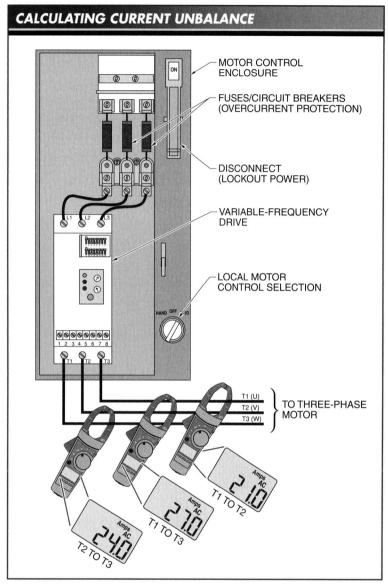

Figure 7-24. Current unbalance is calculated in the same manner as voltage unbalance, except with current measurements.

3. Calculate the average current.
$$I_a = \frac{I_T}{n}$$
$$I_a = \frac{72}{3}$$
$$I_a = 24 \text{ A}$$

4. Calculate the largest current deviation.
$I_d = I_h - I_a$
$I_d = 27 - 24$
$I_d = 3$ A

5. Calculate the current unbalance.

$$I_u = \frac{I_d}{I_a} \times 100$$

$$I_u = \frac{3}{24} \times 100$$

$$I_u = 0.125 \times 100$$

$$I_u = \mathbf{12.5\%}$$

Current unbalance exceeds 10%. For this reason, the circuit must be tested for voltage unbalance.

Electric resistance heating elements are widely used in commercial HVAC systems and may include a wire guard to prevent accidental contact and injury to maintenance personnel.

ELECTRIC HEATING CIRCUITS

Heat is produced any time electric current passes through a conductor that has resistance. A *heating element* is the portion of an electrical component that produces heat by electric current. *Watt density* is the wattage concentration on the surface of a heating element. Watt density is used to indicate the potential of a material to transmit heat. Heating elements are rated according to their total power output (in W), watt density, and/or total wattage. Electric heaters and heating control circuits of a fixed size and output are normally rated by total watt output. For example, heating elements used in commercial hot water heaters are rated in total W or kW. **See Figure 7-25.**

Heating Elements

Heating elements use electricity to heat solid, gas, and liquid materials. In order to effectively and efficiently heat the material, different types of heating elements are used. The two main types of heating elements are open coil and enclosed coil. **See Figure 7-26.**

An *open-coil heating element* is an exposed coiled resistance wire fixed onto a supporting element. An open-coil heating element is the least expensive type of heating element available. Heat is produced when electricity is connected to the resistance wire. Since the wire has resistance, it produces heat. Because the resistance wire is exposed, the heating element must be properly protected so individuals or equipment cannot contact the element. Open-coil heating elements are used to heat air in electric heater applications such as duct heaters or industrial convection ovens.

An *enclosed-coil heating element* is a coiled resistance wire sealed into an enclosure and fixed onto a supporting element. Enclosed-coil heating elements are the safest and most used type of heating elements in commercial and industrial applications. An enclosed-coil heating element has a resistance wire that produces heat surrounded by refractory insulation. *Refractory* is a heat-resistant ceramic material. The refractory insulation is enclosed in a protective metal tube or cover. The amount of heat produced by the heating element is proportional to the resistance of the wire. The lower the resistance, the larger the current flow. The larger the current flow, the higher the temperature of the heating element.

> **TECH FACT**
> *Electric heaters are widely used in moderate climates because the cost of electricity and necessary precautions against excessive heat are offset by their simple control schemes and low installation and maintenance costs.*

ELECTRIC HEATING

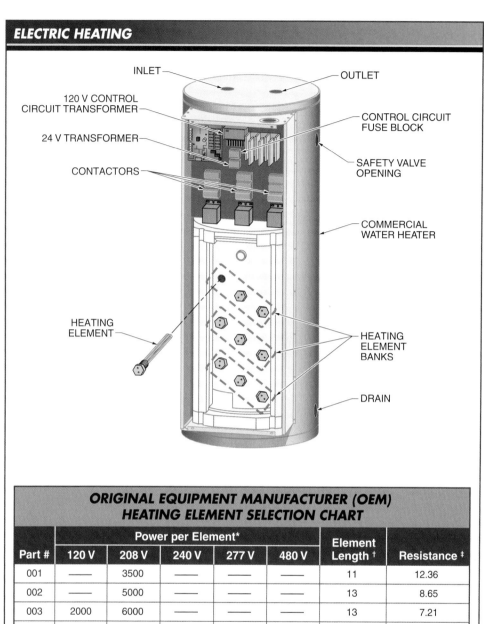

ORIGINAL EQUIPMENT MANUFACTURER (OEM) HEATING ELEMENT SELECTION CHART

Part #	Power per Element*					Element Length †	Resistance ‡
	120 V	208 V	240 V	277 V	480 V		
001	—	3500	—	—	—	11	12.36
002	—	5000	—	—	—	13	8.65
003	2000	6000	—	—	—	13	7.21
004	—	—	3000	4000	—	9	19.20
005	—	3000	4000	—	—	11	14.40
006	—	—	5000	—	—	11	11.52
007	1500	4500	6000	—	—	13	9.60
008	—	500	—	1000	3500	12	47.56
009	500	1500	2000	2500	—	13	28.80
010	—	—	1000	1350	4000	13	41.62
011	—	—	1500	2000	6000	15	27.74

* in W
† in in.
‡ in Ω

Figure 7-25. Electric heating is achieved using heating elements. Original equipment manufacturer (OEM) heating element selection charts provide information on variables such as power, voltage, and resistance.

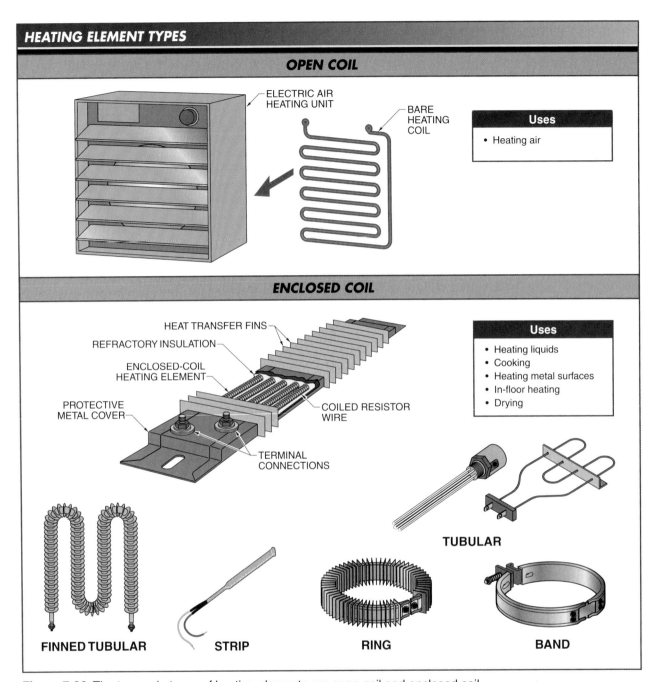

Figure 7-26. The two main types of heating elements are open coil and enclosed coil.

In heating element charts, the OEM normally lists the power (P), voltage (E), and resistance (R) for each heating element. Since heating elements are resistive loads, Ohm's law and the power formula can be applied. The OEM values may be slightly different based on the exact material used and the material resistance. For example, a heating element has OEM resistance ratings of 28.80 Ω. Using Ohm's law and the power formula, the power at each voltage level can be verified and the current can be calculated. Calculating the current is necessary because the circuit conductors, fuses or circuit breakers, and control devices are rated for the amount of current they can safely withstand.

Resistance is calculated using the following formula:

$$R = \frac{E^2}{P}$$

where
R = heating element resistance (in Ω)
E = applied voltage (in V)
P = heating element power (in W)

Example: What is the resistance of a 500 W heating element connected to 120 V?

$$R = \frac{E^2}{P}$$
$$R = \frac{120^2}{500}$$
$$R = \frac{14,400}{500}$$
$$R = \mathbf{28.8\ \Omega}$$

Current is calculated using the following formula:

$$I = \frac{E}{R}$$

where
I = heating element current draw (in A)
E = applied voltage (in V)
R = heating element resistance (in Ω)

Example: What is the current through a heating element with a resistance of 28.8 Ω connected to 120 V?

$$I = \frac{E}{R}$$
$$I = \frac{120}{28.8}$$
$$I = \mathbf{4.17\ A}$$

Note: When a heating element is connected to 240 V rather than 120 V, current doubles from 4.17 A to 8.33 A and power increases four times from 500 W to 2000 W.

Electric heaters that do not have a fixed size normally have a watt density rating. Heating a material to a high temperature quickly is accomplished by increasing the watt density of the heating element. The proper watt density for an application depends on the rate of heat absorption of the material. For example, water conducts (absorbs) heat away from a heating element faster than asphalt. For this reason, heating elements used to heat water may have a higher watt density rating and be smaller in size.

Heating element cables, jackets, and pads used to heat pipes, barrels, totes, and other equipment can be installed in different lengths that can be designed to fit the application. For such applications, the watt density and resistance per foot of the heating element are used to determine operating values. Such heating elements can also be purchased in predetermined sizes with OEM specified ratings. For example, large heating bands or pads used for temperature control of liquid storage containers such as barrels and totes are rated in watts per square foot (W/sq ft). **See Figure 7-27.**

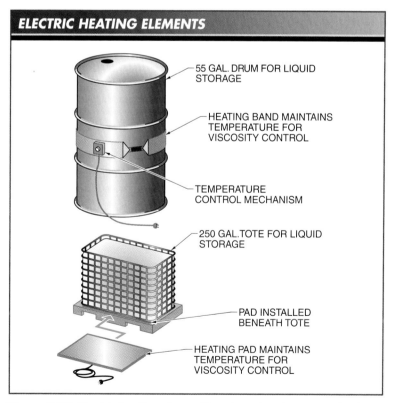

Figure 7-27. Electric heating element bands and pads are used to heat liquid storage containers, such as barrels and totes, to maintain a set temperature for viscosity control.

Preassembled heating pads have a watt density rating of 50 W/sq ft (0.347 W/sq in.). Ohm's law and the power formula can be applied to calculate any unknown quantity of power, voltage, resistance, or current if not provided by the OEM. For example, the power formula can be used when OEM data is not available because a heating element normally lists certain ratings.

Troubleshooting Heating Elements. Problems that can occur in heating circuits include no heat, inadequate heat, excessive heat, or electrical shock. Inadequate or excessive heat normally indicates a control circuit problem or an oversized or undersized heating element. No heat normally indicates a blown fuse, an open circuit breaker, an open heating element, or a control circuit problem. Electrical shock is caused by a high leakage current that occurs when there is a breakdown in insulation. Tests that can be performed to determine a problem in a heating element include an open circuit test, short circuit test, resistance test, insulation breakdown test, and power draw test.

Enclosed-coil heating elements may have an open circuit, short circuit, or change in resistance value due to deterioration, physical damage, or misuse. No heat is produced when a heating element opens. The circuit fuses or circuit breakers open when there is a short circuit. When the heating element resistance increases, it produces less heat. When the heating element resistance decreases, it produces more heat. A DMM set to measure resistance can be used to test the condition of a heating element after power has been completely removed from the circuit. **See Figure 7-28.**

With power off (verified with a DMM set to measure voltage), the heating element is disconnected from the electrical circuit. The new or existing heating element resistance is measured. If OEM data is available, it can be used to determine the condition of the heating element. If the OEM data is not available, Ohm's law and the power formula can be used to calculate the heating element resistance because the heating element includes at least two electrical quantity (voltage, power, or current) ratings. The measured resistance value should be within ±10% of the rated or calculated resistance value.

Heating element insulation can lose its properties because of extreme heat at the heating element or contamination from dust, dirt, oil, vibration, chemicals, and moisture. Deteriorated insulation causes current to leak from the current-carrying conductors inside the heating element to the enclosure cover. Current leakage is a function of insulation resistance. Current leakage increases as insulation resistance decreases.

Although basic DMMs can be used to measure the resistance of a heating element, they cannot be used to measure the condition of the insulation because a basic DMM measures resistance using a low-voltage battery (normally 9 VDC). Insulation resistance must be measured at the operating voltage of the heating element. To test heating element insulation, a megohmmeter must be used. Megohmmeters are designed to measure high resistances at a voltage level equal to or slightly higher than the rated voltage of the device under test. Megohmmeters normally use test voltages of 50 V, 100 V, 250 V, 500 V, or 1000 V.

> **TECH FACT**
> *The U.S. Occupational Safety and Health Administration (OSHA) 29 Code of Federal Regulations (CFR) 1910.306(g)(1)(v)(B) states that induction heating coils must be protected by insulation or refractory materials or both. Induction heating coils heat an electrically conductive material through electromagnetic induction.*

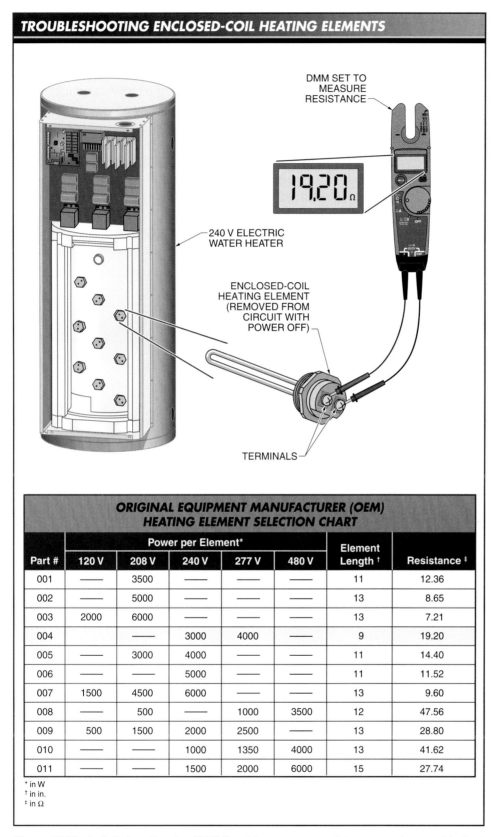

Figure 7-28. A digital multimeter (DMM) set to measure resistance can be used to test the condition of a heating element after all power has been removed from the circuit.

To test the insulation resistance of a heating element, the megohmmeter voltage is set to the next highest setting above the rated heating element voltage. For example, a 120 V, 208 V, or 240 V rated heating element must be tested using the 250 V meter setting. Power is turned off (verified with a DMM set to measure voltage), and the heating element is disconnected from the circuit. The resistance between the heating element enclosure or ground and each current-carrying conductor terminal is tested. A good resistance measurement is normally at least 10 MΩ. A resistance measurement can be less, but never below its recommended minimum resistance value. **See Figure 7-29.**

Heating elements convert electrical energy into heat and are rated in watts. The amount of power a heating element uses is an indication of the condition of the element. Since power equals voltage times current, a power draw test requires first taking a voltage and current measurement and then multiplying the measurement values together to provide an indication of the condition of the heating element.

The advantage of taking a voltage or current measurement is that the heating element does not have to be disconnected from the power source. However, since measurements are taken on a hot circuit, proper PPE must be worn and safety procedures followed.

Electric Heating Circuit Devices

The purpose of an electric heating circuit is to maintain a set temperature of a conditioned medium such as air or water. An electric heating circuit must be able to respond to a change from the set temperature and return the temperature of the conditioned medium to the set temperature. The two main factors in controlling temperature are produced temperature and time. The higher the produced temperature, the faster the temperature rise of the conditioned medium. Likewise, the longer the heating circuit is on, the greater the total amount of heat produced.

Electric heating circuits include a high-voltage power circuit and a low-voltage control circuit.

In the power circuit, the heating elements, fuses or circuit breakers, and contactors are tested/checked when a problem occurs. In the control circuit, the control circuit transformer, control circuit fuses, heating control circuit board, and sensors/switches are tested/checked when a problem occurs. **See Figure 7-30.**

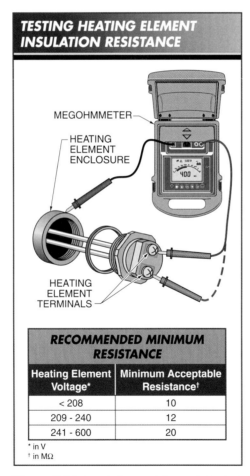

Figure 7-29. Insulation resistance is tested by measuring the resistance between the heating element enclosure or ground and each current-carrying conductor terminal.

TESTING HEATING ELEMENT INSULATION RESISTANCE

RECOMMENDED MINIMUM RESISTANCE

Heating Element Voltage*	Minimum Acceptable Resistance†
< 208	10
209 - 240	12
241 - 600	20

* in V
† in MΩ

ELECTRIC HEATING CIRCUITS

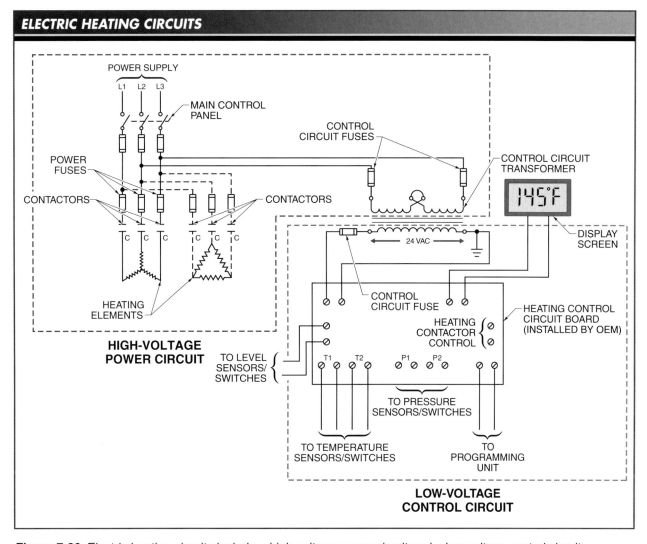

Figure 7-30. Electric heating circuits include a high-voltage power circuit and a low-voltage control circuit.

Troubleshooting Electric Heating Power Circuits. The power circuit is normally the easiest part of an electric heating circuit to test because its fuses or circuit breakers, heating contactors, and heating elements are normally accessible, which allows them to be tested using standard test instruments. Electrical equipment is designed to operate within a set voltage range. When testing electrical equipment, the supply voltage is measured with a DMM set to measure voltage. The voltage should be within +5% to –10% of the equipment voltage rating.

In addition, electrical equipment is designed to draw current in order to produce power. The amount of current that a piece of equipment requires for operation is listed on the equipment nameplate. When testing equipment for current draw, current is best measured with a clamp-on ammeter. Current measurements vary depending on the load on the equipment at any given time but should not exceed the equipment current rating by more than 5%. If a meter has a MIN MAX recording function, the meter should be used to record the current over time for at least one full cycle of the equipment.

A DMM used to test fuses in an energized circuit is the best method to initially test a fuse since the fuse does not have to be removed from the circuit. **See Figure 7-31.** Fuses that are connected into a circuit are tested using the following procedure:

Note: Always wear proper PPE for the location and type of test. Verify the DMM has the minimum CAT III rating for the measuring location.

1. Set the meter to measure voltage. Test the meter on a known, energized source before taking a measurement to verify that the meter is in proper working condition.
2. Measure the voltage into the top of the fuse (line side) by connecting one test lead to the top of the fuse and the other test lead to another power line, neutral, or ground. Connect hot to neutral for 120 VAC and hot to hot for 240 VAC and 480 VAC.
3. Once a voltage measurement is taken from the top of the fuse, move one test lead to the bottom of the fuse. Move the test lead connected to the bottom of the fuse to the bottom of the other fuses, leaving the other test lead connected as is. *Note:* If the fuse is good (closed), voltage going into the fuse is the same coming out of the fuse. If the fuse is bad (open), there is no voltage measured coming out of the fuse.

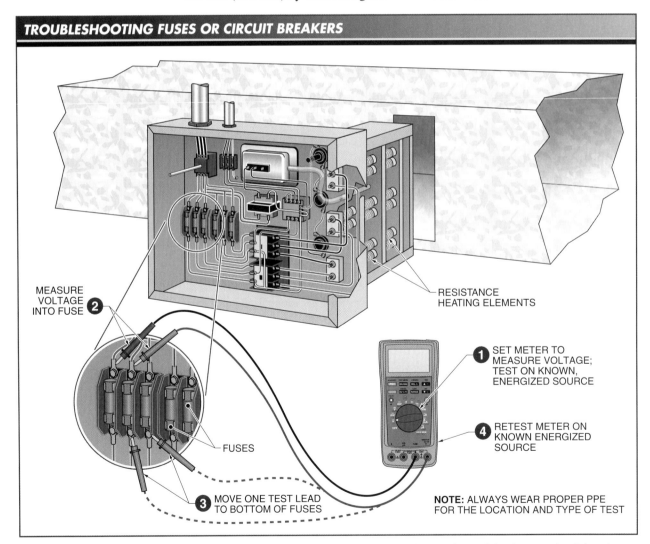

Figure 7-31. Fuses or circuit breakers can be checked by testing the voltage into and out of the fuse or circuit breaker.

4. Retest the meter on a known, energized source to verify that the meter is still working properly.

Fuses can also be tested when they are removed from a circuit. To test fuses removed from a circuit, a DMM set to measure resistance can be used. A good fuse has a measurement of 0 Ω when the DMM leads are connected across the fuse. A bad fuse has a measurement of infinity (OL) when the DMM leads are connected across the fuse. A DMM set to measure resistance should be used only to test fuses that are removed from a circuit.

An open circuit breaker can be identified by the movement of the circuit breaker switch lever away from the ON position. Some circuit breakers also include an indicator window that is red when the circuit breaker is tripped. The indicator window is either white or black when the circuit breaker is not tripped. Circuit breakers that are suspected of having a problem can be tested in the same manner as fuses using a DMM set to measure voltage.

Troubleshooting Electric Heating Control Circuits. Electric heating elements produce heat immediately. The manner in which the heating elements are controlled depends on the application, required temperature accuracy, efficiency, and cost. The two most common types of temperature control circuits are two-position temperature control and proportional temperature control. *Two-position temperature control* is a heating control function in which the heating element can only be ON or OFF. *Proportional temperature control* is a heating control function in which the heating element can produce any amount of heat from 0% to 100% of its rated value. Proportional temperature control is accomplished by varying the amount of voltage applied to the heating element.

Most heating control circuits are located on an OEM-installed circuit board to which external temperature, pressure, and level switches/sensors, overrides, displays, and programming units are connected. External devices can be installed directly in the unit or connected when installing the unit.

An electric heating control circuit is normally tested by observing displayed faults and following OEM guidelines for each displayed fault or following the OEM control circuit troubleshooting instructions in the service manual. Faulty sensors/switches are normally tested and replaced as required.

Troubleshooting Control Circuit Transformers. Electrical equipment uses power. Power is equal to voltage times current. In order to reduce the amount of current required in a power circuit, the voltage is high (120 V, 208 V, 240 V, 480 V, or 600 V). However, because a control circuit does not require a great amount of power, voltage is low (12 V or 24 V). A control circuit transformer is used to lower the voltage from the power circuit to acceptable levels for the control circuit.

Control circuit transformers are tested for proper operation by measuring the voltage in (primary side) and voltage out (secondary side) of the transformer. **See Figure 7-32.** The voltage out should equal +5% to –10% of the transformer secondary (output) rating. If low voltage is present out of a control circuit transformer but not into it, the problem is that the transformer is overloaded or faulty. An overloaded transformer can be replaced by a larger one or the control circuit to the transformer can be checked. A faulty transformer is replaced. If a DMM has a MIN MAX recording function, the meter should be set to record the voltage out of the control circuit transformer over time for at least one complete cycle of the equipment.

TROUBLESHOOTING CONTROL CIRCUIT TRANSFORMERS

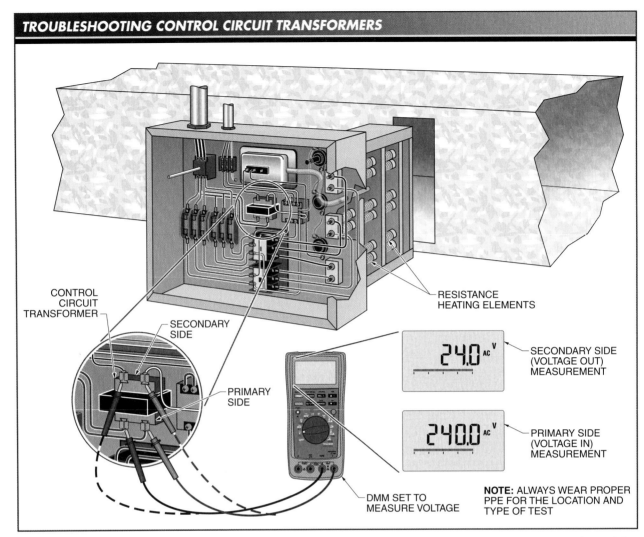

Figure 7-32. Control circuit transformers are tested for proper operation by measuring the voltage in and out of the transformer.

Troubleshooting Contactors. A *contactor* is a control device that uses a small control current to energize and deenergize the load connected to it. A contactor does not include overload protection. In a contactor, a coil is energized and deenergized to move a plunger. The plunger closes and opens a set of contacts. The closing of the contacts allows devices to be controlled from remote locations by a low-voltage circuit. A contactor is used to connect the main power supply to heating elements. The contactor includes high power contacts to withstand the high voltage (120 V, 208 V, 240 V, 480 V, and 600 V) and current of the heating elements. The low-voltage coil (12 V or 24 V) is connected to the control circuit and is used to control the contactor.

To test a contactor, a DMM set to measure voltage is used to measure the voltage into each contactor terminal. There must be voltage to the contactor or there is a problem with the incoming power supply. If there is no voltage to the contactor, the fuses or circuit breakers delivering power to the contactor are tested. If there is voltage at the contactor, the coil voltage is tested. There must be voltage at the coil for the contactor to operate the power contacts. The voltage should be within +5% to –10% of the coil voltage rating. **See Figure 7-33.**

Chapter 7—HVAC Systems 247

TROUBLESHOOTING HEATING CONTACTORS

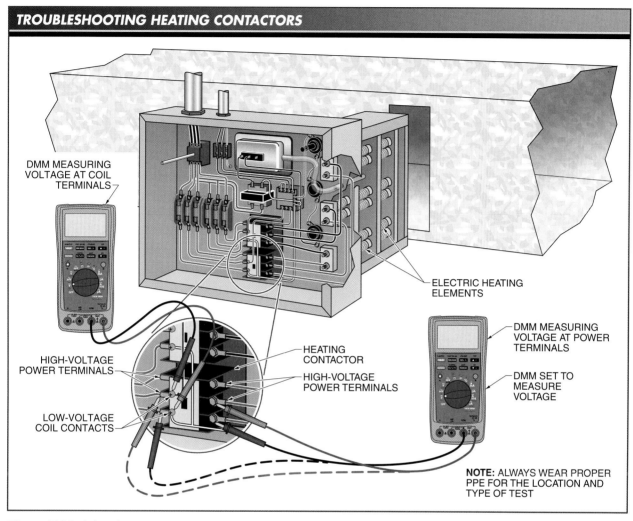

Figure 7-33. A heating contactor is tested by measuring the voltage into and out of each power terminal and coil terminal.

To test contactor contacts, the voltage into and out of the contacts is measured. The voltage in should equal the voltage out. If there is any difference in the voltage measurements, the voltage across the contact when the load (heating element) is on is measured. Even a small voltage drop can cause a heat problem that can damage contacts.

HVAC SYSTEM SOLENOID-OPERATED DIRECTIONAL CONTROL VALVES

A *solenoid* is an electromechanical device that converts electrical energy to linear mechanical force. Solenoids are used in HVAC systems with directional control valves to control the flow of fluids (gas or liquid) within the system. A *directional control valve* is a valve that allows, prevents, or changes fluid flow to specific piping or system actuators. A solenoid includes a coil winding that produces a magnetic field when energized. The magnetic field is used to move an iron shaft that is connected to the valve spool that starts, stops, or changes the direction of the flow of fluid within the valve. Solenoid-operated valves are normally designed to operate on 24 VAC, 24 VDC, or 120 VAC. **See Figure 7-34.**

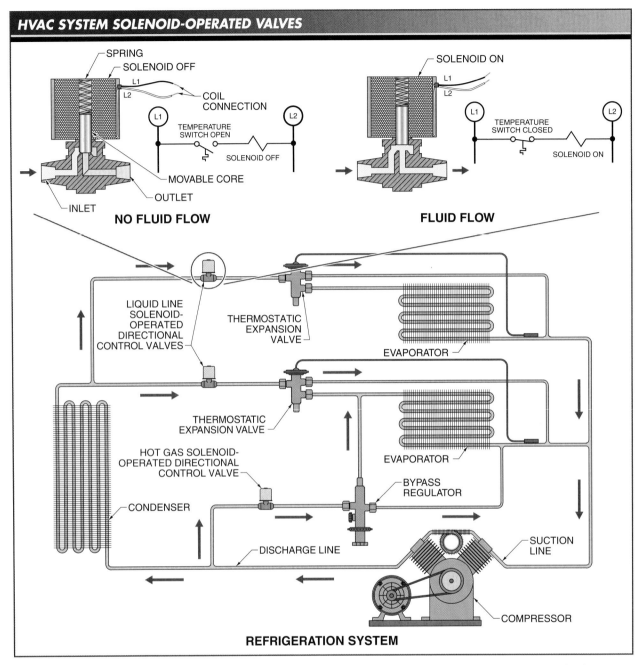

Figure 7-34. Refrigeration systems use solenoid-operated valves to stop, start, and redirect the flow of refrigerant.

Solenoid-operated valves are listed and specified according to the number of positions that they have, if they start, stop, or redirect the fluid flow, and the manner in which they control the flow in each valve position. The terms "positions," "ways," and "normal" are used to describe each valve type. Solenoid-operated valves are normally used with directional control valves. Troubleshooting solenoid-operated directional control valves requires testing the level of voltage to the solenoid.

Positions and Ways

A solenoid-operated valve is placed in different positions to start, stop, or change the direction of fluid flow. A *position* is any of the locations within the valve in which a spool is placed to direct fluid through the valve. A directional control valve normally has two or three positions. Two-position valves are the most common directional control valves used in HVAC systems.

A *way* is the path of a piping port through a directional control valve. Most directional control valves are two-way, three-way, or four-way valves. Two-way and three-way valves are the most common valves used in HVAC systems. A two-way valve has two main ports that allow or stop the flow of fluid. Two-way valves are used as shutoff, check, or quick-exhaust valves. Three- and four-way valves are used to change the direction of fluid flow. **See Figure 7-35.**

Solenoid-operated valves are often used to control the flow of fluid in commercial chiller systems.

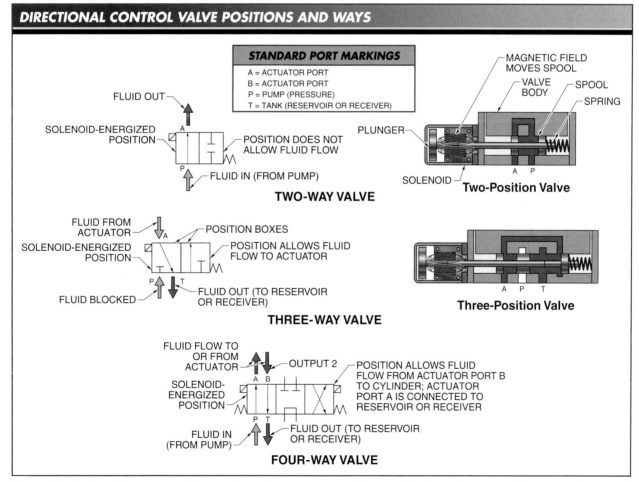

Figure 7-35. A position is any of the locations that a valve spool is capable of moving into, and a way is the path of a piping port in a directional control valve.

Normally Closed and Normally Open Valves

A *normally closed (NC) valve* is a valve that does not allow pressurized fluid out of the valve in the spring-actuated (normal) position. A *normally open (NO) valve* is a valve that allows pressurized fluid to flow out of the valve in the spring-actuated (normal) position. The terms normally closed and normally open are both used when describing fluid power valves and electrical switches. However, they have the opposite meaning for each type of power (fluid power and electrical). In electrical power devices, normally open is used to describe devices that do not have current flow. In fluid power devices, normally open is used to describe devices when there is fluid flow. Normally closed is used to describe electrical power devices when there is current flow and fluid power devices when there is no fluid flow.

Refer to Chapter 7 Quick Quiz® on CD-ROM.

> **TECH FACT**
> When replacing a directional control valve in an HVAC system, it is important to note the flow paths for each position and the number of ways so that it can be replaced with the same directional control valve.

Cleaver-Brooks
Boilers in commercial buildings include solenoid-operated valves that are used to control the flow of gas to the burner nozzle.

Troubleshooting Solenoid-Operated Directional Control Valves

Troubleshooting the electrical portion of a solenoid-operated directional control valve requires verifying that there is the correct level of voltage to the solenoid. The voltage to the solenoid should be within +5% to –10% of the solenoid voltage rating.

If voltage to the solenoid is too high, the solenoid can overheat and destroy coil insulation. If voltage to the solenoid is too low, the solenoid may not be able to move the spool and can draw higher current while attempting to move the spool, which can cause overheating.

For erratically operating valves, the voltage at the solenoid should be recorded over time using a DMM with a MIN MAX function. Erratic solenoid operation can be caused by low voltage at the time of energizing (system voltage sag), too high of system pressure, or faulty mechanical valve problems.

A solenoid coil can also be tested for open or short circuits by taking a resistance measurement with a DMM set to measure resistance when power is off and the solenoid is disconnected from the circuit. **See Figure 7-36.** A functional solenoid has a resistance value that depends on the size of the solenoid coil. The larger the solenoid coil, the larger the conductor size and the lower the resistance. If the coil is open, a DMM displays an overload (OL) when set to measure resistance. If the coil is short-circuited, normally because the conductors to the solenoid have insulation breakdown due to outside moisture, corrosion, or other damage, the DMM displays zero or near zero resistance. Short circuits cause fuses or circuit breakers to open. Any fuse or circuit breaker that has been in a short circuit must be replaced and the circuit tested to determine the cause of the short circuit.

SOLENOID TEST MEASUREMENTS

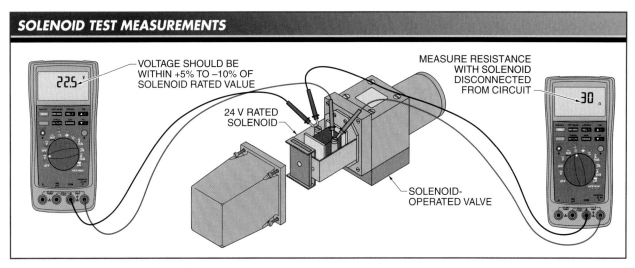

Figure 7-36. Solenoids are tested for proper voltage levels in an operational circuit or for open and short circuits with the power off and the solenoid removed from the circuit.

Case Study—Short Cycling System

The air conditioning system in a building is frequently turning on and off (short cycling). After checking to ensure the differential setting on the pressure switch and thermostat are not set too low and the refrigerant charge is correct, electrical measurements are taken to determine if there is an electrical problem.

The incoming power and control circuit voltage measurements are taken to ensure there is not a low voltage problem. A low voltage problem could cause the compressor motor starter, relays, and other control devices to drop out. The three-phase system is rated at 230 V. The incoming power measurements are as follows:

L1 to L2 = 223.5 V
L2 to L3 = 223.8 V
L1 to L3 = 223.2 V

The voltage is a little low but within the required operating range for the system.

The control circuit voltage measurements are then taken. The control circuit is rated at 120 V. The control circuit voltage measurements are as follows:

H1 to H4 = 223.4 V
X1 to X2 = 116.3 V

The voltage into the control circuit transformer (H1 to H4) is at the required level. The voltage out of the control circuit transformer (X1 to X2) is within the requirements of the 120 VAC rated control devices.

Fluke Corporation

For further verification that there is not a low voltage problem in the control circuit, a DMM is connected to X1 and X2 and set to MIN MAX recording mode. The HVAC system is cycled through a complete on/off cycle and the measurements are checked.

Case Study (continued)

The minimum and maximum voltage measurements are as follows:
control circuit minimum voltage = 114.3 V
control circuit maximum voltage = 117.1 V

The minimum recorded voltage is not low enough to cause control devices to drop out. The current draw of the compressor motor is then measured and compared to the motor nameplate. The motor operating current measurements are as follows:

T1 = 4.25 A
T2 = 4.28 A
T3 = 4.22 A

The current measurements are nearly equal indicating that there is no current unbalance. However, the current measurements are high. The normal current level should be 3.8 A at 230 V when the motor is operating at full power (1 HP). The motor nameplate indicates the motor has a service factor (SF) of 1.25, which means that the motor is designed to produce approximately 25% extra power when required. The service factor amperage (SFA) at 230 V is 4.75 A. This means that the motor is operating within its overload rating, but is operating beyond its normal full-load rating. Since motor overloads are usually rated for the normal full-load rating and not SFA, the motor starter overloads are tripping, leading to short cycling.

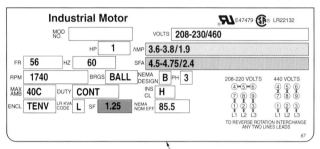
MOTOR NAMEPLATE

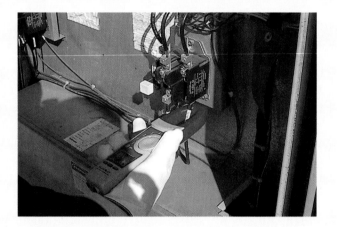

The motor starter overloads could be sized higher to keep the motor operating, but the main cause of the overload condition needs to be determined. The overload condition may be caused by a higher-than-normal torque being placed on the motor shaft due to excessively tight belts or excessive system pressure. It is also possible that the motor was replaced with an undersized motor.

The building maintenance staff notes all checks and measurements made and recommends that a complete follow-up investigation be performed during a scheduled downtime when the system can be shut off for a longer period and the problem investigated further.

Definitions

- An *adjustable motor base* is a mounting base that allows a motor to be easily moved a short distance.
- A *capacitor motor* is a single-phase motor with a capacitor connected in series with the stator windings to produce phase displacement in the starting winding.
- A *capacitor-start motor* is a motor that has the capacitor connected in series with the starting winding.
- *Centrifugal force* is the force that moves rotating bodies away from the center of rotation.
- A *centrifugal switch* is a switch that opens to disconnect the starting winding when the rotor reaches a preset speed and closes to reconnect the starting winding when the rotor speed drops below the preset value.
- A *contactor* is a control device that uses a small control current to energize and deenergize the load connected to it.
- *Current unbalance (imbalance)* is the unbalance that occurs when current on each of the three power lines of a three-phase power supply are not equal.
- A *delta-connected three-phase motor* is a motor in which each phase is connected end to end to form a completely closed-loop circuit.
- *Derating* is the reduction of the total rated operating output of a device to allow for safe operation under abnormal environmental conditions.
- A *dielectric material* is a medium in which an electric field is maintained with little or no outside energy supply.
- A *differential* is the difference between the temperature at which a switch turns on a unit and temperature at which it turns off the unit.
- A *directional control valve* is a valve that allows or prevents fluid flow to specific piping or system actuators.
- A *dual-voltage three-phase motor* is a three-phase motor that operates at more than one voltage level.
- An *electrolyte* is a conducting medium in which the current flow occurs by ion migration.
- An *electrolytic capacitor* is a capacitor formed by winding two sheets of aluminum foil separated by pieces of thin paper impregnated with an electrolyte.
- An *enclosed-coil heating element* is a coiled resistance wire sealed into an enclosure and fixed onto a supporting element.
- A *heating element* is the portion of an electrical component that produces heat by electric current.
- *Inductive reactance* is the opposition to the flow of alternating current in a circuit due to inductance.
- *Locked rotor torque* is the torque a motor produces when the rotor is stationary and full power is applied to the motor.
- A *megohmmeter* is a high-resistance ohmmeter used to measure insulation deterioration on conductors by measuring high resistance values using high-voltage test conditions.
- A *normally closed (NC) valve* is a valve that does not allow pressurized fluid out of the valve in the spring-actuated (normal) position.
- A *normally open (NO) valve* is a valve that allows pressurized fluid to flow out of the valve in the spring-actuated (normal) position.
- An *ohmmeter* is a test instrument that measures resistance.
- An *open-coil heating element* is an exposed coiled resistance wire fixed onto a supporting element.
- *Overcycling* is the process of turning a motor on and off repeatedly.
- An *overload* is the condition that occurs when the load connected to a motor exceeds the full-load torque rating of the motor.

- A *position* is any of the locations within the valve in which a spool is placed to direct fluid through the valve.
- *Proportional temperature control* is a heating control function in which the heating element can produce any amount of heat from 0% to 100% of its rated value.
- *Refractory* is a heat-resistant ceramic material.
- A *shaded-pole motor* is a single-phase AC motor that uses a shaded stator pole for starting.
- A *single-voltage three-phase motor* is a motor that operates at only one voltage level in a three-phase system.
- A *solenoid* is an electromechanical device that converts electrical energy to linear mechanical force.
- A *split-phase motor* is an AC motor that can run on one or more phases.
- A *thermal switch* is a switch that operates its contacts when a preset temperature is reached.
- *Two-position temperature control* is a heating control function in which the heating element can only be ON or OFF.
- *Voltage unbalance (imbalance)* is the unbalance that occurs when voltage at the terminals of an electric motor or other three-phase load are not equal.
- *Watt density* is the wattage concentration on the surface of a heating element.
- A *way* is the path of a piping port through a directional control valve.
- A *wye-connected three-phase motor* is a motor that has one lead end of each of the three phases (phases A, B, and C) internally connected to the other two phases.

Review Questions

1. What are the three main types of electric motors and how are they used in HVAC systems?
2. What are the desired values that should be observed when testing the voltage of electric motors in HVAC systems?
3. What is overcycling and how does it occur in HVAC systems?
4. What tool is used to measure motor insulation failure?
5. What are four common motor environmental and mechanical problems?
6. What are the three different types of single-phase motors?
7. What is the most common problem that occurs within split-phase motors?
8. What is voltage unbalance/imbalance?
9. What is current unbalance/imbalance?
10. How do heating elements generate heat and how are they rated?
11. What are the two types of heating elements?
12. What are three common problems with heating elements?
13. How are control circuit transformers tested?
14. How does a contactor function?
15. How does a solenoid-operated directional control valve operate?

chapter 8
MOTOR CONTROL

Electric motors are used in commercial buildings for circulating heated, cooled, or ventilated air in HVAC systems, operating circulating pumps, and other common applications. For motors to operate safely and efficiently with maximum service life, they must be properly powered and controlled. In order for the work to be performed as designed, each system must have a control circuit. A motor can be controlled manually and/or automatically depending on the application. Magnetic motor starters and motor drives are used for motor control in commercial applications.

OBJECTIVES

- Identify motor power circuits, control circuits, and related devices.
- Identify magnetic motor starters and motor drives.
- Safely perform troubleshooting procedures on motor power circuits, control circuits, and motor drives using test instruments.

MOTOR POWER CIRCUITS

The two major parts of a motor circuit are the motor power circuit and the motor control circuit. A *motor power circuit* is the section of an electrical motor circuit that delivers high voltage or current to an electric motor. A motor power circuit includes a disconnect switch, a control transformer, protective devices (fuses or circuit breakers), a motor starter, and a motor. **See Figure 8-1.**

An electric motor can be DC, single-phase AC, or three-phase AC. Incoming power lines must have a mechanism to switch off and lock out and tag out the circuit power. They must also include fuses or circuit breakers sized to protect the system. A motor can be controlled by a magnetic motor starter or a motor drive.

A motor starter includes devices that are used in the motor power circuit, such as normally open (NO) power contacts and motor overload current monitoring. A motor starter also includes devices that are used in the motor control circuit, such as NO or normally closed (NC) auxiliary contacts, a starter coil or circuit, and NC overload contacts. A magnetic motor starter is a contactor that includes an overload section. Contactors can be used to control single-phase motors that have built-in overload protection. Motor power circuits are classified into different sections for the purpose of troubleshooting.

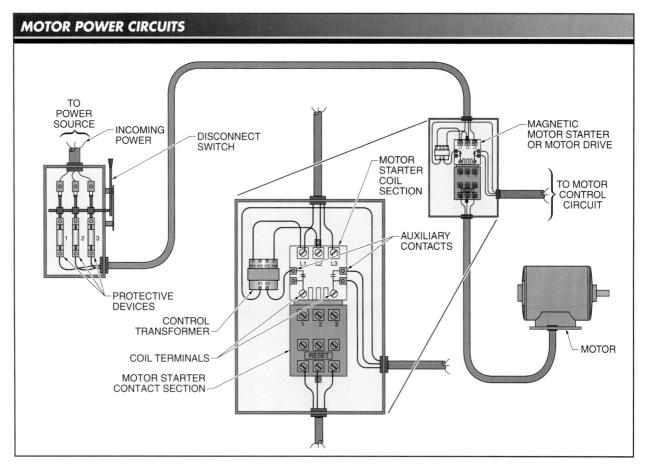

Figure 8-1. A motor power circuit includes a disconnect switch, a control transformer, overload protective devices, a motor starter, and a motor.

Disconnect Switches

A *disconnect switch,* also known as a disconnect, is a switch that disconnects the supply of electric power from electrically powered devices such as motors and machines. Disconnects are used to manually remove power from or apply power to a circuit. The disconnect switch is the point at which the load is connected to the building power distribution system. **See Figure 8-2.**

Disconnect Switches
Media Clip

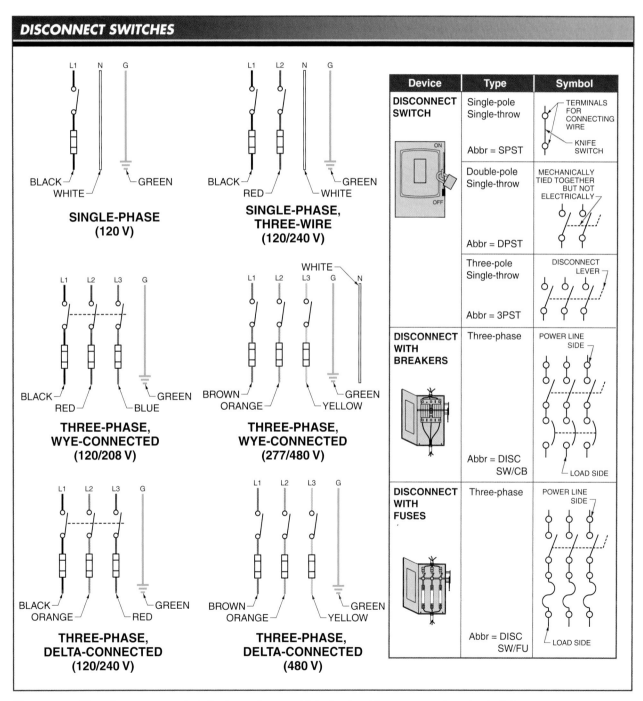

Figure 8-2. The disconnect switch is the point at which the load is connected to the building power distribution system.

Article 430 of the National Electrical Code® (NEC®) covers motors, motor branch-circuit and feeder conductors and their protection, motor overload protection, motor control circuits, and motor control centers.

Overcurrent Protection Media Clip

Disconnect switches include overcurrent protection to protect the load and system from short circuits, faulty ground connections, and excessive current levels. An *overcurrent protective device (OCPD)* is a fuse or circuit breaker that disconnects or discontinues current flow when the amount of current exceeds the design load. A disconnect switch is the point at which a load or energized equipment can be locked out and tagged out to remove power during system maintenance. A disconnect switch enclosure is typically the starting point for troubleshooting a load or energized equipment because it contains fuses and circuit breakers.

Power Circuit Terminal Identification

In a motor power circuit, the terminals and conductors can be identified with different markings depending on the equipment manufacturer and equipment installer. For example, three-phase power lines may be marked as L1, L2, and L3 or R, S, and T. Three-phase motor terminals may be marked T1, T2, and T3 or U, V, and W. Single-phase power lines may be marked L1 and N for 120 VAC circuits or L1 and L2 for 240 VAC circuits.

Single-phase motor terminals may be marked T1 and T2 or with specific manufacturer numbers, such as 1 and 2. Single-phase motor terminals may also be marked with different colors such as blue or black for T1 and white for T2.

DC power lines are normally marked DC+ and DC–. DC motor armature windings are normally marked A1+ and A2–. DC series motor series windings are normally marked S1 and S2. DC shunt motor shunt windings are normally marked F1 and F2. DC compound motors have an armature, series field, and shunt field. **See Figure 8-3.**

Troubleshooting Motor Power Circuits

Troubleshooting is the systematic elimination of the various parts of a system or process to locate a malfunctioning part. When performing troubleshooting tasks, the different parts of a circuit within an electrical system are categorized into sections to aid in determining where to begin the troubleshooting process. For example, an HVAC system can be categorized into a power circuit, control circuit, and interface connecting the power circuit to the control circuit. **See Figure 8-4.**

The power circuit is the high-voltage section of the circuit that includes the incoming power supply, fuses or circuit breakers, motor starter contacts, and motor. In an HVAC circuit, the original equipment manufacturer (OEM) can provide the across-the-line starting circuit. The across-the-line starting circuit is used when the initial inrush current of the compressor motor does not cause a problem, such as a greater than 5% drop in the main voltage supply, when starting.

The control circuit controls the motor starter coils in the power circuit. The control circuit normally operates at a lower voltage than the power circuit. A step-down transformer is the interface used to reduce the voltage from the power circuit to the control circuit.

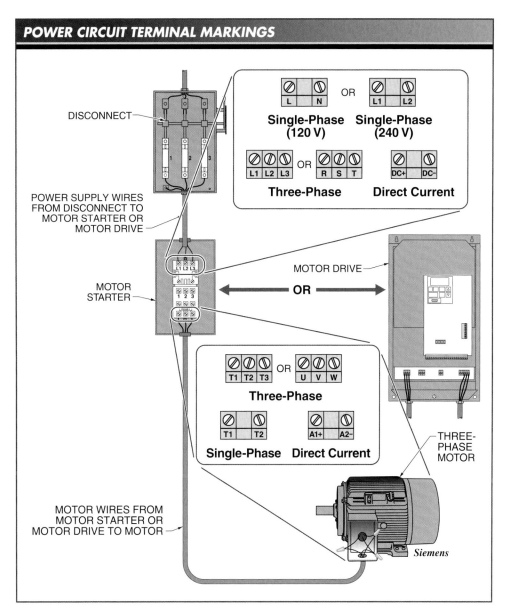

Figure 8-3. In a motor power circuit, the terminals and conductors can be identified with different markings depending on the equipment manufacturer and equipment installer.

When troubleshooting a power circuit, electrical measurements are taken with the appropriate test instruments, such as a DMM. Voltage measurements are the first measurements taken to indicate if power is present. Current measurements must also be taken and compared to nameplate ratings to determine motor loading.

The first voltage measurement taken is the power circuit voltage at the fuses or circuit breakers. Before taking any measurements, it should be ensured that proper PPE is worn, plant and safety procedures are followed, and that the DMM is checked for proper operating condition before and after the voltage measurements.

TECH FACT

The National Electrical Code® (NEC®) requires that the disconnecting means of a motor be properly rated for the corresponding horsepower of the motor and that it be located within sight of the motor.

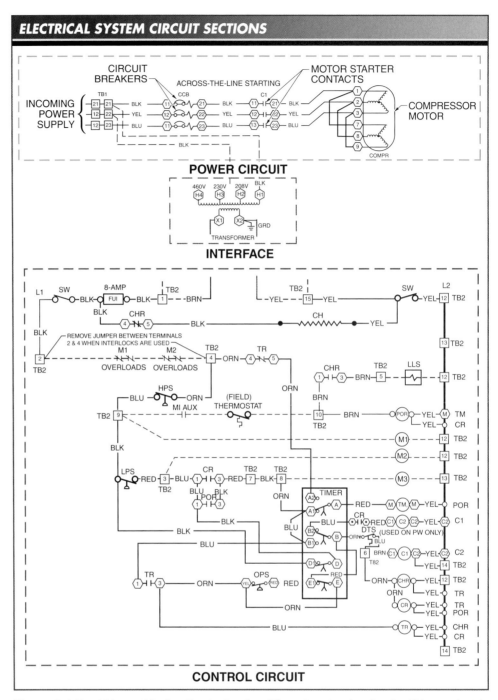

Figure 8-4. To aid in troubleshooting tasks, the different parts of a circuit within an electrical system are categorized into sections, such as the power circuit, control circuit, and interface connecting the power circuit to the control circuit.

When troubleshooting, an electrical print is used as a reference to help identify the components and devices used in a circuit and how they are connected to other components and devices. However, an electrical print does not identify the actual location of the components and devices in the wired panel. The print component layout can be different from the actual component layout. **See Figure 8-5.**

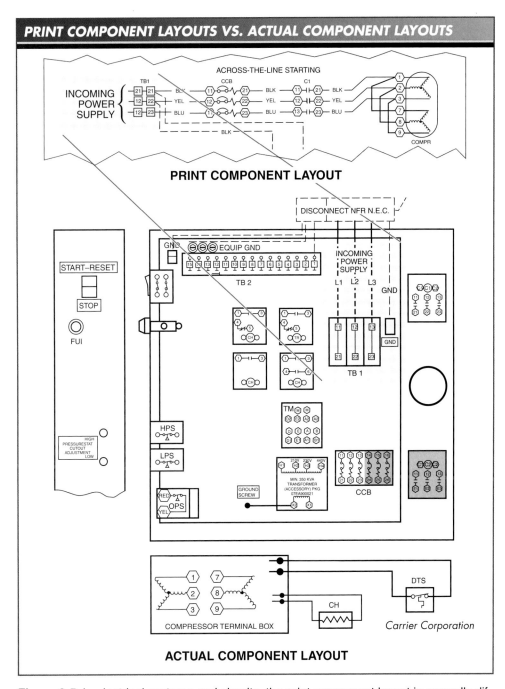

Figure 8-5. In electrical systems and circuits, the print component layout is normally different from the actual component layout.

Troubleshooting the power circuit begins by taking a measurement of the incoming voltage level to verify that the voltage is within +5% to –10% of the equipment voltage rating. Fuses and circuit breakers are also tested for proper operation by taking voltage measurements in and out of each fuse or circuit breaker. A properly operating fuse or circuit breaker must have the same voltage coming out as going in. The voltage in and out of the control transformer must also be tested.

When a motor is running, both voltage and current measurements should be taken. The voltage must be within the +5% to –10% range, and the current should not exceed the motor maximum-rated current as listed on the motor nameplate.

MOTOR CONTROL CIRCUITS

A *motor control circuit* is a circuit that controls a motor as required to start, stop, jog, increase speed, decrease speed, or reverse its direction of rotation. The motor control circuit is normally the low-voltage and current section of a circuit. The motor control circuit controls the motor operation and includes the motor starter coil located in the magnetic motor starter or electronic circuit located in a motor drive.

For safety and code compliancy, motor control circuits normally operate at a lower voltage than the motors they are controlling. Control voltages are normally 120 VAC, 24 VAC, or 24 VDC. A step-down control transformer is used to reduce the voltage between the high-voltage power circuit and the low-voltage control circuit. **See Figure 8-6.**

Most control transformers include a multiple coil or tap primary so the transformer can be connected to different voltage levels. The high-voltage side of a transformer is marked with an "H," and the low voltage side is marked with an "X." Because grounding is not carried through a transformer, the secondary coil of a transformer must be reconnected to ground to re-establish a common ground point in the control circuit.

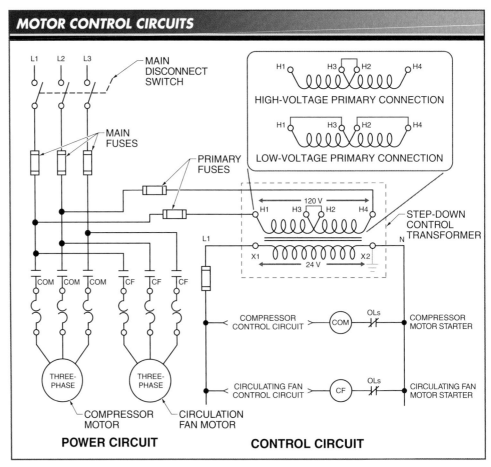

Figure 8-6. In a motor control circuit, a step-down control transformer is used to reduce the voltage between the high-voltage power circuit and the low-voltage control circuit.

The motor control circuit includes the overload contact that stops the motor when the motor draws more than the predetermined current limit. The control circuit also includes control switches, relays, and timers that are used to determine the timing of motor operation. Troubleshooting a control circuit requires using a working print to understand how the circuit must operate and the location of the components and devices to be tested.

Motor control circuits range from basic to complex, depending on the circuit requirements and application. Common motor control circuits include manual start/stop, additional manual start/stop, manual/automatic, and automatic air-compressor control circuits. **See Figure 8-7.** In a manual start/stop control circuit, a start pushbutton is used to start the motor, and a stop pushbutton is used to stop the motor.

When a motor is started and stopped by pushbuttons controlling a motor starter coil, the motor starter coil activates the power circuit to energize and deenergize the motor power circuit. Auxiliary contacts are added in parallel with the start pushbutton to give the circuit memory (holding contact) that enables the motor to remain running after the start pushbutton is pressed and released. Once coil M1 of the magnetic motor starter is energized, it causes coil contacts M1 to close and remain closed (memory) until the coil is deenergized.

A load is often required to be started and stopped from more than one location. In this circuit, the magnetic motor starter may be started or stopped from two locations. Additional stop pushbuttons are connected in series with the existing stop pushbuttons. Additional start pushbuttons are connected in parallel with the existing start pushbuttons.

Pressing any one of the start pushbuttons causes coil M1 to energize. This causes auxiliary contacts M1 to close, adding memory to the circuit until coil M1 is deenergized. Coil M1 may be deenergized by pressing the stop pushbuttons, by an overload that would activate the overload devices, or by a loss of voltage to the circuit. In the case of an overload, the overload must be removed and the circuit overload devices reset before the circuit can return to normal starting condition.

Loads are often required to be controlled manually and/or automatically. In this type of circuit, a three-position selector switch allows the operator to select one of three circuit conditions. For example, a three-position selector switch may be used to place a heating circuit in the manual, automatic, or OFF position. The OFF position is added for safety. In the OFF position, the heating contactor (or other machine being controlled) cannot be energized by the manual or automatic switch. Circuit conditions controlled by three-position selector switches include manual/OFF/automatic, heat/OFF/cool, forward/OFF/reverse, and slow/stop/fast conditions.

An air compressor is a load that is controlled automatically. Air compressors are used to provide pressurized air for pneumatic tools and HVAC controls. A *pressure switch* is an electric switch operated by pressure that acts on a diaphragm or bellows element. Pressure switches start and stop an air-compressor motor based on the pressure in the receiver.

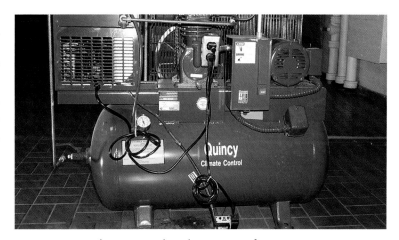

A pressure switch is mounted on the receiver of an air compressor to turn the motor on and off.

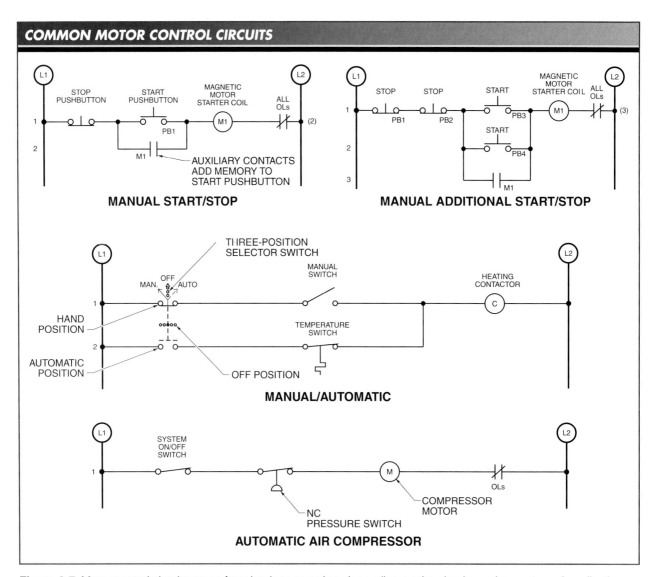

Figure 8-7. Motor control circuits range from basic to complex, depending on the circuit requirements and application.

Control Switches

A *control switch* is a device that regulates the flow of current in a circuit. Control switches can be activated manually, mechanically, or automatically. Manually operated control switches include pushbuttons, selector switches, joysticks, foot switches, or any type of switch that requires an individual to operate it. A limit switch is the only type of mechanically operated switch. Automatically operated switches include temperature switches, pressure switches, liquid-level switches, flow switches, and any other automatic control switch or sensor. Control switch contacts can be either NO, NC, or any combination of NO and NC. The type of contacts used depends on the application. **See Figure 8-8.**

For example, if an NO liquid-level switch is used to control a pump motor, when the liquid level reaches the switch position and closes the switch contacts, the motor turns on. As the liquid level drops below a predetermined level, the switch contacts open, turning off the motor. This type of application is used to keep a liquid level below the switch position. **See Figure 8-9.**

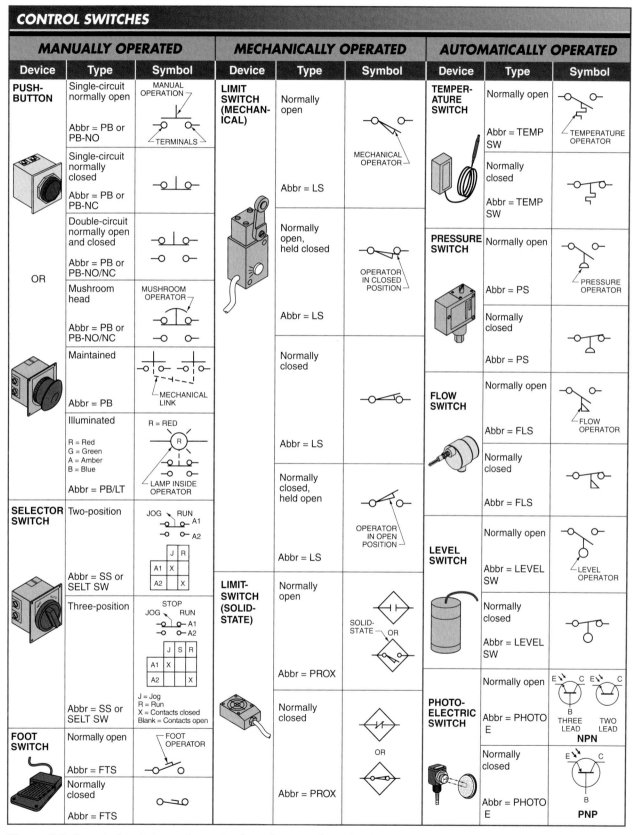

Figure 8-8. A control switch regulates the flow of current in a circuit and can be activated manually, mechanically, or automatically.

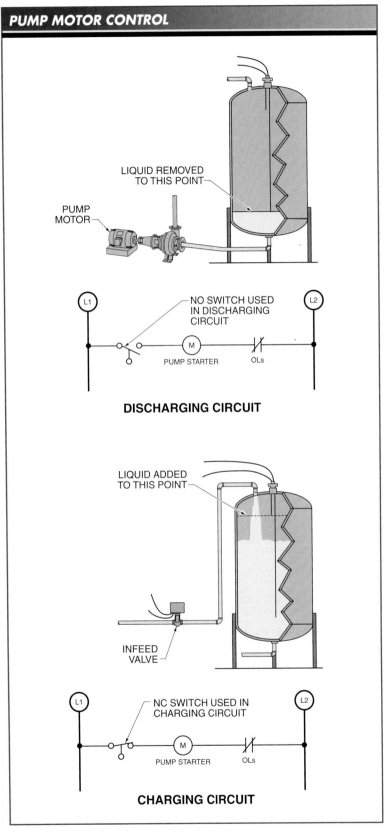

Figure 8-9. NO and NC level switches can be used to control a pump motor to control the liquid level in a tank.

When an NC liquid-level switch is used to control a pump motor, the motor is ON until the liquid level drops below the liquid-level switch position. The motor is OFF as long as the liquid level is above the switch position. This type of application is used to keep the liquid level in a tank above a predetermined switch position.

Single- and Dual-Function Switches. Any type of control switch, such as a pushbutton, limit switch, or pressure switch, can be either a single- or dual-function switch. **See Figure 8-10.** A *single-function switch* is a switch that controls only the turning on or off of a load but not both. To make the control circuit switches operate as single-function switches, an NO motor starter contact (auxiliary contact) must be added into the circuit. The auxiliary contact is placed in parallel with the NO contacts that are used to start the motor.

Once the NO contact is closed and the circuit is completed through the motor starter, the motor starter is energized and the motor starter NO auxiliary contacts close. The motor starter remains on even after the start switch is released. The motor starter is deenergized when any of the NC stop switches are opened. The starter remains off, even when the NC switches return to their closed position.

A *dual-function switch* is a control switch that controls both the on and off control function. When the switch contacts are closed, the load is ON. When the switch contacts are open, the load is OFF.

Basic Air Conditioning System Control Circuit Operation. In a basic air conditioning system, a step-down transformer is used to reduce the incoming voltage to 24 VAC for the control circuit components. A motor starter controls the fan motor and compressor motor. The motor starters apply a high voltage to the motors. A two-position switch allows the system to be placed in the cool or off position. A three-position switch allows the fan motor to be placed in the auto, OFF, or ON position even when the compressor motor is off so that air can continue to be circulated in the building spaces. **See Figure 8-11.**

An NO temperature switch is used to set the temperature at which the cooling system turns on as the temperature in the building space rises. An NC high-pressure switch is connected in series with the compressor motor starter coil and opens if the pressure in the system rises above the safe high-pressure setting. An NO low-pressure switch is connected in series with the compressor motor starter coil and opens if the pressure in the system drops below the safe low-pressure setting.

TECH FACT
A temperature switch may include a thermistor, which is a temperature-sensitive resistor that is placed at the point where temperature is to be measured. The resistance of a thermistor increases with an increase in temperature and decreases with a decrease in temperature, which is sensed by an electronic circuit.

Mechanical and Solid-State Relays

An *electrical relay* is a control device activated by electric power that opens, closes, or activates other mechanisms. Electrical relays control one part of an electrical circuit by opening and/or closing contacts in another circuit. Electrical relays can be either electromechanical or solid-state relays. An *electromechanical relay* is an electric device that uses a magnetic coil to open or close one or more sets of mechanical contacts. A *solid-state relay* is a relay that uses electronic switching devices in place of mechanical contacts. Solid-state relays use an electronic circuit to open and close electronic power switching devices such as SCRs for DC switching and triacs for AC switching.

Electromechanical and solid-state relays are used in most control circuits. For example, an HVAC circuit includes an electromechanical relay "CHR" that controls power to the liquid line solenoid (LLS). The CHR relay coil may require only 20 mA to 100 mA to operate, but the relay contacts can be rated at 10,000 mA or more. **See Figure 8-12.**

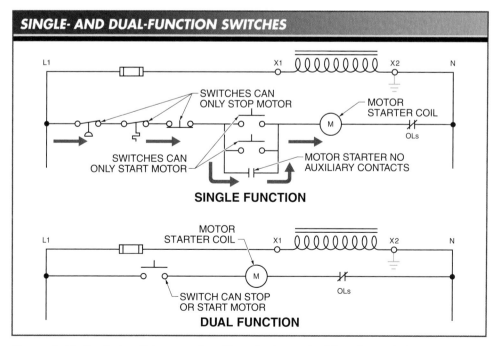

Figure 8-10. Control switches can be either single- or dual-function switches.

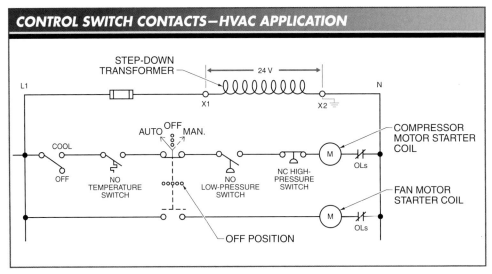

Figure 8-11. A basic HVAC control circuit includes an NO temperature switch, an NO low-pressure switch, and an NC high-pressure switch.

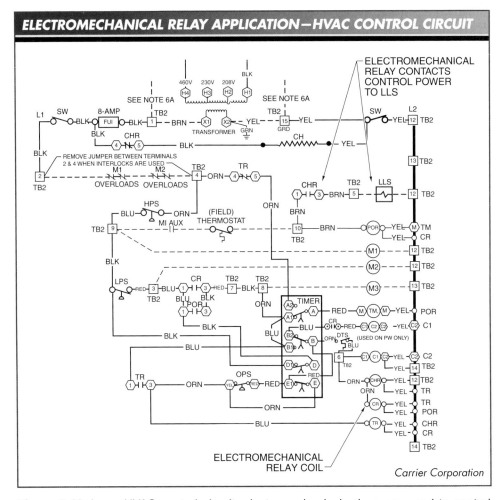

Figure 8-12. In an HVAC control circuit, electromechanical relays are used to control power to solenoids and other devices.

Relays are used to either allow a low voltage to control a high voltage or to multiply the total number of contacts that can be used in a circuit. In a relay, the coil uses low voltage to control contacts that are rated for a high voltage. When energized, a relay coil can open and close multiple contacts. Relays commonly include one to ten sets of contacts. **See Figure 8-13.**

Timers

A *timer* is a control device that uses a preset time period as part of a control function. As with relays, timers can be either mechanical or solid-state. For example, a common commercial control circuit can use a mechanical timer (normally marked "TR") with four contacts. Mechanical and solid-state timers are represented by different schematic symbols on a print. Depending on the type of timer used, the time period can be applied at different intervals of circuit operation. Time-delay relays are available in timing ranges of a few milliseconds to hundreds of hours. The most common timing functions are on-delay and off-delay. **See Figure 8-14.**

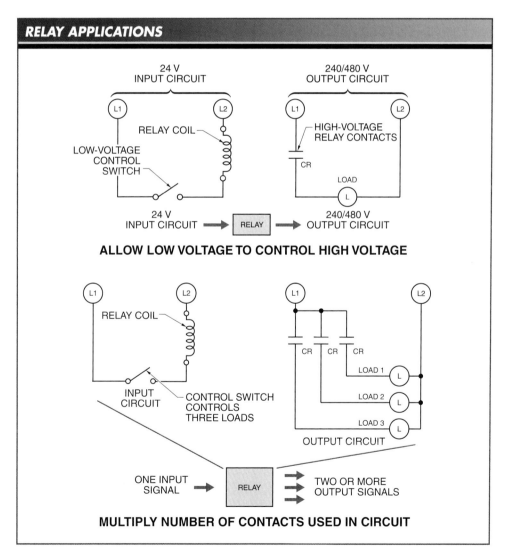

Figure 8-13. Relays are used to either allow a low voltage to control a high voltage or to multiply the total number of contacts that can be used in a circuit.

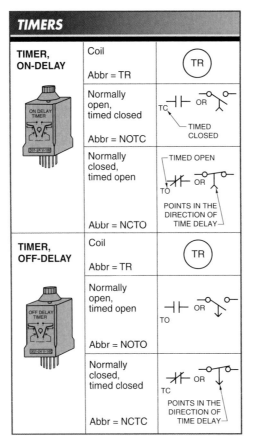

Figure 8-14. In commercial circuits, the most common types of timers are on-delay and off-delay timers.

Other timing functions include one-shot (interval) timers and recycle timers. As with relays, timers can be used to directly switch low-current loads. However, timers are normally used to indirectly control high-current loads by switching relays, contactors, or motor starters.

On-Delay Timers. An *on-delay timer*, also known as a delay-on-operate timer, is a time-based control device that delays for a predetermined time after receiving a signal to activate or turn on. Once activated, an on-delay timer may be used to turn a load on or off, depending on whether NC or NO timing contacts are used. On-delay timer contacts do not change position until the set time passes after the timer receives power. **See Figure 8-15.** The schematic symbol for an on-delay timer is the back end of an arrow pointing upward on the timer contact symbol.

Off-Delay Timers. An *off-delay timer*, also known as a delay-on-release timer, is a timer that does not start its timing function until the power is removed from the timer. **See Figure 8-16.** In a circuit that requires an off-delay timing function, a control switch is used to apply power to the timer. The timer contacts change immediately and the load energizes when power is first applied to the timer. The timer contacts remain in the changed position, and the time period starts only when power is removed from the timer. The schematic symbol for an off-delay timer is the front end of an arrow pointing downward on the timer contact symbol.

Troubleshooting Control Circuits

Troubleshooting a control circuit requires using a print to understand how the circuit must operate and to locate the components to be tested. Troubleshooting starts by testing components at the control circuit incoming power supply and working through the control components to be tested. Troubleshooting control circuits is easily performed when the control circuit is connected to a terminal strip and the terminal numbers are identified on the print. **See Figure 8-17.**

> **TECH FACT**
>
> The four major categories of timers are dashpot, synchronous clock, solid-state, and programmable. Dashpot, synchronous clock, and solid-state timers are stand-alone timers that are installed between the input and output devices. Programmable timers are timing functions that are components of electrical control devices such as programmable logic controllers (PLCs).

ON-DELAY TIMERS

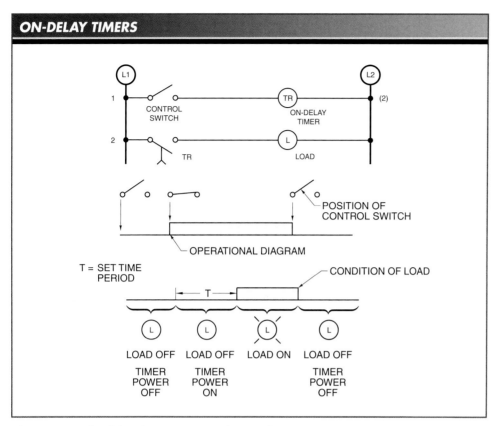

Figure 8-15. On-delay timer contacts do not change position until the set time period passes after the timer receives power.

OFF-DELAY TIMERS

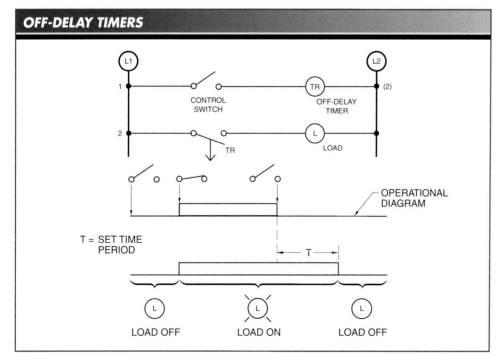

Figure 8-16. An off-delay timer does not start its timing function until the power is removed from the timer.

272 ELECTRICAL SYSTEMS for FACILITIES MAINTENANCE PERSONNEL

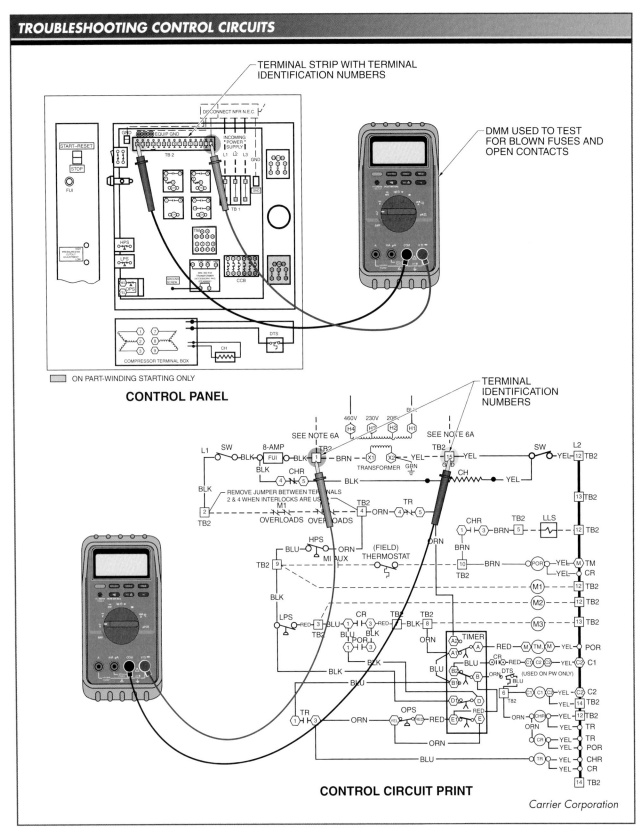

Figure 8-17. Troubleshooting a control circuit requires using a print to understand how the circuit must operate and to locate the components to be tested.

The output voltage of the control transformer must be within +5% to –10% of the control component ratings. For example, in an HVAC circuit, a DMM set to measure voltage is placed at terminals 1 and 15. The common meter lead remains on terminal 15 if the voltage is within specifications at terminals 1 and 15, while the meter lead on terminal 1 is moved across the control circuit to test individual components. If voltage is present at terminal 1, but not at terminal 2, it is an indication that the fuse or switch is open (bad). Also, if voltage is present at terminal 2 but not at terminal 4, it is an indication that a motor overload contact is open.

Note: With experience, troubleshooting a control circuit can be simplified. For example, if a solenoid is not operating, a DMM set to measure voltage can be placed across terminals 1 and 15 to verify control circuit power. Then one meter lead can be moved from terminal 1 to terminal 5 and the other from terminal 15 to terminal 12, so the voltage directly across the solenoid is tested.

MAGNETIC MOTOR STARTERS AND MOTOR DRIVES

A *motor starter* is an electrically operated switch (contactor) that includes motor overload protection. A motor starter is used to turn on and off an electric motor, monitor the motor current draw, and provide additional contacts (auxiliary contacts) that can be used in the control circuit. A motor starter can be a switch that uses mechanical contacts, solid-state switching, or a solid-state motor drive that uses electronic devices to control the flow of current to a motor. **See Figure 8-18.**

Magnetic Motor Starters

A *magnetic motor starter* is a starter that has an electrically operated switch (contactor) and includes motor overload protection. Magnetic motor starters have been the standard method of motor control for most of the last century. Magnetic motor starters are still used for controlling motors in applications in which the only motor control function is to start and stop the motor and saving energy is not a major consideration.

Magnetic motor starters were developed to add control functions, such as controlling a motor from any location, and to allow a low control voltage to control a high motor voltage. Magnetic motor starters perform the following functions:

- They can start a motor from any location when the starter coil is energized. The starter coil closes and opens the power contacts controlling the motor.
- They can stop a motor from any location by deenergizing the starter coil.
- They provide motor overload protection by monitoring the motor current draw through overload devices (heaters) and automatically turn the motor off if the motor is overloaded for an extended period of time.

The length of time that must pass before a motor is turned off (trip time) is determined by the amount of the overload. The greater the overload, the shorter the trip time. Magnetic motor starter trip time is not easily adjustable. The most common method used to adjust trip time on a motor starter is to select one of four different classes of overloads (class 10, 15, 20, or 30) and install the overload into the motor starter. The smaller the trip class number, the shorter the trip time.

In a magnetic motor starter, one power contact is required for every power line that is connected to the motor. Three power contacts are required for three-phase motors. The input to the contacts are marked L1, L2, and L3, and the output of the contacts are marked T1, T2, and T3. A coil closes the contacts when energized and opens the contacts when deenergized.

274 ELECTRICAL SYSTEMS for FACILITIES MAINTENANCE PERSONNEL

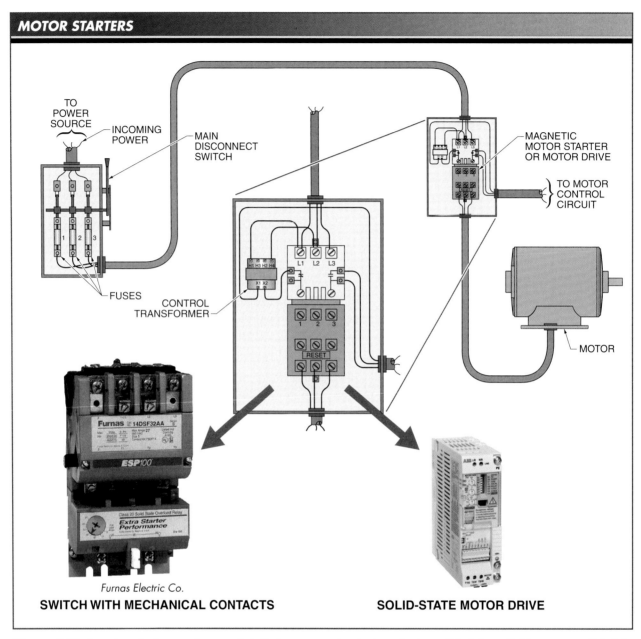

Figure 8-18. A motor starter can be a switch that uses mechanical contacts, solid-state switching, or a solid-state motor drive that uses electronic components to control the flow of current to a motor.

An overload monitoring device is also required for every power line. In a magnetic motor starter, the overload monitoring devices are referred to as heaters and are connected in series with the starter contacts and motor. The overload monitoring devices are referred to as heaters because they increase in temperature as the motor draws more current. **See Figure 8-19.**

Heaters are not fuses or circuit breakers and do not open, but they do open an NC overload (OL) contact that is in series with the motor starter coil. The contact deenergizes the coil and disconnects the power to the motor when the motor draws more current than the heaters are rated for. The OL contact is used in the motor control circuit to remove the motor from the power lines when there is an overload condition.

OVERLOAD CURRENT MONITORING DEVICES

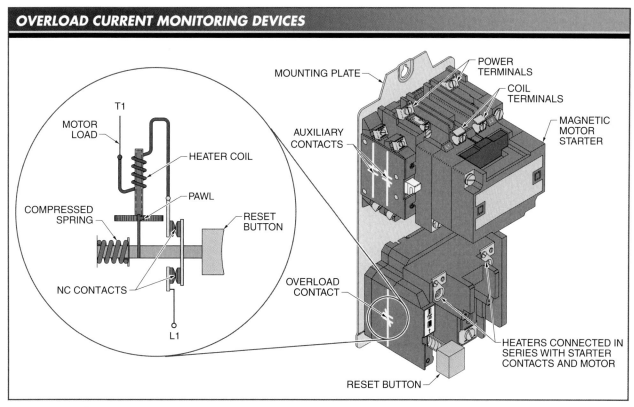

Figure 8-19. Overload current monitoring devices open an NC contact (OL contact) when the motor draws more current than the heater is rated for.

Although motor starters with heaters can indirectly monitor motor current, they are not always accurate because ambient temperatures around the motor starter can affect the trip point. Modern magnetic motor starters use electronic overloads that use current transformers to monitor motor current electronically and an electronic circuit to control the overload contact. A current transformer is similar to a clamp-on ammeter in that they both measure the strength of the magnetic field around a conductor, which is directly proportional to the amount of current flowing in the conductor. **See Figure 8-20.**

Motor Drives

A *motor drive* is an electronic device that controls the direction, speed, torque, and other operating functions of an electric motor, in addition to providing motor protection and monitoring functions.

Motor drives are also referred to as electric motor drives, variable-frequency drives, and adjustable-frequency drives. Certain motor drives are also referred to by the method they use to control a motor, such as inverter drives, vector drives, and direct torque drives.

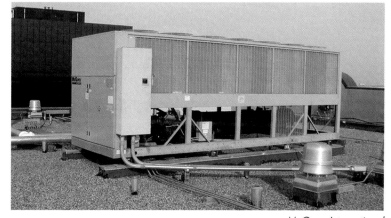

McQuay International

Motor starters are normally located in the electrical cabinet of the equipment they control.

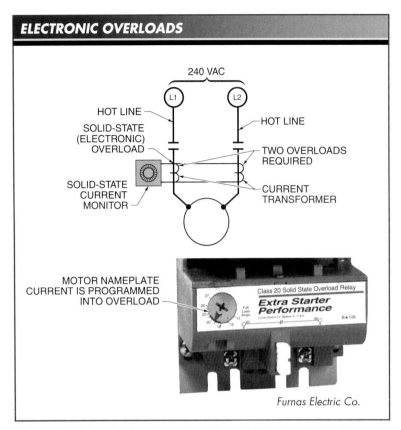

Figure 8-20. An electronic overload has built-in circuitry that senses changes in current and temperature.

Motor drives were developed from magnetic motor starters. Motor drives provide numerous motor control functions and can save energy in most motor applications. Magnetic motor starters are being replaced by motor drives. Most new motor control applications use motor drives. Benefits of motor drives include the following:

- Motor drives can start a motor from any location. Motor drives automatically provide a safe low voltage (12 VDC to 24 VDC) to the drive control circuit terminals. This adds safety and ease in wiring the control circuit. Motor drives do not require a step-down control transformer.
- They can stop a motor from any location. The ability to stop a motor at the drive (as with a manual starter) and away from the drive (as with a magnetic motor starter) provides additional control and safety to the circuit.
- They provide motor overload protection through direct monitoring of the motor current draw. A motor drive turns off the motor when there is an overload. When the drive detects an overload, it records the fault and displays a fault code. Since the amount of current is adjustable, various heater sizes are not required. Also, the trip time is fully adjustable, which allows each drive/motor to be customized to the application requirements.
- They provide a full range of speed control from 0 rpm to full motor speed.
- They control motor acceleration time allowing the motor to start any load without damage.
- They control motor deceleration time and apply a braking force when needed. This can eliminate the need for a separate braking circuit, such as dynamic braking or mechanical shoe brakes, as is required with magnetic motor starter circuits.
- They set minimum and maximum speed limits on the motor so the motor operates within a certain speed range.
- They connect and interface with PCs, which allows drive parameters and drive faults to be displayed or output to a printer.

When motor drives are used to control motors, motors are protected from an overload by programming the motor nameplate-rated current into the drive. On basic motor drives, the motor overload protection is programmed with an adjustment dial on the motor drive. On advanced motor drives, motor overload protection is programmed into the motor drive using a keypad. **See Figure 8-21.**

MOTOR DRIVE OVERLOAD PROGRAMMING

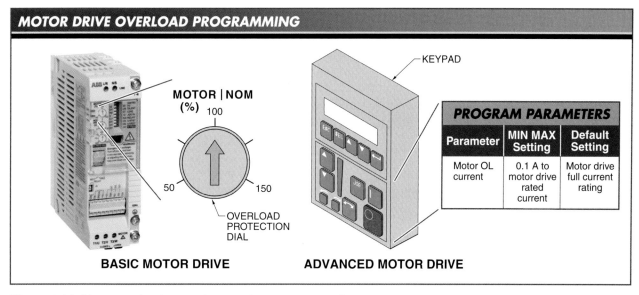

Figure 8-21. Motor overload protection may be programmed into a motor drive by setting a dial or using a keypad.

The three main sections of a motor drive are the converter section, DC bus section, and inverter section. The converter section (rectifier) receives the incoming AC voltage and changes the voltage to DC. AC input voltage that is different from the AC output voltage requires that the converter section first step up or step down the AC voltage to the proper level. **See Figure 8-22.**

The DC bus section filters the voltage and maintains the proper DC voltage level. The DC bus section delivers the DC voltage to the inverter section for conversion back to AC voltage. The inverter section controls the speed of a motor by controlling the power frequency and controls motor torque by controlling the voltage sent to the motor.

MOTOR DRIVE SECTIONS

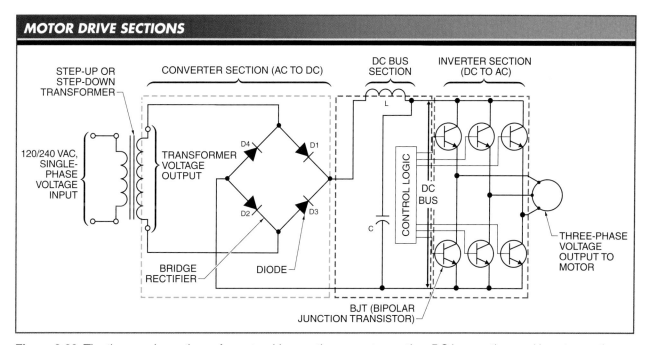

Figure 8-22. The three main sections of a motor drive are the converter section, DC bus section, and inverter section.

When magnetic motor starters are used to control motors, the current and voltage levels have to match the nameplate rating of the motor. For example, a three-phase, 240 V rated motor controlled by a magnetic motor starter must be supplied by a three-phase, 240 V power supply. When certain motor drives are used to control motors, the type and level of the input supply voltage does not have to match the rated voltage type and level of the motor. For example, a motor drive supplied with 120 V single-phase power can be used to control a 240 V three-phase motor. This allows the use of energy-efficient three-phase motors where only single-phase power is available, such as in small commercial HVAC systems. **See Figure 8-23.**

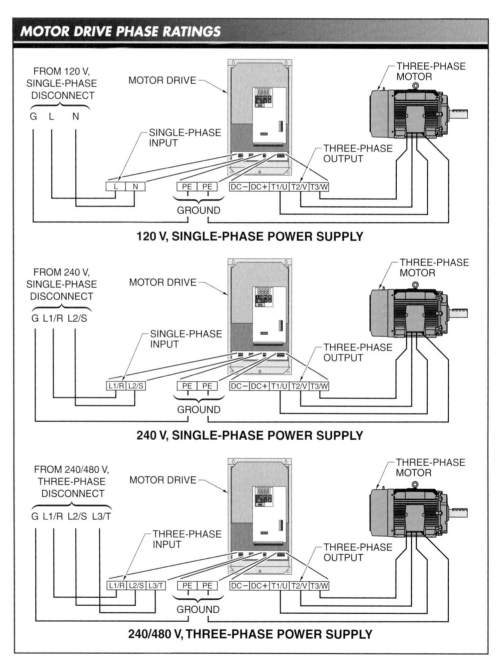

Figure 8-23. The input voltage type (single-phase or three-phase) and voltage level (120 V, 240 V, or 480 V) to a motor drive can be different from the output of the drive.

Motor Circuit Connections

When a magnetic motor starter is used to control a motor, a step-down transformer, a motor starting coil, auxiliary contacts, and an overload contact are required in the control circuit, in addition to the external switches. However, when a motor drive is used to control a motor, only the external switch requires connection to the motor drive control circuit terminals.

When using a motor drive, a step-down transformer is not required because the motor drive steps down the control voltage to the terminals to which the switches are connected. The control voltage at motor drive control terminals is normally 12 VDC or 24 VDC. A motor starter coil, auxiliary contacts, and overload contacts are also not required with a motor drive because they are replaced with an electronic circuit inside the drive. The electronic circuit is mainly used for motor speed control and is programmed through use of a keypad on the motor drive faceplate.

Most motor drives also include additional terminals for connecting other inputs or outputs, such as error reset switches, motor at speed sensors, and drive fault indicators. Motor drives can also have a potentiometer (internal or external) with an external signal activated by liquid level, temperature, pressure, resistance, voltage, or current that is used to control motor speed. **See Figure 8-24.**

Troubleshooting Motor Drives

Troubleshooting motor drives and electrical equipment is inherently dangerous because it normally requires removing motor drive housings, which exposes energized internal parts with dangerous voltages present. The motor and driven load may also be running, which exposes individuals working on or near the equipment to machine hazards. In addition, unexpected events can occur during troubleshooting, such as a motor drive stopping unexpectedly due to a drive undervoltage fault. The following safety guidelines must be followed when troubleshooting motor drives:

- Only qualified individuals (those who have special knowledge, training, and experience in the installation, programming, maintenance, and troubleshooting of electric motors and drives) should troubleshoot motor drives.
- Recommendations and instructions from the motor drive manufacturer and applicable federal, state, and local regulations should be referred to. Failure to follow OEM recommendations can result in serious physical injury and property damage.
- The equipment and processes that the motor drive controls and the consequences of unexpected starting, stopping, and running of a motor should be understood.
- Using two-way radios within the vicinity of motor drives should be avoided. Motor drives are susceptible to radio frequency interference (RFI), and using two-way radios in close proximity to a motor drive can result in the drive unexpectedly starting and stopping.
- Proper PPE must be used. For example, rubber insulating gloves, leather protector gloves, and rubberized insulated matting are used to provide maximum insulation from electrical shock hazards.
- It should be verified that the motor drive and motor operate as designed for the application once a problem has been identified and corrected.

TECH FACT
Motor drives can be used to control the flow rate of pumps and fans in HVAC applications, thereby allowing less energy to be used during periods of low demand for flow. These energy savings help reduce the utility bills for commercial buildings and help manage the quality of indoor air.

280 ELECTRICAL SYSTEMS for FACILITIES MAINTENANCE PERSONNEL

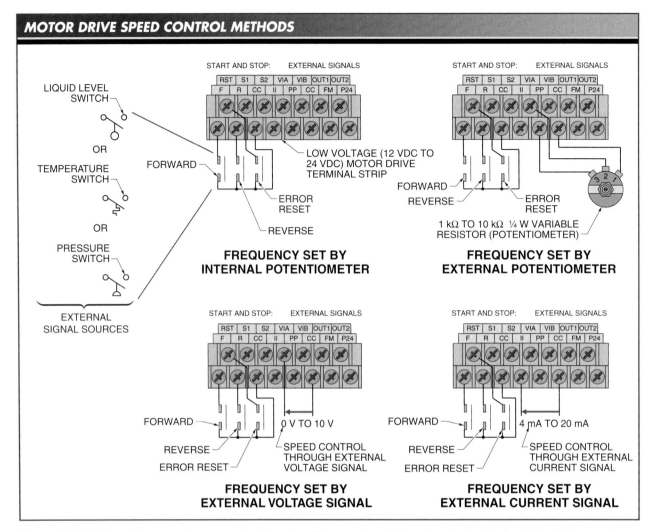

Figure 8-24. Motor drives can have a potentiometer with an external signal activated by liquid level, temperature, pressure, resistance, voltage, or current that is used to control motor speed.

Saftronics Inc.

Some motor drives may be removed from their housings so that benchtop troubleshooting can be performed away from energized electrical parts.

Gathering Information. The initial task when troubleshooting a motor drive is to gather information about the problem. Individuals are sent to unfamiliar locations to troubleshoot motor drive problems without the aid of engineering or maintenance shop records. Equipment operators are valuable sources of information concerning motor drive problems. Individuals can gather useful troubleshooting information through the following questions:

- What function was the motor drive performing when it failed, such as accelerating, decelerating, or running a motor at speed?

- Did the motor drive display a fault code or error message? If so, what was the fault code?
- How long has the problem been occurring?
- Does the problem occur constantly, at a particular time, or randomly?
- Is the problem linked to a time of day, a specific event, or a specific process?
- Has maintenance been performed on the motor drive or motor recently? If so, what type of maintenance was performed and who performed the work? Did the problem begin after the maintenance was completed?
- Have there been recent changes to the load, system, or motor drive programming?

Individuals must obtain appropriate OEM-provided motor drive manuals and programming parameters. Motor drive manuals include installation, operation, and troubleshooting procedures. OEM manuals also contain motor drive schematic diagrams, fault code explanations, and parameter descriptions. Troubleshooting a motor drive without OEM manuals is extremely difficult.

Motor drive parameters are saved as hard copies or as electronic files that are downloaded to personal computers (PCs) to display text. Motor drive parameter recording systems guarantee that motor drive parameters are not lost or destroyed when a motor drive is reset to OEM default settings. A *dual inline-package (DIP) switch* is a group of several identical manual electric switches that are packaged together in a standard switch body. DIP switches are designed to be installed on a printed circuit board (PCB) with other electronic devices and are normally used to program the operation of an electronic device for specific applications. Certain motor drives include DIP switches that are used to program selected drive parameters.

Inspecting Motor Drive Applications. After gathering information, the motor drive application must be inspected. An inspection provides familiarity with the physical layout and operation of an application. In addition to inspecting the motor and drive, the disconnect switch, motor coupling, and load can also be inspected. Inspections normally provide reasons as to the cause of the problem. **See Figure 8-25.** A motor drive application is inspected using the following procedure:

1. Verify that power disconnect switches are ON.
2. Access the fault history of the motor drive for information on possible causes and record the software version number.
3. Inspect the motor drive for physical damage and signs of overheating or fire.
4. Record the model number, serial number, input voltage, input current, output current, and horsepower rating from the motor drive nameplate.
5. Inspect the exterior of the motor and the area adjacent to the motor for proper clearance to provide proper ventilation to cool the motor.
6. Verify that the motor power rating corresponds to the motor drive power rating.
7. Verify that the motor is correctly aligned with the driven load and that the coupling or other connection method between the motor and driven load is not loose or broken.
8. Verify that the motor and the driven load are securely fastened in place and that no object is preventing the motor or load from rotating.
9. Determine if any special equipment is required for maintenance on the motor drive application.

282 ELECTRICAL SYSTEMS for FACILITIES MAINTENANCE PERSONNEL

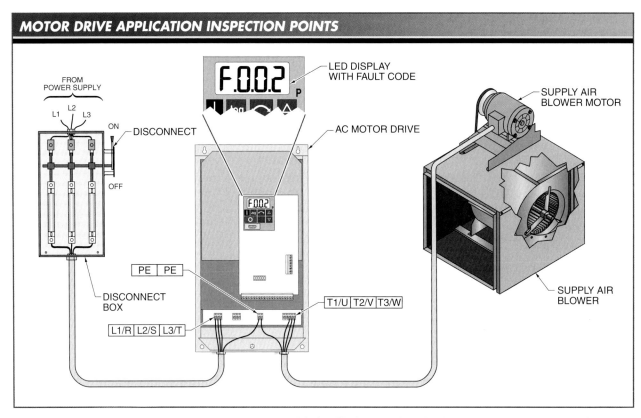

Figure 8-25. Motor drive application inspections provide familiarity with the physical layout and operation of a motor and motor drive application.

Troubleshooting Incoming Power. In order for a motor drive to function properly, the incoming power must be on, be within the voltage operating range of the drive, have sufficient voltage and current capacity, and be free of power quality problems. Common incoming power problems are high input voltage, low input voltage, no input voltage, voltage unbalance, improper grounding, and harmonic distortion. Momentary power problems that may also occur are voltage sags or voltage swells. Individuals must frequently test the incoming line voltage to motor drives. Incoming line voltage tests are performed at the power terminal strip of a motor drive using a DMM set to measure AC voltage. **See Figure 8-26.** Incoming line voltage is tested using the following procedure:

1. Wear proper PPE for the procedure, verify that disconnect switches are ON, and verify that fuses or circuit breakers are operational.
2. Verify that line conductors from the local disconnect switch terminate at the correct location on the motor drive.
3. Verify that line conductors are either shielded or in separate metal conduits with no other conductors and are the proper AWG size.
4. Verify that the grounding conductor is the proper AWG size and terminates at the correct location.
5. Verify that each connection at the power supply terminal strip is tight.

TECH FACT

In addition to obtaining motor drive manuals from the equipment manufacturers, developing a troubleshooting checklist will help individuals gather information and maintain a written record of the results of their inspections.

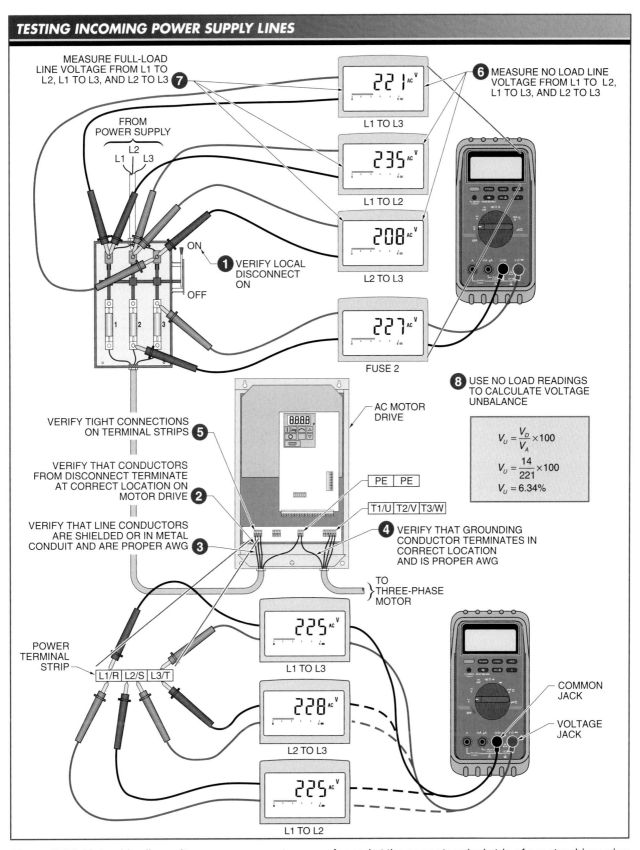

Figure 8-26. Motor drive line voltage measurements are performed at the power terminal strip of a motor drive using a DMM set to measure AC voltage.

6. Measure and record the line voltage with no load from L1 to L2, L1 to L3, and L2 to L3. Verify that the voltage is within the operating range of the motor drive. *Note:* When a measurement of 0 VAC is present, additional electrical distribution system tests must be performed.
7. Measure and record the line voltage under full-load operating conditions from L1 to L2, L1 to L3, and L2 to L3. Compare full-load readings with the no-load readings taken in step 6. *Note:* A voltage difference greater than 3% between no load and full load indicates that the motor drive does not have sufficient capacity or that the drive is overloaded.
8. Use the readings from the no load measurements in step 6 to calculate voltage unbalance. *Note:* A value greater than 2% is not acceptable and requires further testing.

A reading of 0 VAC can be an indication that the local disconnect switch is OFF, a fuse is blown, or a circuit breaker is tripped. **See Figure 8-27.** If no voltage is measured in step 6 of the incoming line voltage test, the cause is determined using the following procedure:

1. Wear proper PPE and stand to the side of the disconnect switch and motor drive when energizing to protect against hazards created by equipment failure or arcing. Turn the local disconnect switch ON if it is in the OFF position.
2. With local disconnect switch in the ON position, use a DMM set to measure voltage to verify that voltage is present at the disconnect switch.
3. Open the disconnect switch cover and measure the line voltage from L1 to L2, L1 to L3, and L2 to L3. *Note:* If any of the measurements are 0 VAC or significantly less than the known line voltage, a problem exists in the electrical distribution system. **Warning:** Do not turn disconnect on while the internal cover of the disconnect switch is removed.
4. Use a DMM set to measure voltage to test fuses and circuit breakers. Verify that fuses or circuit breakers have the correct voltage rating, current rating, and trip characteristic for the motor drive in use. Replace any blown fuses or reset any tripped breakers. Do not remove or install fuses with the disconnect switch ON.
5. Close the disconnect switch cover. Wear proper PPE and stand to the side of the disconnect switch and motor drive when energizing to protect against hazards created by equipment failure or arcing. Turn the local disconnect switch ON. Turn the local disconnect switch OFF if a fuse blows or circuit breaker trips again when power is applied to the motor drive. Lock out and tag out the disconnect switch.
6. Use a DMM set to measure voltage to verify that the AC line voltage is not present at the motor drive power terminal strip. Disconnect the line conductors at the power terminal strip of the motor drive and insulate the conductors.
7. Replace any blown fuses or reset any tripped circuit breakers. Remove lockout and tagout, and turn the local disconnect switch ON.

If fuses do not blow or circuit breakers do not trip, the problem is with the motor drive. If fuses blow or circuit breakers trip, the cause of the problem is with the wiring and connections to the motor drive.

When a problem with the incoming power is identified, the appropriate solution is applied. The solution may require modifications to the power source that supplies the motor drive or troubleshooting the motor drive. **See Appendix.**

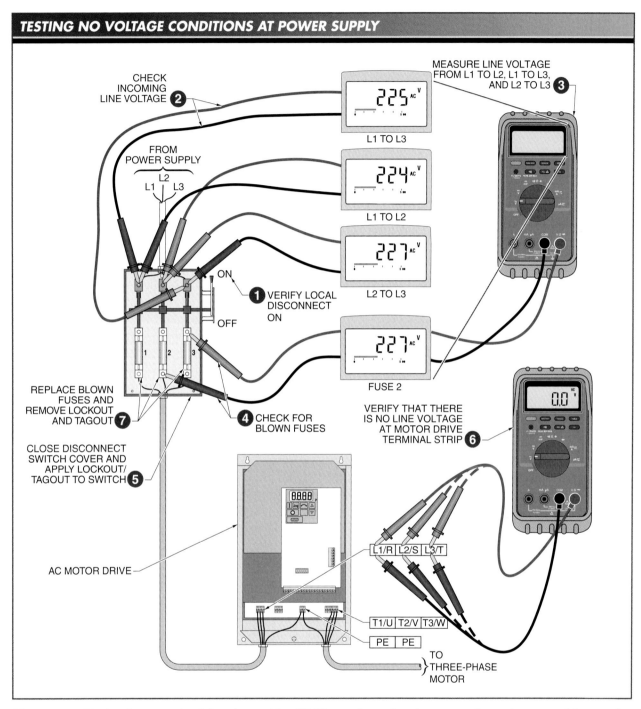

Figure 8-27. Testing fuses or circuit breakers with a DMM can determine the cause of no voltage conditions at the power supply.

Motor Drive Tests. When the incoming power is eliminated as the source of the problem, the motor drive is tested. A series of tests are used to eliminate the motor drive, drive parameters, input and output devices, motor, and load as problems. Common motor drive problems include component failure, incorrect parameter settings, and input and output devices that are incorrectly connected or that have failed. Motor problems and load problems are often mistaken for motor drive problems.

Motor drive fault codes aid in identifying problems. Tests must be performed in the proper sequence to identify a problem correctly in the least amount of time. A *motor drive initial test* is a test that verifies if a motor drive is operational. A partial failure of a motor drive is uncommon. Motor drives are normally completely operational or completely not operational. A motor drive set to OEM default settings and controlled by an integral keypad must be tested with the motor disconnected.

When a problem is first noticed, parameter settings, inputs, outputs, the motor, and the load are not tested as the source of the problem. When the control mode is sensorless vector control or closed-loop vector control, it may not be possible to run the vector control drive with the motor disconnected. When possible, the control mode should be changed to constant torque or variable torque in order to perform a motor drive initial test. **See Figure 8-28.** A motor drive is tested using the following procedure:

1. Wear proper PPE for the procedure and press the stop (O) button on the motor drive keypad if the motor drive is on.
2. Turn the disconnect switch OFF and apply lockout and tagout devices. Wait for the DC bus capacitors to discharge. Do not manually discharge the capacitors by shorting + to –. Remove the motor drive cover. Use a DMM set to measure voltage to verify that the AC line voltage is not present. Use a DMM to verify that the DC bus capacitors have discharged. Do not rely on the DC bus charged LED lamps because LED lamps can burn out, which could indicate a false reading.
3. Disconnect the load conductors from the motor drive power terminal strip. Make notation of the location of the motor wires in order to maintain the correct motor rotation on reconnection.
4. Remove the lockout and tagout from the local disconnect switch and reinstall the motor drive cover.
5. Stand to the side of the disconnect switch and motor drive when energizing to protect against hazards created by equipment failure or arcing. Turn the local disconnect switch ON. *Note:* The start (I) button should not be pushed. The motor drive LED display or clear text display will activate. The motor drive cooling fans may or may not start when power is applied, depending on the drive model. If the fans do not start, it should be verified that they start when the start (I) button is pressed. If there are any loud noises, smoke, or explosions, the local disconnect switch should be turned OFF immediately and the motor drive components should be tested.
6. Record or download the motor drive parameter values. Reset parameters to OEM default settings.
7. Program the input mode to control the motor drive by the integral keypad, the speed reference to internal, and the display mode to display motor drive output frequency in hertz (Hz).
8. Stand to the side of the motor drive when pushing the start (I) button to protect against hazards created by equipment failure or arcing. Press the start (I) button. *Note:* The cooling fans should start if they did not start when power was first applied. The LED display should gradually increase to a low speed. If the LED display shows 0 Hz, the ramp up (Δ) button should be pressed until 5 Hz is displayed.
9. Press the ramp up (Δ) button until 60 Hz is displayed.
10. Program the display mode to display the motor drive output voltage. *Note:* The motor drive output voltage must be approximately the same as the 60 Hz drive input voltage.
11. Press the stop (O) button. *Note:* The voltage should decrease to 0 VAC.

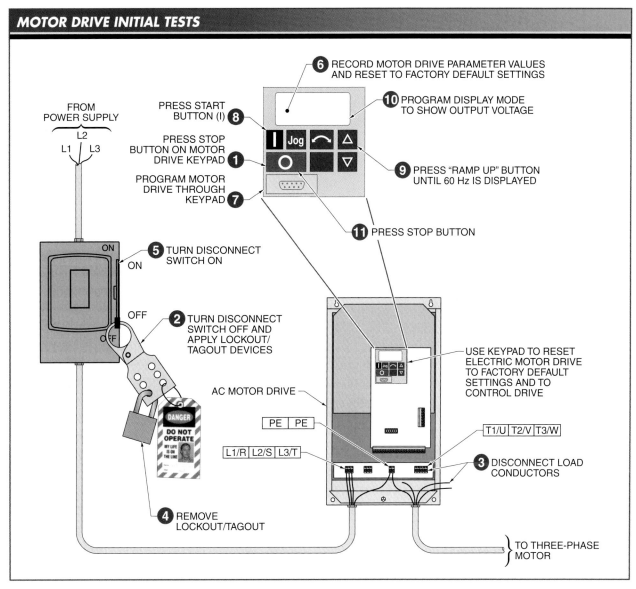

Figure 8-28. A motor drive initial test is performed to verify if a motor drive is operational.

The motor drive is not the source of the problem if it operates properly. Proceed to the motor drive, motor, and load tests if the motor drive does not operate properly.

A *motor drive, motor, and load test* is a test used to verify that a motor drive and motor properly rotate the driven load. A motor drive set to OEM defaults and controlled by the integral keypad is tested with the motor connected. Inputs and outputs are not tested as the source of the problem. If the control mode was changed to perform the motor drive test, return the control mode to its original setting. **See Figure 8-29.** A motor drive, motor, and load test is performed using the following procedure:

1. Wear proper PPE for the procedure and press the stop (O) button on the motor drive keypad if the motor drive is on.
2. Turn the disconnect switch OFF. Lock out and tag out the disconnect switch.

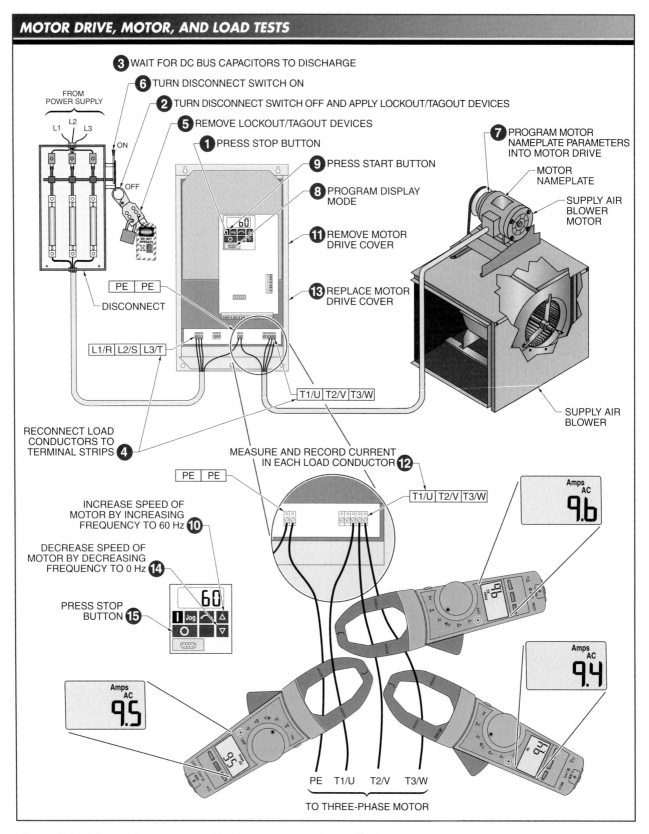

Figure 8-29. Motor drive, motor, and load tests are used to verify that a motor drive and motor properly rotate the driven load.

3. Wait for the DC bus capacitors to discharge. Do not manually discharge the capacitors by shorting + to –. Remove the motor drive cover. Use a DMM set to measure voltage to verify that the AC line voltage is not present. Use a DMM to verify that the DC bus capacitors have discharged. Do not rely on the DC bus charged LED lamps because LED lamps can burn out, which could indicate a false reading.
4. Reconnect the load conductors to their previous locations (from step 3 of the motor drive initial test) on the power terminal strip to maintain correct motor rotation because incorrect motor rotation causes damage in certain applications. Reinstall the motor drive cover.
5. Remove the lockout and tagout from the local disconnect switch.
6. Stand to the side of the disconnect switch and motor drive when energizing to protect against hazards created by equipment failure or arcing. Turn the disconnect switch ON. Do not push the start (I) button. The motor drive LED display or "clear text" display activates.
7. Program the appropriate parameters from the motor nameplate data into the motor drive.
8. Program the display mode to display the motor drive output frequency.
9. Stand to the side of the disconnect switch and motor drive when pressing the start button to protect against hazards created by equipment failure or arcing. Press the start (I) button. Do not start the motor drive until it is verified that personnel are not at risk from the driven load. The LED display on the motor drive must indicate a gradual increase to a low speed such as 5 Hz. If 0 Hz is displayed, press the ramp up (Δ) button until 5 Hz is displayed. *Note:* Motor drives control motor speed by changing the frequency. With a motor drive, motor speed is displayed as a percentage or in Hz but never in rpm.
10. Increase the speed of the motor to 60 Hz using the ramp up (Δ) button. The motor and driven load must accelerate smoothly to 60 Hz. *Note:* Any unusual noises or vibrations must be recorded and the frequency of the occurrence recorded. Unusual noises or vibrations indicate alignment problems or require the use of the skip frequency parameter to avoid unwanted mechanical resonance.
11. Remove the motor drive cover. **Warning:** Dangerous voltage levels exist when the motor drive cover is removed and the drive is energized. Exercise extreme caution and use the appropriate PPE.
12. Measure and record the current in each of the three load conductors using a true-rms clamp-on ammeter. *Note:* True-rms clamp-on ammeters are required because the current waveform of a motor drive is nonsinusoidal. Current readings are taken at 60 Hz because motor nameplate current is normally based on 60 Hz. Current readings of the three load conductors must be equal or within ±1% of each other. A problem with the load conductors or motor is present if the current readings of the load conductors are not within the specified range. An *overloaded motor* is a motor with a current reading greater than 105% of the nameplate current rating. There is a problem with the motor or the load if the current readings are greater than 105% of the nameplate current rating.
13. Replace the motor drive cover.
14. Decrease the speed of the motor to 0 Hz using the ramp down (∇) button. The motor and driven load should decelerate smoothly to 0 Hz. *Note:* Unusual noises or vibrations and the

Refer to Chapter 8 Quick Quiz® on CD-ROM.

frequency of the occurrence must be recorded. Unusual noises or vibrations indicate alignment problems or can require the use of the skip frequency parameter to avoid unwanted mechanical resonance.

15. Press the stop (O) button. *Note:* If the motor drive, motor, and load performed without any problems, the drive, motor, and load are not the source of the problem. There is a problem if the motor drive, motor, and load did not perform correctly. The problem is the motor drive parameters, motor, or load because the motor drive was eliminated as the problem in the initial test.

Case Study—Motor Drive Service Call

A pump system controlled by a motor drive has a pressure surge problem that is causing noise and mechanical problems in the system. The system needs to be checked to ensure the drives operating parameters are correctly set for the application.

To prevent overstressing the motor and instantaneous pressure surges, a motor drive is used because the motor acceleration time (ramp up time) and deceleration time (ramp down time) can be adjusted for a longer, smoother starting and stopping of pump motors. Before making any changes, the maintenance staff checks the drive service manual to determine the parameter number used for the acceleration and deceleration setting.

Baldor

The factory default setting shows a 10 sec ramp up time and a 10 sec ramp down time. This time may be too short for the system pumps and should be changed to 15 sec.

The maintenance staff must wear the proper PPE for the service location and use a DMM and current clamp for taking electrical measurements to verify the changes and system are operating properly.

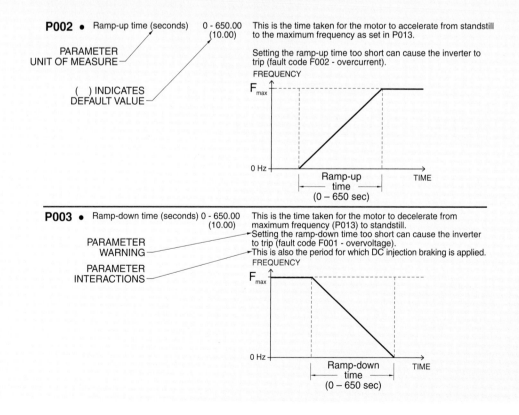

Case Study—Motor Drive Service Call (continued)

After changing the acceleration and deceleration times to 15 sec, voltage measurements to the drive are taken and monitored. When a drive is used to control a motor, the voltage applied to the motor increases as the drive speed increases. By varying the applied voltage to the motor at different speeds, proper motor torque is maintained.

The voltage applied to the motor is good, so the current draw of the motor is checked. Motor current draw should not exceed the current rating listed on the nameplate of the motor.

The current draw of the motor is within the nameplate rating. The added acceleration and deceleration time has fixed the problem and the system is monitored for several days.

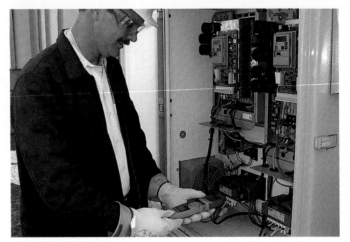

Definitions

- A *control switch* is a device that regulates the flow of current in a circuit.
- A *disconnect switch*, also known as a disconnect, is a switch that disconnects the supply of electric power from electrically powered devices such as motors and machines.
- A *dual-function switch* is a control switch that controls both the on and off control function.
- A *dual inline-package (DIP) switch* is a group of several identical manual electric switches that are packaged together in a standard switch body.
- An *electrical relay* is a control device activated by electric power that opens, closes, or activates other mechanisms.
- An *electromechanical relay* is an electric device that uses a magnetic coil to open or close one or more sets of contacts.
- A *magnetic motor starter* is a starter that has an electrically operated switch (contactor) and includes motor overload protection.
- A *motor control circuit* is a circuit that controls a motor as required to start, stop, jog, increase speed, decrease speed, or reverse its direction of rotation.
- A *motor drive* is an electronic device that controls the direction, speed, torque, and other operating functions of an electric motor, in addition to providing motor protection and monitoring functions.
- A *motor drive initial test* is a test that verifies if a motor drive is operational.
- A *motor drive, motor, and load test* is a test used to verify that a motor drive and motor properly rotate the driven load.
- A *motor power circuit* is the section of an electrical motor circuit that delivers high voltage or current to an electric motor.
- A *motor starter* is an electrically operated switch (contactor) that includes motor overload protection.
- An *off-delay timer*, also known as a delay-on-release timer, is a timer that does not start its timing function until the power is removed from the timer.
- An *on-delay timer*, also known as a delay-on-operate timer, is a time-based control device that delays for a predetermined time after receiving a signal to activate or turn on.
- An *overcurrent protection device (OCPD)* is a fuse or circuit breaker that disconnects or discontinues current flow when the amount of current exceeds the design load.
- An *overloaded motor* is a motor with a current reading greater than 105% of the nameplate current rating.
- A *pressure switch* is an electric switch operated by pressure that acts on a diaphragm or bellows element.
- A *single-function switch* is a switch that controls only the turning on or off of a load but not both.
- A *solid-state relay* is a relay that uses electronic switching devices in place of mechanical contacts.
- A *timer* is a control device that uses a preset time period as part of a control function.
- *Troubleshooting* is the systematic elimination of the various parts of a system or process to locate a malfunctioning part.

Review

1. What is the purpose of a motor power circuit, and what are the components included?
2. What is the purpose of a disconnect switch?
3. What tests should be performed when troubleshooting a power circuit?
4. At what voltage level do motor control circuits operate compared to the motors they control, and what control voltages are normally used?
5. How and why should a control transformer be grounded?
6. How does a manual start/stop control circuit start a motor once the start pushbutton is used?
7. What is the purpose and function of a pressure switch?
8. What are the possible ways that control switches can be activated?
9. How are control circuit switches made to operate as single-function switches?
10. How do basic HVAC control circuits operate?
11. What are the types of electrical relays and how do they function?
12. What are the differences between on-delay timers and off-delay timers?
13. What is the purpose of a magnetic motor starter?
14. What is the most common method used to adjust trip time on a motor starter?
15. What are four benefits of using motor drives?
16. What safety concerns must be considered when troubleshooting motor drives?
17. What are common incoming power problems regarding motor drives?
18. What is the purpose of a motor drive, motor, and load test?

APPENDIX

CONTENTS

Electrical Symbols	296
HVAC Symbols	298
Electrical/Electronic Abbreviations/Acronyms	299
Locking Wiring Devices	300
Nonlocking Wiring Devices	301
Motor Drive Incoming Power Troubleshooting Guide	302
Annual Motor Maintenance Checklist	303
Single-Phase Full-Load Currents	304
Three-Phase Full-Load Currents	304
Exit Lamp Specifications	304
Heater Selections	305
Recommended Ballast Output Voltage Limits	306
Recommended Short-Circuit Current Test Limits	306
Voltage Conversions	306
Incandescent Lamp Troubleshooting Guide	307
Fluorescent Lamp Troubleshooting Guide	307
HID Lamp Troubleshooting Guide	309
Hazardous Locations – Article 500	310

ELECTRICAL SYMBOLS . . .

Lighting Outlets

Symbol Name	
OUTLET BOX AND INCANDESCENT LIGHTING FIXTURE	CEILING / WALL
INCANDESCENT TRACK LIGHTING	
BLANKED OUTLET	B
DROP CORD	D
EXIT LIGHT AND OUTLET BOX. SHADED AREAS DENOTE FACES.	
OUTDOOR POLE-MOUNTED FIXTURES	
JUNCTION BOX	J
LAMPHOLDER WITH PULL SWITCH	L_{PS}
MULTIPLE FLOODLIGHT ASSEMBLY	
EMERGENCY BATTERY PACK WITH CHARGER	
INDIVIDUAL FLUORESCENT FIXTURE	
OUTLET BOX AND FLUORESCENT LIGHTING TRACK FIXTURE	
CONTINUOUS FLUORESCENT FIXTURE	
SURFACE-MOUNTED FLUORESCENT FIXTURE	

Panelboards

FLUSH-MOUNTED PANELBOARD AND CABINET	
SURFACE-MOUNTED PANELBOARD AND CABINET	

Convenience Outlets

Symbol Name	
SINGLE RECEPTACLE OUTLET	
DUPLEX RECEPTACLE OUTLET	
TRIPLEX RECEPTACLE OUTLET	
SPLIT-WIRED DUPLEX RECEPTACLE OUTLET	
SPLIT-WIRED TRIPLEX RECEPTACLE OUTLET	
SINGLE SPECIAL-PURPOSE RECEPTACLE OUTLET	
DUPLEX SPECIAL-PURPOSE RECEPTACLE OUTLET	
RANGE OUTLET	R
SPECIAL-PURPOSE CONNECTION	DW
CLOSED-CIRCUIT TELEVISION CAMERA	
CLOCK HANGER RECEPTACLE	C
FAN HANGER RECEPTACLE	F
FLOOR SINGLE RECEPTACLE OUTLET	
FLOOR DUPLEX RECEPTACLE OUTLET	
FLOOR SPECIAL-PURPOSE OUTLET	
UNDERFLOOR DUCT AND JUNCTION BOX FOR TRIPLE, DOUBLE, OR SINGLE DUCT SYSTEM AS INDICATED BY NUMBER OF PARALLEL LINES	

Busducts and Wireways

SERVICE, FEEDER, OR PLUG-IN BUSWAY	B B B
CABLE THROUGH LADDER OR CHANNEL	C C C
WIREWAY	W W W

Switch Outlets

SINGLE-POLE SWITCH	S
DOUBLE-POLE SWITCH	S_2
THREE-WAY SWITCH	S_3
FOUR-WAY SWITCH	S_4
AUTOMATIC DOOR SWITCH	S_D
KEY-OPERATED SWITCH	S_K
CIRCUIT BREAKER	S_{CB}
WEATHERPROOF CIRCUIT BREAKER	S_{WCB}
DIMMER	S_{DM}
REMOTE CONTROL SWITCH	S_{RC}
WEATHERPROOF SWITCH	S_{WP}
FUSED SWITCH	S_F
WEATHERPROOF FUSED SWITCH	S_{WF}
TIME SWITCH	S_T
CEILING PULL SWITCH	S
SWITCH AND SINGLE RECEPTACLE	S
SWITCH AND DOUBLE RECEPTACLE	S
A STANDARD SYMBOL WITH AN ADDED LOWERCASE SUBSCRIPT LETTER IS USED TO DESIGNATE A VARIATION IN STANDARD EQUIPMENT	a,b

Appendix

...ELECTRICAL SYMBOLS

Commercial and Industrial Systems

Device	Symbol
PAGING SYSTEM DEVICE	◇
FIRE ALARM SYSTEM DEVICE	□
COMPUTER DATA SYSTEM DEVICE	◀
PRIVATE TELEPHONE SYSTEM DEVICE	◁
SOUND SYSTEM	▽
FIRE ALARM CONTROL PANEL	FACP

Signaling System Outlets for Residential Systems

PUSHBUTTON	•
BUZZER	⌐⌐
BELL	◐
BELL AND BUZZER COMBINATION	◐⌐
COMPUTER DATA OUTLET	◀
BELL RINGING TRANSFORMER	BT
ELECTRIC DOOR OPENER	D
CHIME	CH
TELEVISION OUTLET	TV
THERMOSTAT	T

Underground Electrical Distribution or Electrical Lighting Systems

MANHOLE	M
HANDHOLE	H
TRANSFORMER-MANHOLE OR VAULT	TM
TRANSFORMER PAD	TP
UNDERGROUND DIRECT BURIAL CABLE	———
UNDERGROUND DUCT LINE	⊏⊐
STREET LIGHT STANDARD FED FROM UNDERGROUND CIRCUIT	⊗

Above-Ground Electrical Distribution or Lighting Systems

POLE	○
STREET LIGHT AND BRACKET	⊗
PRIMARY CIRCUIT	———
SECONDARY CIRCUIT	- - -
DOWN GUY	→—
HEAD GUY	—•—
SIDEWALK GUY	—○→
SERVICE WEATHERHEAD	⌒

Panel Circuits and Miscellaneous

Device	Symbol
LIGHTING PANEL	■
POWER PANEL	▨
WIRING – CONCEALED IN CEILING OR WALL	⌒
WIRING – CONCEALED IN FLOOR	- - -
WIRING EXPOSED	- - -
HOME RUN TO PANEL BOARD — Indicate number of circuits by number of arrows. Any circuit without such designation indicates a two-wire circuit. For a greater number of wires indicate as follows: —///— (3 wires) —////— (4 wires), etc.	⌒➤
FEEDERS — Use heavy lines and designate by number corresponding to listing in feeder schedule	———
WIRING TURNED UP	—○
WIRING TURNED DOWN	—●
GENERATOR	G
MOTOR	M
INSTRUMENT (SPECIFY)	I
TRANSFORMER	T
CONTROLLER	⊠
EXTERNALLY-OPERATED DISCONNECT SWITCH	▯
PULL BOX	▨

HVAC SYMBOLS

Equipment Symbols	Ductwork	Heating Piping
EXPOSED RADIATOR	DUCT (1ST FIGURE, WIDTH; 2ND FIGURE, DEPTH) — 12 X 20	HIGH-PRESSURE STEAM — HPS
RECESSED RADIATOR	DIRECTION OF FLOW →	MEDIUM-PRESSURE STEAM — MPS
FLUSH ENCLOSED RADIATOR	FLEXIBLE CONNECTION	LOW-PRESSURE STEAM — LPS
PROJECTING ENCLOSED RADIATOR	DUCTWORK WITH ACOUSTICAL LINING	HIGH-PRESSURE RETURN — HPR
UNIT HEATER (PROPELLER) – PLAN	FIRE DAMPER WITH ACCESS DOOR — FD / AD	MEDIUM-PRESSURE RETURN — MPR
UNIT HEATER (CENTRIFUGAL) – PLAN	MANUAL VOLUME DAMPER — VD	LOW-PRESSURE RETURN — LPR
UNIT VENTILATOR – PLAN	AUTOMATIC VOLUME DAMPER	BOILER BLOW OFF — BD
STEAM	EXHAUST, RETURN OR OUTSIDE AIR DUCT – SECTION — 20 X 12	CONDENSATE OR VACUUM PUMP DISCHARGE — VPD
DUPLEX STRAINER	SUPPLY DUCT – SECTION — 20 X 12	FEEDWATER PUMP DISCHARGE — PPD
PRESSURE-REDUCING VALVE	CEILING DIFFUSER SUPPLY OUTLET — 20" DIA CD / 1000 CFM	MAKEUP WATER — MU
AIR LINE VALVE	CEILING DIFFUSER SUPPLY OUTLET — 20 X 12 CD / 700 CFM	AIR RELIEF LINE — V
STRAINER	LINEAR DIFFUSER — 96 X 6-LD / 400 CFM	FUEL OIL SUCTION — FOS
THERMOMETER	FLOOR REGISTER — 20 X 12 FR / 700 CFM	FUEL OIL RETURN — FOR
PRESSURE GAUGE AND COCK	TURNING VANES	FUEL OIL VENT — FOV
RELIEF VALVE	FAN AND MOTOR WITH BELT GUARD	COMPRESSED AIR — A
AUTOMATIC 3-WAY VALVE		HOT WATER HEATING SUPPLY — HW
AUTOMATIC 2-WAY VALVE		HOT WATER HEATING RETURN — HWR
SOLENOID VALVE	LOUVER OPENING — 20 X 12-L / 700 CFM	

Air Conditioning Piping	
REFRIGERANT LIQUID	RL
REFRIGERANT DISCHARGE	RD
REFRIGERANT SUCTION	RS
CONDENSER WATER SUPPLY	CWS
CONDENSER WATER RETURN	CWR
CHILLED WATER SUPPLY	CHWS
CHILLED WATER RETURN	CHWR
MAKEUP WATER	MU
HUMIDIFICATION LINE	H
DRAIN	D

ELECTRICAL/ELECTRONIC ABBREVIATIONS/ACRONYMS

Abbr/Acronym	Meaning	Abbr/Acronym	Meaning	Abbr/Acronym	Meaning
A	Ammeter; Ampere; Anode; Armature	FU	Fuse	PNP	Positive-Negative-Positive
AC	Alternating Current	FWD	Foward	POS	Positive
AC/DC	Alternating Current; Direct Current	G	Gate; Giga; Green; Conductance	POT.	Potentiometer
A/D	Analog to Digital	GEN	Generator	P-P	Peak-to-Peak
AF	Audio Frequency	GRD	Ground	PRI	Primary Switch
AFC	Automatic Frequency Control	GY	Gray	PS	Pressure Switch
Ag	Silver	H	Henry; High Side of Transformer; Magnetic Flux	PSI	Pounds Per Square Inch
ALM	Alarm			PUT	Pull-Up Torque
AM	Ammeter; Amplitude Modulation	HF	High Frequency	Q	Transistor
AM/FM	Amplitude Modulation; Frequency Modulation	HP	Horsepower	R	Radius; Red; Resistance; Reverse
		Hz	Hertz	RAM	Random-Access Memory
ARM.	Armature	I	Current	RC	Resistance-Capacitance
Au	Gold	IC	Integrated Circuit	RCL	Resistance-Inductance-Capacitance
AU	Automatic	INT	Intermediate; Interrupt	REC	Rectifier
AVC	Automatic Volume Control	INTLK	Interlock	RES	Resistor
AWG	American Wire Gauge	IOL	Instantaneous Overload	REV	Reverse
BAT.	Battery (electric)	IR	Infrared	RF	Radio Frequency
BCD	Binary Coded Decimal	ITB	Inverse Time Breaker	RH	Rheostat
BJT	Bipolar Junction Transistor	ITCB	Instantaneous Trip Circuit Breaker	rms	Root Mean Square
BK	Black	JB	Junction Box	ROM	Read-Only Memory
BL	Blue	JFET	Junction Field-Effect Transistor	rpm	Revolutions Per Minute
BR	Brake Relay; Brown	K	Kilo; Cathode	RPS	Revolutions Per Second
C	Celsius; Capacitance; Capacitor	L	Line; Load; Coil; Inductance	S	Series; Slow; South; Switch
CAP.	Capacitor	LB-FT	Pounds Per Foot	SCR	Silicon Controlled Rectifier
CB	Circuit Breaker; Citizen's Band	LB-IN.	Pounds Per Inch	SEC	Secondary
CC	Common-Collector Configuration	LC	Inductance-Capacitance	SF	Service Factor
CCW	Counterclockwise	LCD	Liquid Crystal Display	1 PH; 1φ	Single-Phase
CE	Common-Emitter Configuration	LCR	Inductance-Capacitance-Resistance	SOC	Socket
CEMF	Counter Electromotive Force	LED	Light Emitting Diode	SOL	Solenoid
CKT	Circuit	LRC	Locked Rotor Current	SP	Single-Pole
CONT	Continuous; Control	LS	Limit Switch	SPDT	Single-Pole, Double-Throw
CPS	Cycles Per Second	LT	Lamp	SPST	Single-Pole, Single-Throw
CPU	Central Processing Unit	M	Motor; Motor Starter; Motor Starter Contacts	SS	Selector Switch
CR	Control Relay			SSW	Safety Switch
CRM	Control Relay Master	MAX.	Maximum	SW	Switch
CT	Current Transformer	MB	Magnetic Brake	T	Tera; Terminal; Torque; Transformer
CW	Clockwise	MCS	Motor Circuit Switch	TB	Terminal Board
D	Diameter; Diode; Down	MEM	Memory	3 PH; 3φ	Three-Phase
D/A	Digital to Analog	MED	Medium	TD	Time Delay
DB	Dynamic Braking Contactor; Relay	MIN	Minimum	TDF	Time Delay Fuse
DC	Direct Current	MN	Manual	TEMP	Temperature
DIO	Diode	MOS	Metal-Oxide Semiconductor	THS	Thermostat Switch
DISC.	Disconnect Switch	MOSFET	Metal-Oxide Semiconductor Field-Effect Transistor	TR	Time Delay Relay
DMM	Digital Multimeter			TTL	Transistor-Transistor Logic
DP	Double-Pole	MTR	Motor	U	Up
DPDT	Double-Pole, Double-Throw	N; NEG	North; Negative	UCL	Unclamp
DPST	Double-Pole, Single-Throw	NC	Normally Closed	UHF	Ultrahigh Frequency
DS	Drum Switch	NEUT	Neutral	UJT	Unijunction Transistor
DT	Double-Throw	NO	Normally Open	UV	Ultraviolet; Undervoltage
DVM	Digital Voltmeter	NPN	Negative-Positive-Negative	V	Violet; Volt
EMF	Electromotive Force	NTDF	Nontime-Delay Fuse	VA	Volt Amp
F	Fahrenheit; Fast; Field; Forward; Fuse	O	Orange	VAC	Volts Alternating Current
FET	Field-Effect Transistor	OCPD	Overcurrent Protection Device	VDC	Volts Direct Current
FF	Flip-Flop	OHM	Ohmmeter	VHR	Very High Frequency
FLC	Full-Load Current	OL	Overload Relay	VLF	Very Low Frequency
FLS	Flow Switch	OZ/IN.	Ounces Per Inch	VOM	Volt-Ohm-Milliammeter
FLT	Full-Load Torque	P	Peak; Positive; Power; Power Consumed	W	Watt; White
FM	Frequency Modulation	PB	Pushbutton	w/	With
FREQ	Frequency	PCB	Printed Circuit Board	X	Low Side of Transformer
FS	Float Switch	PH; φ	Phase	Y	Yellow
FTS	Foot Switch	PLS	Plugging Switch	Z	Impedance

LOCKING WIRING DEVICES

2-Pole, 3-Wire

Wiring Diagram	NEMA ANSI	Receptacle Configuration	Rating
	ML2 C73.44		15 A 125 V
	L5-15 C73.42		15 A 125 V
	L5-20 C73.72		20 A 125 V
	L6-15 C73.74		15 A 250 V
	L6-20 C73.75		20 A 250 V
	L6-30 C73.76		30 A 250 V
	L7-15 C73.43		15 A 277 V
	L7-20 C73.77		20 A 277 V
	L8-20 C73.79		20 A 480 V
	L9-20 C73.81		20 A 600 V

3-Pole, 4-Wire

Wiring Diagram	NEMA ANSI	Receptacle Configuration	Rating
	L14-20 C73.83		20 A 125/250 V
	L14-30 C73.84		30 A 125/250 V
	L15-20 C73.85		20 A 3φ 250 V
	L15-30 C73.86		30 A 3φ 250 V
	L16-20 C73.87		20 A 3φ 480 V
	L16-30 C73.88		30 A 3φ 480 V
	L17-30 C73.89		30 A 3φ 600 V

3-Pole, 3-Wire

Wiring Diagram	NEMA ANSI	Receptacle Configuration	Rating
	ML3 C73.30		15 A 125/250 V
	L10-20 C73.96		20 A 125/250 V
	L10-30 C73.97		30 A 125/250 V
	L11-15 C73.98		15 A 3φ 250 V
	L11-20 C73.99		20 A 3φ 250 V
	L12-20 C73.101		20 A 3φ 480 V
	L12-30 C73.102		30 A 3φ 480 V
	L13-30 C73.103		30 A 3φ 600 V

4-Pole, 4-Wire

Wiring Diagram	NEMA ANSI	Receptacle Configuration	Rating
	L18-20 C73.104		20 A 3φ Y 120/208 V
	L18-30 C73.105		30 A 3φ Y 120/208 V
	L19-20 C73.106		20 A 3φ Y 277/480 V
	L20-20 C73.108		20 A 3φ Y 347/600 V

4-Pole, 5-Wire

Wiring Diagram	NEMA ANSI	Receptacle Configuration	Rating
	L21-20 C73.90		20 A 3φ Y 120/208 V
	L22-20 C73.92		20 A 3φ Y 277/480 V
	L23-20 C73.94		20 A 3φ Y 347/600 V

Appendix

NONLOCKING WIRING DEVICES

2-Pole, 3-Wire

Wiring Diagram	NEMA ANSI	Receptacle Configuration	Rating
	5-15 C73.11		15 A 125 V
	5-20 C73.12		20 A 125 V
	5-30 C73.45		30 A 125 V
	5-50 C73.46		50 A 125 V
	6-15 C73.20		15 A 250 V
	6-20 C73.51		20 A 250 V
	6-30 C73.52		30 A 250 V
	6-50 C73.53		50 A 250 V
	7-15 C73.28		15 A 277 V
	7-20 C73.63		20 A 277 V
	7-30 C73.64		30 A 277 V
	7-50 C73.65		50 A 277 V

4-Pole, 4-Wire

Wiring Diagram	NEMA ANSI	Receptacle Configuration	Rating
	18-15 C73.15		15 A 3φ Y 120/208 V
	18-20 C73.26		20 A 3φ Y 120/208 V
	18-30 C73.47		30 A 3φ Y 120/208 V
	18-50 C73.48		50 A 3φ Y 120/208 V
	18-60 C73.27		60 A 3φ Y 120/208 V

3-Pole, 3-Wire

Wiring Diagram	NEMA ANSI	Receptacle Configuration	Rating
	10-20 C73.23		20 A 125/250 V
	10-30 C73.24		30 A 125/250 V
	10-50 C73.25		50 A 125/250 V
	11-15 C73.54		15 A 3φ 250 V
	11-20 C73.55		20 A 3φ 250 V
	11-30 C73.56		30 A 3φ 250 V
	11-50 C73.57		50 A 3φ 250 V

3-Pole, 4-Wire

Wiring Diagram	NEMA ANSI	Receptacle Configuration	Rating
	14-15 C73.49		15 A 125/250 V
	14-20 C73.50		20 A 125/250 V
	14-30 C73.16		30 A 125/250 V
	14-50 C73.17		50 A 125/250 V
	14-60 C73.18		60 A 125/250 V
	15-15 C73.58		15 A 3φ 250 V
	15-20 C73.59		20 A 3φ 250 V
	15-30 C73.60		30 A 3φ 250 V
	15-50 C73.61		50 A 3φ 250 V
	15-60 C73.62		60 A 3φ 250 V

MOTOR DRIVE INCOMING POWER TROUBLESHOOTING GUIDE

Symptom/Fault Code	Problem	Cause	Solution
Motor drive overvoltage faults. Blown converter (rectifier) semiconductor	High input voltage/ voltage swell	Switching of power factor correction capacitors	Stop switching power factor correction capacitors. Install motor drive on another feeder
		Utility switching transformer taps for load adjustment	Install line reactor, or install motor drive on another feeder
		Proximity to low-impedance voltage source	Install line reactor
		Transformer secondary voltage is high	Adjust taps on transformer
Motor drive undervoltage faults. Difference between no load and full load voltage is greater than 3%	Low input voltage/ voltage sag	Power source for motor drive unable to deliver enough current	Increase kVA source rating, or install motor drive on another feeder
		Low transformer secondary voltage	Adjust secondary taps on supply transformer
		Large starting load(s)	Increase kVA source rating, or install motor drive on another feeder
		Motor drive is overloaded	Troubleshoot motor or load
	Voltage sine wave flat-topping at motor drive	Power source for motor drive unable to deliver enough current	Increase kVA source rating, or install motor drive on another feeder
Motor drive does not turn on. Blown fuse or tripped circuit breaker	No input voltage	Incorrect fuse or circuit breaker	Install correct fuse or circuit breaker
		Conductors feeding motor drive shorted or have ground fault	Repair or replace conductors
		Problem with motor drive	Troubleshoot motor drive
Motor drive overload faults	Voltage unbalance greater than 2%	Unbalance from utility	Contact utility
		Single-phase loads on the same feeder as the motor drive switching ON and OFF	Install motor drive on separate feeder
Intermittent motor drive faults and/or motor drive intermittently does not operate per design	Improper grounding	Undersized grounds or no grounds	Install proper size ground
		Loose ground connections	Tighten ground connections
	Intermittent fault/ erratic operation	Electric noise, EMI/RFI	Troubleshoot motor drive
Harmonics	Harmonics present on electrical distribution system	Motor drive or existing nonlinear loads can be source	Install input reactor

ANNUAL MOTOR MAINTENANCE CHECKLIST

Motor File #: _____   Serial #: _____

Date Installed: _____   Motor Location: _____

_____

MFR: _____   Type: _____   Frame: _____

HP: _____   Volts: _____   Amps: _____

RPM: _____   Date Serviced: _____

Step	Operation	Mechanic
1	Turn OFF and lock out all power to the motor and its control circuit.	
2	Clean motor exterior and all ventilation ducts.	
3	Uncouple motor from load and disassemble.	
4	Clean inside of motor.	
5	Check centrifugal switch assemblies.	
6	Check rotors, armatures, and field windings.	
7	Check all peripheral equipment.	
8	Check bearings.	
9	Check brushes and commutator.	
10	Check slip rings.	
11	Reassemble motor and couple to load.	
12	Flush old bearing lubricant and replace.	
13	Check motor's wire raceway.	
14	Check drive mechanism.	
15	Check motor terminations.	
16	Check capacitors.	
17	Check all mounting bolts.	
18	Check and record line-to-line resistance.	
19	Check and record megohmmeter resistance from T1 to ground.	
20	Check and record insulation polarization index.	
21	Check motor controls.	
22	Reconnect motor and control circuit power supplies.	
23	Check line-to-line voltage for balance and level.	
24	Check line current draw against nameplate rating.	
25	Check and record inboard and outboard bearing temperatures.	

NOTES:

SINGLE-PHASE FULL-LOAD CURRENTS*						
kVA	120 V	208 V	240 V	277 V	380 V	480 V
0.050	0.4	0.2	0.2	0.2	0.1	0.1
0.100	0.8	0.5	0.4	0.3	0.2	0.2
0.150	1.2	0.7	0.6	0.5	0.4	0.3
0.250	2.0	1.2	1.0	0.9	0.6	0.5
0.500	4.2	2.4	2.1	1.8	1.3	1.0
0.750	6.3	3.6	3.1	2.7	2.0	1.6
1.0	8.3	4.8	4.2	3.6	2.6	2.1
1.5	12.5	7.2	6.2	5.4	3.9	3.1
2.0	16.7	9.6	8.3	7.2	5.2	4.2
3.0	25	14.4	12.5	10.8	7.9	6.2
5.0	41	24	20.8	18	13.1	10.4
7.5	62	36	31	27	19.7	15.6
10	83	48	41	36	26	20.8
15	125	72	62	54	39	31

* in A

THREE-PHASE FULL-LOAD CURRENTS*				
kVA	208 V	240 V	480 V	600 V
3	8.3	7.2	3.6	2.9
4	12.5	10.8	5.4	4.3
6	16.6	14.4	7.2	5.8
9	25	21.6	10.8	8.6
15	41	36	18	14.4
22	62	54	27	21.6
30	83	72	36	28
45	124	108	54	43

* in A

EXIT LAMP SPECIFICATIONS			
Typical Specification	Incandescent	Fluorescent	LED
Operating Hours	24/7	24/7	24/7
Total Bulb Wattage to Light Sign	24	8	2.5
Cost per kWh	$0.12	$0.12	$0.12
Annual Energy Cost (per sign)	$25.22	$8.40	$2.63
Energy Cost (100 signs)	$2522.00	$840.00	$263.00
Average bulb life	1000 hr	11,000 hr	60,000 hr

HEATER SELECTIONS

Heater number	Full-load current (A)*					Heater number	Full-load current (A)*				
	Size 0	Size 1	Size 2	Size 3	Size 4		Size 0	Size 1	Size 2	Size 3	Size 4
10	0.20	0.20	—	—	—	47	6.68	6.68	—	—	—
11	0.22	0.22	—	—	—	48	7.21	7.21	—	—	—
12	0.24	0.24	—	—	—	49	7.81	7.81	7.89	—	—
13	0.27	0.27	—	—	—	50	8.46	8.46	8.57	—	—
14	0.30	0.30	—	—	—	51	9.35	9.35	9.32	—	—
15	0.34	0.34	—	—	—	52	10.00	10.00	10.1	—	—
16	0.37	0.37	—	—	—	53	10.7	10.7	11.0	12.2	—
17	0.41	0.41	—	—	—	54	11.7	11.7	12.0	13.3	—
18	0.45	0.45	—	—	—	55	12.6	12.6	12.9	14.3	—
19	0.49	0.49	—	—	—	56	13.9	13.9	14.1	15.6	—
20	0.54	0.54	—	—	—	57	15.1	15.1	15.5	17.2	—
21	0.59	0.59	—	—	—	58	16.5	16.5	16.9	18.7	—
22	0.65	0.65	—	—	—	59	18.0	18.0	18.5	20.5	—
23	0.71	0.71	—	—	—	60	—	19.2	20.3	22.5	23.8
24	0.78	0.78	—	—	—	61	—	20.4	21.8	24.3	25.7
25	0.85	0.85	—	—	—	62	—	21.7	23.5	26.2	27.8
26	0.93	0.93	—	—	—	63	—	23.1	25.3	28.3	30.0
27	1.02	1.02	—	—	—	64	—	24.6	27.2	30.5	32.5
28	1.12	1.12	—	—	—	65	—	26.2	29.3	33.0	35.0
29	1.22	1.22	—	—	—	66	—	27.8	31.5	36.0	38.0
30	1.34	1.34	—	—	—	67	—	—	33.5	39.0	41.0
31	1.48	1.48	—	—	—	68	—	—	36.0	42.0	44.5
32	1.62	1.62	—	—	—	69	—	—	38.5	45.5	48.5
33	1.78	1.78	—	—	—	70	—	—	41.0	49.5	52
34	1.96	1.96	—	—	—	71	—	—	43.0	53	57
35	2.15	2.15	—	—	—	72	—	—	46.0	58	61
36	2.37	2.37	—	—	—	73	—	—	—	63	67
37	2.60	2.60	—	—	—	74	—	—	—	68	72
38	2.86	2.86	—	—	—	75	—	—	—	73	77
39	3.14	3.14	—	—	—	76	—	—	—	78	84
40	3.45	3.45	—	—	—	77	—	—	—	83	91
41	3.79	3.79	—	—	—	78	—	—	—	88	97
42	4.17	4.17	—	—	—	79	—	—	—	—	103
43	4.58	4.58	—	—	—	80	—	—	—	—	111
44	5.03	5.03	—	—	—	81	—	—	—	—	119
45	5.53	5.53	—	—	—	82	—	—	—	—	127
46	6.08	6.08	—	—	—	83	—	—	—	—	133

*Full-load current (A) does not include FLC × 1.15 or 1.25.

RECOMMENDED BALLAST OUTPUT VOLTAGE LIMITS

Ballast	Lamp Size		rms Voltage (Volts)
	Wattage	ANSI Number	
Low-pressure sodium	18	L69	300–325
	35	L70	455–505
	55	L71	455–505
	90	L72	455–525
	135	L73	645–715
	180	L74	645–715
Mercury-vapor	50	H46	225–255
	75	H43	225–255
	100	H38	225–255
	175	H39	225–255
	250	H37	225–255
	400	H33	225–255
	700	H35	405–455
	1000	H36	405–455
Metal-halide	70	M85	210–250
	100	M90	250–300
	150	M81	220–260
	175	M57	285–320
	250	M80	230–270
	250	M58	285–320
	400	M59	285–320
	1000	M47	400–445
	1500	M48	400–445
High-pressure sodium	35	S76	110–130
	50	S68	110–130
	70	S62	110–130
	100	S54	110–130
	150	S55	110–130
	150	S56	200–250
	200	S66	200–230
	250	S50	175–225
	310	S67	155–190
	400	S51	175–225
	1000	S52	420–480

RECOMMENDED SHORT-CIRCUIT CURRENT TEST LIMITS

Ballast	Lamp Size		Short-Circuit Current (Amps)
	Wattage	ANSI Number	
Low-pressure sodium	18	L69	0.30–0.40
	35	L70	0.52–0.78
	55	L71	0.52–0.78
	90	L72	0.8–1.2
	135	L73	0.8–1.2
	180	L74	0.8–1.2
Mercury-vapor	50	H46	0.85–1.15
	75	H43	0.95–1.70
	100	H38	1.10–2.00
	175	H39	2.0–3.6
	250	H37	3.0–3.8
	400	H33	4.4–7.9
	700	H35	3.9–5.85
	1000	H36	5.7–9.0
Metal-halide	70	M85	0.85–1.30
	100	M90	1.15–1.76
	150	M81	1.75–2.60
	175	M57	1.5–1.90
	250	M80	2.9–4.3
	250	M58	2.2–2.85
	400	M59	3.5–4.5
	1000	M47	4.8–6.15
	1500	M48	7.4–9.6
High-pressure sodium	35	S76	0.85–1.45
	50	S68	1.5–2.3
	70	S62	1.6–2.9
	100	S54	2.45–3.8
	150	S55	3.5–5.4
	150	S56	2.0–3.0
	200	S66	2.50–3.7
	250	S50	3.0–5.3
	310	S67	3.8–5.7
	400	S51	5.0–7.6
	1000	S52	5.5–8.1

VOLTAGE CONVERSIONS

To Convert	To	Multiply By
rms	Average	0.9
rms	Peak	1.414
Average	rms	1.111
Average	Peak	1.567
Peak	rms	0.707
Peak	Average	0.637
Peak	Peak-to-Peak	2

INCANDESCENT LAMP TROUBLESHOOTING GUIDE

Problem	Possible Cause	Corrective Action
Short lamp life	Voltage higher than lamp rating	Ensure that voltage is equal to or less than lamp's rated voltage; lamp life is decreased when the voltage is higher than the rated voltage; a 5% higher voltage shortens lamp life by 40%
	Lamp exposed to rough service conditions or vibration	Replace lamp with one rated for rough service or resistant to vibration
Lamp does not turn ON after new lamp is installed	Fuse blown, poor electrical connection, faulty control switch	Check circuit fuse or CB; check electrical connections and voltage out of control switch; replace switch when voltage is present into, but not out of the control switch

FLUORESCENT LAMP TROUBLESHOOTING GUIDE . . .

Problem	Circuit Type	Possible Cause	Corrective Action
Lamp blinks and has shimmering effect during lighting period	All types	Depletion of emission material on electrodes	Replace with new lamp
New lamp blinks	All types	Loose lamp connection	Reseat lamp in socket securely; ensure lamp holders are rigidly mounted and properly spaced
		Low voltage applied to circuit	Ensure that voltage is within ±7% of rated voltage
		Cold area or draft hitting lamp	Protect lamp with enclosure; use a low temperature-rated ballast
	Preheat	Defective or old starter	Replace starter; lamp life is reduced when starter is not replaced
Lamp does not light or is slow starting	All types	Lamp failure	Replace lamp; test faulty lamp in another fixture. Check circuit fuse or CB; check voltage at fixture
		Loss of power to the fixture or low voltage	Troubleshoot fluorescent fixture when voltage is present at the fixture; replace broken or cracked holder; check for poor wire connection
	Preheat	Normal end of starter life	Replace starter
	Rapid-start	Failed capacitor in ballast	Replace ballast
		Lamp not seated in holder	Seat lamp properly in holder; in rapid-start circuits, the holder includes a switch that removes power when the lamp is removed due to high voltage present
Bulb ends remain lighted after switch is turned OFF	Preheat	Starter contacts stuck together	Replace starter; lamp life is reduced when starter is not replaced

...FLUORESCENT LAMP TROUBLESHOOTING GUIDE

Problem	Circuit Type	Possible Cause	Corrective Action
Short lamp life	All types	Frequent turning ON and OFF of lamps	Normal operation is based on one start per three hour period of operation time; short lamp life must be expected when frequent starting cannot be avoided
		Supply voltage excessive or low	Check the supply voltage against the ballast rating; short lamp life must be expected when supply voltage is not within ±7% of lamp rating
		Low ambient temperature; low temperature causes a slow start	Protect lamp with enclosure; use a low temperature-rated ballast
	Instant-start	One lamp burned out and other burning dimly due to series-start ballast circuit	Replace burned-out lamp; ballast is damaged when lamp is not replaced
		Wrong lamp type; may be using rapid-start or preheat lamp instead of instant-start	Replace lamp with correct type
Light output decreases	All types	Light output decreases over first 100 hours of operation	Rated light output is based on output after 100 hours of lamp operation; before 100 hours of operation, light output may be as much as 10% higher than normal
		Low circuit voltage	Check the supply voltage against the ballast rating; short lamp life and low light output must be expected when supply voltage is not within ±7% of lamp rating
		Dirt build-up on lamp and fixture	Clean bulb and fixture

Appendix

HID LAMP TROUBLESHOOTING GUIDE

Problem	Possible Cause	Corrective Action
Lamp does not start	Normal end of life operating characteristic	Replace with new lamp
	Loose lamp connection	Re-seat lamp in socket securely. Ensure lamp holder is rigidly mounted and properly spaced
	Defective photocell used for automatic turn ON	Replace defective photocell
	Low line voltage applied to circuit	Ensure that voltage is within ±7% of rated voltage
	Cold area or draft hitting lamp	Protect lamp with enclosure. Use a low temperature-rated ballast
	Defective ballast	Replace ballast
Lamp cycles ON and OFF or flickers when ON	Normal end of life operating characteristic	Replace with new lamp
	Poor electrical connection or loose bulb	Check electrical connections and socket contacts
	Line voltage variations	Ensure that voltage is within ±7% of rated voltage. Move lamps to separate circuit when lamps are on same circuit as high-power loads. High-power loads cause a voltage dip when turned ON. This voltage dip may cause the lamp to turn OFF
Short lamp life	Wrong wattage lamp or ballast	HID lamps and ballast must be matched in size. Lamp life is shortened when a large-wattage lamp is used. The same size and type ballast must be installed when replacing a ballast
	Power tap set too low for line voltage	Check ballast taps and ensure they are set for the correct supply voltage
	Defective sodium ballast starter	Replace starter
Low light output	Normal end of life operating characteristic	Replace with new lamp
	Dirty lamp or lamp fixture	Keep lamp fixture clean
	Early blackening of bulb caused by incorrect lamp size or ballast	Ensure that lamp size, lamp type, and ballast match

HAZARDOUS LOCATIONS—ARTICLE 500

Hazardous Location – A location where there is an increased risk of fire or explosion due to the presence of flammable gases, vapors, liquids, combustible dusts, or easily-ignitable fibers or flyings.

Location – A position or site.

Flammable – Capable of being easily ignited and of burning quickly.

Gas – A fluid (such as air) that has no independent shape or volume but tends to expand indefinitely.

Vapor – A substance in the gaseous state as distinguished from the solid or liquid state.

Liquid – A fluid (such as water) that has no independent shape but has a definite volume. A liquid does not expand indefinitely and is only slightly compressible.

Combustible – Capable of burning.

Ignitable – Capable of being set on fire.

Fiber – A thread or piece of material.

Flyings – Small particles of material.

Dust – Fine particles of matter.

Classes	Likelihood that a flammable or combustible concentration is present
I	Sufficient quantities of flammable gases and vapors present in air to cause an explosion or ignite hazardous materials
II	Sufficient quantities of combustible dust are present in air to cause an explosion or ignite hazardous materials
III	Easily ignitable fibers or flyings are present in air, but not in a sufficient quantity to cause an explosion or ignite hazardous materials

Divisions	Location containing hazardous substances
1	Hazardous location in which hazardous substance is normally present in air in sufficient quantities to cause an explosion or ignite hazardous materials
2	Hazardous location in which hazardous substance is not normally present in air in sufficient quantities to cause an explosion or ignite hazardous materials

Class I Division I:
- Spray booth interiors
- Areas adjacent to spraying or painting operations using volatile flammable solvents
- Open tanks or vats of volatile flammable liquids
- Drying or evaporation rooms for flammable vents
- Areas where fats and oils extraction equipment using flammable solvents is operated
- Cleaning and dyeing plant rooms that use flammable liquids that do not contain adequate ventilation
- Refrigeration or freezer interiors that store flammable materials
- All other locations where sufficient ignitable quantities of flammable gases or vapors are likely to occur during routine operations

Class II Division I:
- Grain and grain products
- Pulverized sugar and cocoa
- Dried egg and milk powders
- Pulverized spices
- Starch and pastes
- Potato and wood flour
- Oil meal from beans and seeds
- Dried hay
- Any other organic materials that may produce combustible dusts during their use or handling

Class III Division I:
- Portions of rayon, cotton, or other textile mills
- Manufacturing and processing plants for combustible fibers, cotton gins, and cotton seed mills
- Flax processing plants
- Clothing manufacturing plants
- Woodworking plants
- Other establishments involving similar hazardous processes or conditions

HAZARDOUS LOCATIONS

Class	Group	Material
I	A	Acetylene
	B	Hydrogen, butadiene, ethylene oxide, propylene oxide
	C	Carbon monoxide, ether, ethylene, hydrogen sulfide, morpholine, cyclopropane
	D	Gasoline, benzene, butane, propane, alcohol, acetone, ammonia, vinyl chloride
II	E	Metal dusts
	F	Carbon black, coke dust, coal
	G	Grain dust, flour, starch, sugar, plastics
III	No groups	Wood chips, cotton, flax, and nylon

GLOSSARY

A

abbreviation: A letter or combination of letters that represents a word or words.

AC voltage: Voltage that reverses its direction of flow at regular intervals.

adjustable motor base: A mounting base that allows a motor to be easily moved a short distance.

ambient temperature: The temperature of the air surrounding a piece of equipment.

ampere: The amount of electrons passing a predetermined point in a circuit in 1 sec.

analog display: An electromechanical device that displays measurements using the mechanical motion of a pointer.

analog multimeter: A meter that can measure two or more electrical properties and displays the measured properties using a pointer that moves along several calibrated scales.

apparent power (P_A): The product of the voltage and current in a circuit calculated without considering the phase shift that can be present between the voltage and current in the circuit.

arc blast: An explosion that occurs when the air surrounding electrical equipment becomes ionized and conductive.

arc blast hood: An eye and face protection device that consists of a flame-resistant hood and face shield.

arc flash: An extremely high-temperature discharge produced by an electrical fault in the air.

arc tube: The light-producing element of an HID lamp.

assembly drawing: A drawing that shows an entire machine or device with all detailed parts.

audible signal: Any sound that can be heard. Sound is produced by a series of pressure vibrations originating from a vibrating object.

automatic input: A switch independent of an individual or object that enters information or controls a circuit.

automatic transfer switch: An electrical device that transfers the load of a building from public utility circuits to the output of a standby generator during a power failure.

average voltage (V_{avg}): The mathematical mean of instantaneous voltage values in the sine wave; equal to 0.637 of the peak value.

B

ballast: An autotransformer that delivers a specific voltage and limits the electric current supplied to a lamp.

bar graph: A graph composed of segments that function similar to an analog pointer.

bayonet base: A lamp base that has two pins located on opposite sides.

branch circuit: The part of an electrical wiring system between the final set of circuit breakers or fuses and the fixtures and receptacles.

brush: The sliding contact that rides against the slip rings; used to connect the armature to the external circuit (load).

busway: A grounded metal enclosure that contains factory-mounted bare or insulated conductors.

C

cable: Two or more conductors grouped together within a common protective cover and used to connect individual components.

capacitance (C): The ability of a component or circuit to store energy in the form of an electric charge.

capacitive leakage current: Leakage current that flows through conductor insulation due to a capacitive effect.

capacitive reactance (X_C): The opposition to current flow by a capacitor.

capacitor: An electric device that stores electrical energy by means of an electrostatic field.

capacitor motor: A single-phase motor with a capacitor connected in series with the stator windings to produce phase displacement in the starting winding.

capacitor-start motor: A motor that has the capacitor connected in series with the starting winding.

carrier frequency: The frequency that controls the number of times the solid-state switches in the inverter section of a drive switch on and off.

caution: A signal word used to indicate a potentially hazardous situation that, if not avoided, may result in minor or moderate injury.

centrifugal force: The force that moves rotating bodies away from the center of rotation.

centrifugal switch: A switch that opens to disconnect the starting winding when the rotor reaches a preset speed and closes to reconnect the starting winding when the rotor speed drops below the preset value.

circuit breaker: An overcurrent protection device with a mechanical mechanism that may manually or automatically open a circuit when an overload condition or short circuit occurs.

clamp-on ammeter: A test instrument that measures current in a circuit by measuring the strength of the magnetic field around a conductor.

closed transition switch: An automatic transfer switch that closes one power source before opening another power source.

color rendering: The appearance of a color when illuminated by a light source.

color temperature: A description of the coolness or warmth of a light source; measured in Kelvin (K).

compact fluorescent lamp (CFL): A fluorescent lamp that has a smaller diameter than a conventional fluorescent lamp and a folded bulb configuration.

conductive leakage current: The small amount of current that normally flows through the insulation of a conductor.

conductor: A material that has little resistance and permits electrons to move through it easily.

conduit: A tube or pipe used to support and protect electrical conductors.

contactor: A control device that uses a small control current to energize or deenergize the load connected to it.

control-circuit transformer: A transformer that is used to step down the voltage to the control circuit of a system or machine.

control switch: A device that regulates the flow of current in a circuit.

copper-clad aluminum: A conductor that has copper bonded to an aluminum core.

cord: Two or more conductors in one cover used to deliver power to a load by means of a plug.

current (I): The amount of electrons flowing through an electrical circuit.

current unbalance (imbalance): The unbalance that occurs when current on each of the three power lines of a three-phase power supply are not equal.

D

danger: A signal word used to indicate an imminently hazardous situation that, if not avoided, will result in death or serious injury.

DC voltage: Voltage that flows in one direction only.

decision: An electronic determination of how an electrical circuit is to operate with a predetermined set of known conditions.

delta-connected three-phase motor: A motor in which each phase is connected end to end to form a completely closed-loop circuit.

delta connection: A connection that has each coil connected end to end to form a closed loop.

derating: The reduction of the total rated operating output of a device to allow for safe operation under abnormal environmental conditions.

dielectric material: A medium in which an electric field is maintained with little or no outside energy supply.

differential: The difference between the temperature at which a switch turns on a unit and temperature at which it turns off the unit.

digital display: An electronic device that displays measurements as numerical values.

digital multimeter (DMM): A meter that can measure two or more electrical properties and display the measured properties as numerical values.

diode: A semiconductor device that offers high opposition to current flow in one direction and low opposition to current flow in the opposite direction.

directional control valve: A valve that allows or prevents fluid flow to specific piping or system actuators.

disconnect switch: A switch that disconnects the supply of electric power from electrically powered devices such as motors and machines.

distribution substation: An outdoor facility located close to the point of electrical service use; used to change voltage levels, provide a central place for system switching, monitoring, and protection, and redistribute power.

double-pole (DP) switch: A control device that consists of two switches in one and is used for controlling two separate loads.

dual-function switch: A control switch that controls both the on and off control function.

dual inline-package (DIP) switch: A group of several identical manual electric switches that are packaged together in a standard switch body.

dual-voltage three-phase motor: A three-phase motor that operates at more than one voltage level.

duplex receptacle: An electrical contact device containing two receptacles.

E

electrical relay: A control device activated by electric power that opens, closes, or activates other mechanisms.

electrical test instrument: A device used to measure electrical quantities such as voltage (V), current (A), resistance (Ω), power (W), capacitance (f), and frequency (Hz).

electric arc: A discharge of electric current across an air gap.

electric discharge lamp: A lamp that produces light by means of an arc discharged between two electrodes.

electrolyte: A conducting medium in which the current flow occurs by ion migration.

electrolytic capacitor: A capacitor formed by winding two sheets of aluminum foil separated by pieces of thin paper impregnated with an electrolyte.

electromagnet: A device consisting of a core and coil that generates magnetism only when an electric current is present in the coil.

electromagnetic induction: The production of electricity by the interaction of a conductor cutting through a magnetic field.

electromagnetism: The magnetism produced when electric current passes through a conductor.

electromechanical relay: An electric device that uses a magnetic coil to open or close one or more sets of contacts.

elevation drawing: An orthographic view of a vertical surface without allowance for perspective.

emergency exit lighting system: A lighting system that is activated when a power failure occurs and provides adequate lighting for egress from a building.

enclosed-coil heating element: A coiled resistance wire sealed into an enclosure and fixed onto a supporting element.

equipment grounding conductor (EGC): An electrical conductor that provides a low-impedance ground path between electrical equipment and enclosures within a distribution system.

exit lighting system: Illuminated signage that indicates the safest and quickest path out of a building.

F

face shield: Any eye and face protection device that covers the entire face with a plastic shield and is used for protection from flying objects.

fault current: Any current that travels a path other than the normal operating path for which a system is designed.

feeder: A conductor between the service equipment and the last branch-circuit OCPD.

feeder panel: A junction box used to house electrical circuit breakers and main connections.

ferromagnetic metal: A material that has extremely high and variable magnetic susceptibility and is strongly attracted to a magnetic field.

field winding: A group of wires used to produce the magnetic field in the stator of a generator.

filament: A conductor with a resistance high enough to cause the conductor to heat.

first aid: Help for an injured individual immediately after an injury and before professional medical help arrives.

floor plan: A plan view of a structure that shows the arrangement of walls and partitions as they appear in an imaginary section taken horizontally and approximately 5′-0″ above floor level.

fluorescent lamp: A low-pressure discharge lamp in which ionization of mercury vapor transforms ultraviolet energy generated by the discharge into light.

footcandle (fc): The amount of light produced by a lamp (lumens) divided by the area that is illuminated.

four-way switch: A control device that is used in combination with two three-way switches to allow control of a load from three locations.

fuse: An overcurrent protection device with a fusible link that melts and opens the circuit when an overload condition or short circuit occurs.

G

generator: A device that converts mechanical energy into electrical energy by means of electromagnetic induction.

ground: An intentional or accidental electrical connection between a circuit or equipment and the earth or other conducting member.

grounded conductor: A conductor that has been intentionally grounded; typically known as the neutral conductor.

ground fault: Any amount of current above a certain level that may deliver a dangerous shock.

ground-fault circuit interrupter (GFCI): A fast-acting electrical device that automatically deenergizes a circuit by opening the circuit in response to the grounded current exceeding a predetermined value.

grounding: The connection of all exposed non-current-carrying metal parts to the earth.

grounding electrode conductor (GEC): A conductor that connects grounded parts of a power distribution system to an NEC® approved earth grounding system.

H

harmonic filter: A device used to reduce harmonic frequencies and total harmonic distortion in a power distribution system.

heating element: The portion of an electrical component that produces heat by electric current.

high-intensity discharge (HID) lamp: A lamp that produces light from an arc tube.

high-pressure sodium lamp: An HID lamp that produces light when current flows through sodium vapor under high pressure and high temperature.

I

ignitor: A component inside a lamp ballast that produces a high starting voltage.

illumination: The effect that occurs when light contacts a surface.

impulse transient voltage: A short-duration transient voltage commonly caused by a lightning strike, which results in a short, unwanted voltage placed on the power distribution system.

incandescent lamp: An electric lamp that produces light by means of the flow of current through a tungsten filament inside a gas-filled, sealed glass bulb.

inductance (L): The property of a circuit that causes it to oppose a change in current due to energy stored in a magnetic field.

inductive reactance (X_L): The opposition to the flow of alternating current in a circuit due to inductance.

input (switch): A device or component that controls the flow of electricity or information into a circuit.

instant-start circuit: A fluorescent lamp starting circuit that provides sufficient voltage to strike an arc instantly.

insulating matting: A floor covering that provides personnel with protection from electrical shock when working on live electrical circuits.

interface: Any device that allows two different types of components, voltage levels, voltage types, or systems to be interconnected.

invisible light: The portion of the electromagnetic spectrum on either side of the visible light spectrum.

isolated-ground receptacle: An electrical connection device in which the grounding terminal is isolated from the device yoke or strap.

K

K factor: The measure of the extra heat caused by harmonic distortion in a transformer.

knuckle thread: A rounded thread normally rolled from sheet metal and used in various forms for electric bulbs and bottle caps.

K-rated transformer: A transformer designed to withstand the extra heating effects caused by harmonic distortion.

K rating: A listed rating on a transformer that represents the ability of the transformer to operate properly with certain loads.

L

lamp: An output device that converts electrical energy into light.

lateral service: An electrical service in which service-entrance conductors are run underground from the utility service to the building.

leakage current: Current that leaves the normal path of current and flows through a ground wire.

light: The portion of the electromagnetic spectrum that produces radiant energy.

light-emitting diode (LED): A diode that emits light when forward direct current is applied.

light level sensor: A control device that measures the amount of ambient light and activates a control switch at a preset light level.

line diagram: A drawing that has a series of single lines (rungs) that indicate the logic of the control circuit and how the control devices are interconnected.

linear load: Any load in which current increases proportionately as voltage increases and current decreases proportionately as voltage decreases.

linear scale: A scale that is divided into equally spaced segments.

locked rotor torque: The torque a motor produces when the rotor is stationary and full power is applied to the motor.

lockout: The process of removing the source of electrical power and installing a lock that prevents the power from being turned on.

low-pressure sodium lamp: An HID lamp that operates at a low vapor pressure and uses sodium as the vapor.

lumen (lm): The unit used to measure the total amount of light produced by a light source.

M

magnet: A substance that produces a magnetic field and attracts iron.

magnetic motor starter: A starter that has an electrically operated switch (contactor) and includes motor overload protection.

magnetism: A force by which materials exert an attraction or repulsion on other materials.

main bonding jumper (MBJ): A connection at the service equipment that connects the equipment grounding conductor, grounding electrode conductor, and grounded conductor (typically the neutral conductor).

manual input: A switch that requires an individual to enter information or control a circuit.

mechanical input: A switch that detects the physical presence of an object.

megohmmeter: A high-resistance ohmmeter used to measure insulation deterioration on conductors by measuring high resistance values during high-voltage test conditions.

mercury-vapor lamp: An HID lamp that produces light by means of an electrical discharge through mercury vapor.

metal-halide lamp: An HID lamp that produces light by means of an electrical discharge through mercury vapor and metal halide in the arc tube.

motion sensor: A sensing device that detects the movement of a temperature variance and automatically switches when the movement is detected.

motor control center (MCC): A central location for troubleshooting and servicing motor control circuits.

motor control circuit: A circuit that controls a motor as required to start, stop, jog, increase speed, decrease speed, or reverse its direction of rotation.

motor drive: An electronic device that controls the direction, speed, torque, and other operating functions of an electric motor, in addition to providing motor protection and monitoring functions.

motor drive initial test: A test that verifies if a motor drive is operational.

motor drive, motor, and load test: A test used to verify that a motor drive and motor properly rotate the driven load.

motor power circuit: The section of an electrical motor circuit that delivers high voltage or current to an electric motor.

motor starter: An electrically operated switch (contactor) that includes motor overload protection.

multimeter: A portable test instrument that is capable of measuring two or more electrical quantities.

N

National Fire Protection Association (NFPA): A national organization that provides guidance in assessing the hazards of the products of combustion.

nonlinear load: Any load in which the current is not a pure proportional sine wave because current is drawn in short pulses rather than a smooth even sine wave.

nonlinear scale: A scale that is divided into unequally spaced segments.

normally closed (NC) valve: A valve that does not allow pressurized fluid out of the valve in the spring-actuated (normal) position.

normally open (NO) valve: A valve that allows pressurized fluid to flow out of the valve in the spring-actuated (normal) position.

O

Occupational Safety and Health Administration (OSHA): A federal agency that requires all employers to provide a safe working environment for their employees.

off-delay timer: A timer that does not start its timing function until the power is removed from the timer.

ohmmeter: A test instrument that measures resistance.

Ohm's law: The relationship between voltage (E), current (I), and resistance (R) in a circuit.

oil capacitor: A capacitor that uses oil or oil-impregnated paper as a dielectric.

on-delay timer: A time-based control device that delays it operation for a predetermined time after receiving a signal to activate or turn on.

open-coil heating element: An exposed coiled resistance wire fixed onto a supporting element.

open transition switch: An automatic transfer switch that opens one power source before closing another power source.

oscillatory transient voltage: A transient voltage that includes both positive and negative polarity values.

output (load): An electrical device that converts electrical energy into some other form of energy, such as light, heat, sound, linear motion, or rotary motion.

overcurrent: Electrical current in excess of the equipment limit, total amperage load of the circuit, or conductor or equipment rating.

overcurrent protective device (OCPD): A fuse or circuit breaker (CB) that disconnects or discontinues current flow when the amount of current exceeds the design load.

overcycling: The process of turning a motor on and off repeatedly.

overhead service: An electrical service in which service-entrance conductors are run from the utility pole through the air and to the building.

overload: The condition that occurs when the load connected to a motor exceeds the full-load torque rating of the motor.

overloaded motor: A motor with a current reading greater than 105% of nameplate current rating.

overvoltage: A voltage increase more than 10% above the normal rated line voltage for a period of longer than 1 min.

P

panelboard: A wall-mounted distribution cabinet containing a group of OCPDs and short-circuit protection devices for lighting, appliance, or power distribution branch circuits.

parallel connection: A connection with two or more devices connected so that there is more than one path for current flow.

peak-to-peak voltage (V_{p-p}): The voltage measured from the maximum positive alternation to the maximum negative alternation of a sine wave.

peak value: The maximum instantaneous value of a sine wave.

peak voltage (V_{max}, V_p): The maximum value of either the positive or negative alternation of a sine wave.

permanent magnet: A magnet that holds its magnetism for a long period of time.

personal protective equipment (PPE): Clothing and/or equipment worn by an individual to reduce the possibility of injury in the workplace.

phase shift: The state when voltage and current in a circuit do not reach their maximum amplitude and zero level simultaneously.

pictorial diagram: An electrical wiring sketch showing actual positions of system devices and wiring.

plot plan: A drawing that shows property lines of a building lot, elevation information, compass directions, lengths of property lines, and locations of structures to be built on the lot.

polarity: The positive (+) or negative (−) state of an object.

pole: The number of completely isolated circuits that a switch can switch.

position: Any of the locations within the valve in which a spool is placed to direct fluid through the valve.

power (P): The rate of doing work or using energy.

power factor (PF): The ratio of true power used in an AC circuit to apparent power delivered to the circuit.

power formula: The relationship between power (P), voltage (E), and current (I) in an electrical circuit.

preheat circuit: A fluorescent lamp starting circuit that heats the cathode before an arc is created.

pressure switch: An electric switch operated by pressure that acts on a diaphragm or bellows element.

primary division: A division of an analog display scale with a listed value.

proportional temperature control: A heating control function in which the heating element in a heating control circuit can produce any amount of heat from 0% to 100% of its rated value.

R

rapid-start circuit: A fluorescent lamp starting circuit that brings the lamp to full brightness in about 2 sec.

reactive power (VAR): Power supplied to a reactive load, such as a transformer coil, motor winding, or solenoid coil.

receptacle (outlet): A device used to connect equipment with a cord and plug to an electrical system.

rectifier: A device that converts AC voltage to DC voltage by allowing the voltage and current to flow in one direction only.

refractory: A heat-resistant ceramic material.

resistance (R): The opposition to the flow of electrons in a circuit; measured in ohms (Ω).

resistive circuit: A circuit that contains components that produce only resistance, such as heating elements and incandescent lamps.

root-mean-square voltage (V_{rms}): The AC voltage value of a sine wave that produces the same amount of heat in a pure resistive circuit as DC voltage of the same value.

S

safety glasses: An eye protection device with special impact-resistant glass or plastic lenses, reinforced frames, and side shields.

safety label: A label that indicates areas or tasks that can pose a hazard to personnel and/or equipment.

schematic diagram: A drawing that shows electrical system circuitry with symbols that depict electrical devices and lines representing the conductors.

secondary division: A division of an analog display scale that divides the primary divisions into halves, thirds, fourths, fifths, etc.

separately derived system (SDS): A system that supplies a facility with electrical power derived, or taken, from a transformer, storage batteries, a photovoltaic system, or a generator.

series connection: A connection that has two or more devices connected so there is only one path for current flow.

series-parallel circuit: A combination of series and parallel-connected components.

shaded-pole motor: A single-phase AC motor that uses a shaded stator pole for starting.

short circuit: An overcurrent that leaves the normal current-carrying path by going around the load and back to the power source or ground.

single-function switch: A switch that controls only the turning on or off of a load but not both.

single-line diagram: An electrical drawing that uses a single line and basic symbols to illustrate the current path, voltage values, circuit disconnects, OCPDs, transformers, and panelboards for a power distribution system.

single-pole (SP) switch: An electrical control device used to turn lights or appliances on and off from a single location.

single-voltage three-phase motor: A motor that operates at only one voltage level in a three-phase system.

sinusoidal waveform: A waveform that is not distorted during its rise from zero to peak value and back down to zero.

slip ring: A metallic ring connected to the end of an armature loop and is used to connect the induced voltage to a brush.

solenoid: An electromechanical device that converts electrical energy to linear mechanical force.

solid-state relay: A relay that uses electronic switching devices in place of mechanical contacts.

sound: A series of pressure variations that produce an audible signal.

split-phase motor: An AC motor that can run on one or more phases.

subdivision: A division of an analog display scale that divides the secondary divisions into halves, thirds, fourths, fifths, etc.

surface leakage current: Current that flows from areas on conductors where insulation has been removed to allow electrical connections.

switchboard: A panel or an assembly of panels containing electrical switches, meters, busses, and other overcurrent protection devices (OCPDs).

switchgear: A high-power electrical device that switches or interrupts devices or circuits in a building distribution system.

symbol: A graphic element that represents a quantity, unit, or component.

T

tagout: The process of placing a danger tag on the source of electrical power to indicate that the equipment may not be operated until the danger tag is removed by the individual who installed the tag.

tap: A connection point provided along a transformer coil.

temperature rise: The increase of equipment temperature over ambient temperature after the equipment is energized and loaded.

temporary magnet: A magnet that loses its magnetism as soon as the magnetizing force is removed.

thermal switch: A switch that operates its contacts when a preset temperature is reached.

three-way switch: A single-pole, double-throw (SPDT) switch.

throw: The number of closed switch positions per pole.

timer: A control device that uses a preset time period as part of a control function.

transformer: An electrical device that uses electromagnetism to change voltage from one level to another or to isolate one voltage from another.

transient voltage: A short, temporary, undesirable voltage in an electrical circuit, also known as a voltage spike.

transmission line: An aerial conductor that carries large amounts of electrical power at high voltages over long distances.

transmission substation: An outdoor facility located along a utility system that is used to change voltage levels, provide a central place for system switching, monitoring, and protection, and redistribute power.

troubleshooting: The systematic elimination of the various parts of a system or process to locate a malfunctioning part.

true power (P_T): The actual power used in an electrical circuit to produce work, such as heat, light, sound, and motion.

tungsten-halogen lamp: An incandescent lamp filled with a halogen gas such as iodine or bromine.

two-position temperature control: A heating control function in which the heating element can only be on or off.

two-way switch: A single-pole, single-throw (SPST) switch.

U

undervoltage: A voltage decrease more than 10% below the normal rated line voltage for a period of longer than 1 min.

utility: An organization that installs, operates, or maintains the electrical, communication, gas, water, or related services in a community.

utility plan: A drawing that indicates the location and intended path of utility lines such as electrical, water, sewage, gas, and communication cables.

V

visible light: The portion of the electromagnetic spectrum to which the human eye responds.

voltage (E): The amount of electrical pressure in a circuit; measured in volts (V).

voltage tester: An electrical test instrument that indicates the approximate voltage amount and type (AC or DC) in a circuit using a movable pointer or vibration.

voltage unbalance (imbalance): The unbalance that occurs when voltage at the terminals of an electric motor or other three-phase load are not equal.

W

warning: A signal word used to indicate a potentially hazardous situation that, if not avoided, could result in death or serious injury.

watt (W): The unit of electrical power equal to the power produced by a current of 1 A across a potential difference of 1 V.

watt density: The wattage concentration on the surface of a heating element.

way: The path of a piping port through a directional control valve.

wire: Any individual conductor.

wiring diagram: A drawing that indicates the connections of all devices and components in an electrical system.

wye-connected three-phase motor: A motor that has one lead end of each of the three phases (phases A, B, and C) internally connected to the other two phases.

wye connection: A connection that has one end of each coil connected together and the other end of each coil left open for external connections.

INDEX

Numbers in italics refer to figures.

A

abbreviations, *101*, 101
acceleration times, 212, *213*
AC generators, *3*, 3–5, *4*, *5*
across-the-line starting circuits, 258
actual component layouts, *261*
AC voltage, *68*, 68–70, *69*
adjustable motor bases, *217*, 217
advanced multimeters, *100*, 100
air conditioning system control circuits, 266–267
alternating current (AC), 75
ambient temperature, 147
amperage ratings, receptacle, *33*, 33
amperes, 74
analog displays, *105*, 105
analog multimeters, *104*, 104–105
analog scales, 105, *106*
apparent power (P_A), *80*, 80–81
arc blast hoods, 52, *53*
arc blasts, 48
arc flashes, 48
arc tubes, 188
assembly drawings, 36, *37*
audible signals, 161
automatic inputs, 93
automatic transfer switches, 22–23, *23*
average (AVE) modes, 144
average voltage (V_{avg}), 68, *69*

B

ballasts, 180
bar graphs, 106–107, *107*
basic first aid, 55–57
bayonet bases, 176, *177*
belt tension adjustment, *217*, 217
branch circuits, 26–31, *27*
break-before-make switches, 23
brushes, 2
building power distribution systems, 16–38, *17*
   automatic transfer switches, 22–23, *23*
   branch circuits, 26–31, *27*
   busways, *24*, 24
   conduit, 23, *24*
   diagrams, 18–20, *19*, 34, 36–37, *37*
   feeder panels, 23
   monitoring, 20, *22*, 22
   motor control centers (MCCs), 24–25, *25*, *26*
   panelboards, 26–31, *27*
   power requirements, 16
   protection, 20, 28, *29*
   receptacles, 31–34, *32*, *33*, *35*
   switchboards, 16
   switchgears, 16–18, *18*
   transformers, 18–20, *21*
   voltage-level requirements, 16
burns, 57
busways, *24*, 24
bypass (automatic transfer) switches, 22–23

C

cables, 29, *30*
capacitance (C), 77
capacitive leakage current, *120*, 120
capacitive loads, *81*
capacitive reactance (X_C), 77, *79*
capacitor motors, 223–226, *225*
capacitors, 77, *79*, 120
capacitor-start motors, *223*, 223–226
cardiopulmonary resuscitation (CPR), 56
carrier frequencies, 161
CAT ratings, 122–123, *124*
caution signal words, 44, *45*
centrifugal force, 218
centrifugal switches, 218–219
CFLs, 182
circuit breakers, 28, *29*, 61–63, *62*, 145
   measuring current of, 143–144, *144*
   testing, 146–147, *147*, 261–262
   troubleshooting, *244*, 245
circuit loading, 141, 143–144, *144*
circuit measurements, *75*, 75

circuits, 83–88, 141, 143–144
　measuring loads of, 141, 143–144, *144*
　parallel, 85–88, *86, 87*
　series, 83–85, *84*
　series-parallel, 88–89, *89*
　testing current of, 143–144, *144*
circuit sections, *91,* 91–93, 258, *260*
clamp-on ammeters, 111–114, *112, 113, 115, 143*
clamp-on meters, *100,* 100
closed transition switches, 23
color rendering, 188
color temperature, 180
compact fluorescent lamps (CFLs), 182
conductive leakage current, *120,* 120
conductors, 29–31, *30, 32*
conduit, 23, *24*
contact bounce, 107
contactors, 246–247, *247*
continuity test functions, 108–109, *110*
control circuits. *See* motor control circuits
control-circuit transformers, 161, *162,* 165, 245
　testing, 165, *166*
　troubleshooting, 245, *246*
control switches, 264–267, *265, 268*
copper-clad aluminum, 31
cords, 29, *30*
CPR, 56
current (I), 74–76
　alternating current (AC), 75
　direct current (DC), 75
　measurement and testing procedure, 143–144, *144*
　measurements, *75,* 75
　in parallel circuits, *87,* 88
　problems, 75–76, 153–155, *154*
　relationships, *8*
　in series circuits, *84,* 85
current imbalance (unbalance), 234–236, *235*
current measurements, 111–117, *112*
　clamp-on ammeters, *113,* 113–114, *115*
　in-line ammeters, 114, *116,* 116–117
current unbalance (imbalance), 153–155, *154,* 234–236, *235*

D

danger signal words, 44, *45*
danger tags, 53, *54*
DC voltage, *68,* 68
deceleration times, 212
decisions, *91,* 93
delay-on-operate timers, 270
delay-on-release timers, 270
delta-connected three-phase motors, 227–228, *229*
delta connections, *5,* 5

derating, 216
diagrams, 34, 36–37, *37*
dielectric material, 224
differential, 213
digital displays, 106, *107*
digital multimeters (DMMs), *100,* 100, *104,* 105–107, *143. See also* multimeters
diodes, 179
DIP switches, 281
direct current (DC), 75
directional control valves, 247
disconnects, *257,* 257–258
disconnect switches, *257,* 257–258
distribution lines, 8
distribution panel noise and vibration, 161
distribution panels, 141, 143–144, *144*
distribution substations, *6,* 8, *9*
distribution systems. *See* building power distribution systems; utility power transmission and distribution systems
DMMs. *See* digital multimeters (DMMs); multimeters
double-insulated electric safety tools, 59–61
drawings, 34, 36–37, *37*
dual-function switches, 266, *267*
dual inline-package (DIP) switches, 281
dual-voltage, delta-connected three-phase motors, 230–231, *231*
dual-voltage, wye-connected three-phase motors, 229–230, *230*
dual-voltage three-phase motors, 228
duplex receptacles, 31, *32, 131*

E

effective voltage, 3
EGCs, 141
electrical balance, 11, *13*
electrical measurement and test instruments, *100,* 100–101, 141, *143*
electrical prints, 34, 260
electrical relays, 267
electrical services, 9–15, *10*
　single-phase, three-wire, 9, *11*
　three-phase, four-wire, 10–13, *12, 14, 15*
electrical shock, 46–48, *47*
electrical system circuit sections, *91,* 91–93, 258, *260*
electrical test instrument terminology, *101,* 101
electrical warnings, 45, *46*
electric arcs, 48
electric circuit sections, *91,* 91–93
electric discharge lamps, 188
electric heating, 236–247, *237, 239*
　devices, 242–247, *243*
　heating elements. *See* separate entry

electric panel noise, 161
electrolytes, 224
electrolytic capacitors, 224
electromagnetic induction, 2
electromagnetic spectrum, *172,* 172
electromagnetism, 90, *91*
electromagnets, 2
electromechanical relays, 267, *268*
electronic overloads, 275, *276*
elevation drawings, 36, *37*
emergency exit lighting systems, 194–195, *195*
enclosed-coil heating elements, 236, 240, *241*
equipment grounding conductors (EGCs), 141
equipment safety. *See* property and equipment safety
exit lighting systems, *194,* 194
explosion warnings, 45, *46*
eye protection, 52–53, *53*

F

face shields, 52, *53*
fault current, 48
feeder panels, 23
feeders, 23
ferromagnetic metals, 89
fiberglass ladders. *See* nonconductive ladders
field windings, 2
filaments, 174
fire extinguishers, *58,* 58
first aid, 55–57
flame-resistant (FR) clothing, 51–53, *52*
floor plans, 36, *37*
floor receptacles, *131*
fluorescent lamps, 180–182, *181, 182, 183*
  compact fluorescent lamps (CFLs), 182
  troubleshooting, *183,* 183–188
footcandles (fc), 173
FR clothing, 51–53, *52*
fuses, 28, *29,* 61–63, *62,* 145
  measuring current of, 143–144, *144*
  testing, 145–146, *146,* 261–262
  troubleshooting, *244,* 244–245

G

GECs, 141
generators, 2–5, *3, 4, 5*
GFCIs, *32,* 33
grids, 8
ground, 140
grounded conductors, 141
ground-fault circuit interrupter (GFCI) receptacles, *131,* 132–134, *133*
ground-fault circuit interrupters (GFCIs), *32,* 33
ground faults, 33
grounding, 48, *49,* 58, *59*
grounding electrode conductors (GECs), 141

H

harmonic distortion, 155–161, *159*
  electric panel noise, 161
  harmonic filters, 159–160, *160*
  overheating of neutral conductors, 158, *159*
harmonic filters, 159–160, *160*
heat, 215
heating elements, 236–242
  calculations for, 238–239
  troubleshooting, 240–242, *241, 242*
  types of, 236, *238*
high-intensity discharge (HID) lamps, 188–193, *189*
high-pressure sodium lamp, *193,* 193
HVAC control circuits, 267–269, *268*

I

IEC standard 61010, 122–123, *124*
ignitors, 193
illumination, 173
impedance
  loads, *79*
  and Ohm's law, 78
improper phase sequences, 155, *156*
impulse transient voltage, 73, *74*
incandescent lamps, 174–178, *175, 177, 178, 179*
incoming power supply lines, 282–285, *283*
inductance (L), 77
inductive loads, *81*
inductive reactance (XL), 77, *79,* 218
infrared (IR) thermometers, *148,* 148
in-line ammeters, 114, *116,* 116–117
inputs. *See* switches (inputs)
instant-start circuits, 184–186, *186*
insulating matting, 53
interfaces, *91,* 93
invisible light, 172
isolated-ground receptacles, 31, *32,* 130, *131, 132*
I-trap filters. *See* harmonic filters

K

K factors, 162
knuckle threads, 176
K-rated transformers, 162–163, *163*
K ratings, 162

L

ladder (line) diagrams, 36, *37*
ladders, 59–61, *60*
lamps, 71, *72*, 173
lateral service, 9, *10*
leakage current, 119–121, *120*
LED lamps, 179, *180*
light, 172–174, *173*
light-emitting diodes (LEDs), 179, *180*
lighting transformers, 161, *162*
linear loads, 155, *159*
linear scales, *105*, 105
line diagrams, 36, *37*
load balancing, 11, *13*
load centers (panelboards), 26–31, *27*
loads (outputs), *91*, 92, 141, 143–144, *144*
   linear, 155, *159*
   nonlinear, 156–157, *157*, *159*
locked rotor torque, 223
locking receptacles, 134–135, *135*
lockout, 53, *54*, *56*
lockout/tagout procedures, 53–55, *54*, *56*
low-pressure sodium lamps, 190–191
lumens (lm), 173

M

magnetic motor starters, 273–275, *274*, *278*
magnetism, 89–90
magnets, 89
main bonding jumpers (MBJs), 141
make-before-break switches, 23
manual inputs, 92
maximum (MAX) modes, 144
MBJs, 141
MCCs, 24–25, *25*, *26*
measurement procedures
  for clamp-on ammeters, 114, *115*
  current, 114–117, *115*, *116*, 143–144, *144*
  for in-line ammeters, *116*, 116–117
  for megohmmeters, 121, *122*
  resistance, 110–111, *111*
  voltage, 117, *118*
  for voltage testers, 102–103, *103*. *See also* meter reading
mechanical inputs, 92
megohmmeters, 118–119, *119*, 121, *122*
  for motor insulation, 213
  usage, *100*, 100
mercury-vapor lamps, 191, *192*
metal-halide lamps, 191, *192*
meter reading, *105*, 105, *106*, 106, *107*
minimum (MIN) modes, 144
MIN MAX recording modes, 144
modules, 26
momentary power interruptions, *149*
monitoring devices, 20, *22*, 22
motor circuit connections, 279
motor control centers (MCCs), 24–25, *25*, *26*, *62*, 63
motor control circuits, *262*, 262–273, *264*
  control switches, 264–267, *265*, *268*
  relays, 267–269, *268*, *269*
  timers, 269–270, *270*, *271*
  troubleshooting, 270–273, *272*
motor drive initial tests, 286–287, *287*
motor drive, motor, and load tests, 287, *288*
motor drives, 275–278, *277*, *278*. *See also* troubleshooting: motor drives
motor drive speed control, 279, *280*
motor drive tests, 285–290, *287*, *288*
motor insulation, 213, *214*, *215*
motor overload protection, 276, *277*
motor power circuits, *256*, 256–262
  disconnect switches (disconnects), *257*, 257–258
  power circuit terminal identification, 258, *259*
  troubleshooting, 258–262
motors in HVAC systems, 210–236, *211*, *212*
  derating, 216, *217*
  environmental considerations for, 215–217
  malfunctions, 210–213
  mounting and positioning, *217*, 217
  temperature ratings for, *216*
  types of. *See* single-phase motors; three-phase motors
motor starters, 273–275, *274*, *278*
motor voltage variations, 72, *73*
motor winding resistance measurements, 108–109, *109*, 118–119, *119*
multimeters, *100*, 100, 103–117, *104*, *143*
  abbreviations, *101*, 101
  advanced, *100*, 100
  analog, *104*, 104–105, *106*
  clamp-on current probe accessories, *112*, 112–113
  current measurements, 111–117, *112*
    with clamp-on ammeters, *112*, *113*, 113–114, *115*, 143
    with in-line ammeters, 112, 114, *116*, 116–117
  digital (DMMs), *100*, 100, *104*, 105–107
  measurement categories, 122–123, *124*
  resistance measurements, 108–111, *109*, *110*, *111*
  symbols, 101, *102*
  voltage measurements, 117, *118*

N

National Electrical Code® (NEC®), 50
National Fire Protection Association (NFPA), 50
NC valves, 250
NEC®, 50
neutral conductors, 158, *159*, *160*

Index

neutral-to-ground voltage troubleshooting, 140–141, *142*
NFPA, 50
NFPA 70E®, 50
nonconductive ladders, 59–61, *60*
noncontact thermometers, 141, *143*
nonlinear loads, 156–157, *157, 159*
nonlinear scales, *105,* 105
nonlocking receptacles, 134–135, *135*
normally closed (NC) valves, 250
normally open (NO) valves, 250
NO valves, 250

O

Occupational Safety and Health Administration (OSHA), 50
OCPDs, 28, 258
off-delay timers, 270, *271*
ohmmeters, 110, 213
Ohm's law, *77,* 77–78
oil capacitors, 224
on-delay timers, 270, *271*
open-coil heating elements, 236
open transition switches, 23
oscillatory transient voltage, 73, *74*
OSHA, 50
outlets. *See* receptacles
outputs. *See* loads (outputs)
overcurrent, 28
overcurrent and overloads, 75
overcurrent protective devices (OCPDs), 28, 258, 293
overcycling, 212–213
overhead service, 9, *10*
overloaded motors, 289
overload monitoring devices, 274, *275*
overload protection, 61–63, *62*
overloads, 28, 215–216, *216,* 275, *276*
overvoltage, 70, *149*

P

panelboards, 26–31, *27*
parallel circuits, 85–88, *86, 87*
parallel connections, 85, *86*
peak-to-peak voltage ($V_{p\text{-}p}$), 68, *69*
peak value, 3
peak voltage (V_{max}, V_p), 68
permanent magnets, 89, *90*
personal protective equipment (PPE), 48–53, *50*
personal safety, 44–57
 arc flashes and arc blasts, 48
 electrical shock, 46–48, *47*
 first aid, 55–57
 lockout/tagout procedures, 53–55, *54, 56*
 NFPA 70E®, 50

personal protective equipment (PPE), 48–53, *50*
 safety labels, 44–45, *45, 46*
phase sequences, 155, *156*
phase shifts, *81,* 81–82
pictorial diagrams, 36, *37*
plot plans, 36, *37*
plug-in units, *26*
plugs, *33,* 34, *35*
polarity, 68
positions, 248–249, *249*
power (P), *8,* 8, 78–82, *80*
 in parallel circuits, *87,* 88
 in series circuits, *84,* 85
power circuits. *See* motor power circuits
power circuit terminal identification, 258, *259*
power distribution systems. *See* building power distribution systems; utility power transmission and distribution systems
power factor (PF), 81, *82*
power formulas, 82, *83*
power interruptions, *149*
power lines, *7,* 7–8
power panels, 141, 143–144, 161
power production, 2–5
power protection devices, 20, 28, *29*
power quality meters, 141, *143*
power quality problems, 148–161, *149*
 current unbalance, 153–155, *154*
 harmonic distortion, 155–161, *159*
  electric panel noise, 161
  overheating of neutral conductors, 158, *159*
 improper phase sequences, 155, *156*
 voltage drop, *149,* 150, *151*
 voltage unbalance, *152,* 152–153, *153*
power ratings, *2,* 2
power requirements, 16
power supply lines, 282–285, *283*
power transformers, 161, *162*
power transmission systems, 5–8, *6*
PPE. *See* personal protective equipment (PPE)
preheat circuits, 184, *185*
pressure switches, 263
primary divisions, 105, *106*
print component layouts, *261*
property and equipment safety, 57–63
 circuit breakers, 61–63, *62*
 double-insulated electric safety tools, 59–61
 fire extinguishers, *58,* 58
 fuses, 61–63, *62*
 grounding, 58, *59*
 overload protection, 61–63, *62*
proportional temperature control, 245
pump motor control, 264, *266,* 266

R

rapid-start circuits, 186–188, *187*
reactive power (VAR), *80,* 80
receptacles, 31–34, *32,* 130–141, *131*
    120/208 V, 137–139, *138*
    120 V, 134–135, *135*
    208 V, 136–137, *137*
    configurations, *33, 35*
    ground-fault circuit interrupter (GFCI), *131,* 132–134, *133*
    isolated-ground, 130, *131, 132*
    locking, 134–135, *135*
    nonlocking, 134–135, *135*
    single-phase, 134–135, *135,* 136–139, *137, 138*
    special-use, *136,* 136
    standard, 130, *131, 132*
    three-phase, 139, *140*
rectifiers, 70
reed switches, *90*
refractory, 236
relays, 267–269, *268, 269*
resistance (R), 76–77, 108–111, *109, 110, 111*
    in parallel circuits, 86–87, *87*
    in series circuits, 83, *84*
resistance heating element systems, 61, *62,* 70–71, *71*
resistive circuits, 76–77
resistive loads, 79, *81*
root-mean-square voltage (V_{rms}), 3, 68, *69*
rubber insulating gloves, 50–51, *51, 54*

S

safety. *See* personal safety; property and equipment safety
safety circuits, 185
safety glasses, 52, *53*
safety labels, 44–45, *45, 46*
schematic diagrams, 36, *37*
SDSs, 140
secondary divisions, 105, *106*
separately derived systems (SDSs), 140
series circuits, 83–85, *84*
series connections, *83,* 83
series-parallel circuit applications, 88–89
series-parallel circuits, 88–89, *89*
services. *See* electrical services
shaded-pole motors, 218
shock, 56–57. *See also* electrical shock
short circuits, 28, 76
single-function switches, 266, *267*
single-line diagrams, 18–19, *19,* 36, *37*
single-phase, three-wire services, 9, *11*
single-phase AC generators, *3,* 3, *4*
single-phase AC voltage, 68, *69*

single-phase motors, 218–226
    capacitor, 223–226, *225*
    capacitor-start, *223,* 223–226
    shaded-pole, 218
    split-phase, 218–223, *219, 221*
single-phase receptacles, 134–135, *135,* 136–139, *137, 138*
single receptacles, *131*
single-voltage three-phase motors, 227
sinusoidal waveforms, 155
slip rings, 2
solenoid-operated directional control valves, 247–251, *248*
solenoids, 247, *251*
solid-state relays, 267
sound, 161
special-purpose receptacles, *131*
special-use receptacles, *136,* 136
split-phase motors, 218–223, *219, 221*
split-wired duplex receptacles, *131*
standard receptacles, 31, *32,* 130, *131, 132*
step-up/step-down transformers, 5, *6, 7,* 161, *162*
subdivisions, 105, *106*
substations, 6–7, *8, 9*
surface leakage current, *120,* 120–121
surges. *See* transient voltage
sustained power interruptions, *149*
switchboards, 16
switches (inputs), *91,* 92–93
switchgears, 16–18, *18*
symbols, 34, 101, *102*

T

tagout, 53, *54*
tagout/lockout procedures. *See* lockout/tagout procedures
taps, 70
temperature problems, 147–148
temperature rise, 147–148
temporary magnets, 90
temporary power interruptions, *149*
testing procedures
    circuit breakers, 146–147, *147*
    current unbalance, *154,* 154–155
    fuses, 145–146, *146*
    ground-fault circuit interrupter (GFCI) receptacles, *133,* 133–134
    harmonic distortion, 158
    phase sequences, 155, *156*
    receptacles. *See specific types under* receptacles
    transformers, 164–165
    voltage drop, 150, *151*
test instruments. *See* electrical measurement and test instruments

thermal imagers, *148,* 148
thermal switches, 219–220, *220, 222,* 222–223
three-phase, four-wire services, 10–13, *12, 14, 15*
three-phase AC generators, *5,* 5
three-phase AC voltage, 68, *69*
three-phase motors, 226–236, *227*
  delta-connected, 227–228, *229*
  dual-voltage, 228
  dual-voltage, delta-connected, 230–231, *231*
  dual-voltage, wye-connected, 229–230, *230*
  single-voltage, 227
  troubleshooting, 231–236, *232*
  wye-connected, 227, *228*
three-phase receptacles, 139, *140*
timers, 269–270, *270, 271*
tools, 59–61
transfer switches, 22–23, *23*
transformers, 6–7, *7,* 18–20, 161–165, *162*
  connection types, 20, *21*
  isolation, 18
  K-rated, 162–163, *163*
  overloading, 163, *164*
  single-line diagrams for, 18–19, *19*
  testing, 163–165, *166*
  troubleshooting, 245, *246*
transient voltage, 72–73, *74, 149*
transmission lines, *7,* 7–8
transmission line voltages, *8,* 8
transmission substations, *6,* 6–7
transmission systems, 5–8, *6*
trap filters. *See* harmonic filters
troubleshooting, 258
  control circuits, 270–273, *272*
  motor drives, 279–290
    circuit breakers, 284, *285*
    fuses, 284, *285*
    gathering information, 280–281
    incoming power supply lines, 282–285, *283*
    initial tests, 286–287, *287*
    inspections, 281, *282*
    motor drive, motor, and load tests, 287, *288*
    tests, 285–290, *287, 288*
    voltage conditions, 284, *285*
  motor power circuits, 258–262
true power (P_T), 79, *80*
tungsten-halogen lamps, 178–179, *179*
two-position temperature control, 245

U

undervoltage, 70, *149*
utilities, 5
utility plans, 36, *37*

utility power transmission and distribution systems, 5–15, *6*
  distribution lines, 8
  distribution substations, 8, *9*
  services. *See* electrical services
  transmission lines, 7–8
  transmission substations, 6–7

V

visible light, *172,* 172
voltage (E), 68–73
  alternating current (AC), *68,* 68–70, *69*
  direct current (DC), *68,* 68
  measurements, *75,* 75
  overvoltage, 70, *149*
  in parallel circuits, *87,* 87
  problems, 70–73
  relationships, *8,* 8
  in series circuits, *84,* 84
  transient voltage, 72–73, *74, 149*
  troubleshooting, 140–141, *142*
  undervoltage, 70, *149*
voltage drop, *149,* 150, *151*
voltage fluctuations, *149*
voltage imbalance (unbalance), 232–234, *233, 234*
voltage-level requirements, 16
voltage measurements, 117, *118*
voltage sags, *149*
voltage spikes, 72–73, *74, 149*
voltage swells, *149*
voltage testers, *100,* 100–103
voltage unbalance (imbalance), *152,* 152–153, 232–234, *233*
  in motors, 232–234, *233, 234*
  in power distribution systems, *152,* 152–153, *153*

W

warning signal words, 44, *45*
watt (W), 2, 178, 208
watt density, 236
ways, *249,* 249
wires, 29, *30*
wiring diagrams, 36, *37*
wiring methods, 158, *160*
wye-connected three-phase motors, 227, *228*
wye connections, *5,* 5

USING THE *ELECTRICAL SYSTEMS FOR FACILITIES MAINTENANCE PERSONNEL* INTERACTIVE CD-ROM

Before removing the Interactive CD-ROM from the protective sleeve, please note that the book cannot be returned for refund or credit if the CD-ROM sleeve seal is broken.

Windows System Requirements

To use this CD-ROM on a Windows® system, your computer must meet the following minimum system requirements:
- Microsoft® Windows® 7, Windows Vista®, or Windows® XP operating system
- Intel® 1.3 GHz processor (or equivalent)
- 128 MB of available RAM (256 MB recommended)
- 335 MB of available hard disk space
- 1024 × 768 monitor resolution
- CD-ROM drive (or equivalent optical drive)
- Sound output capability and speakers
- Microsoft® Internet Explorer® 6.0 or Firefox® 2.0 web browser
- Active Internet connection required for Internet links

Macintosh System Requirements

To use this CD-ROM on a Macintosh® system, your computer must meet the following minimum system requirements:
- Mac OS X 10.5 (Leopard) or 10.6 (Snow Leopard)
- PowerPC® G4, G5, or Intel® processor
- 128 MB of available RAM (256 MB recommended)
- 335 MB of available hard disk space
- 1024 × 768 monitor resolution
- CD-ROM drive (or equivalent optical drive)
- Sound output capability and speakers
- Apple® Safari® 2.0 web browser or later
- Active Internet connection required for Internet links

Opening Files

Insert the Interactive CD-ROM into the computer CD-ROM drive. Within a few seconds, the home screen will be displayed allowing access to all features of the CD-ROM. Information about the usage of the CD-ROM can be accessed by clicking on Using This Interactive CD-ROM. The Quick Quizzes®, Illustrated Glossary, Flash Cards, Virtual Meters, Troubleshooting Simulations, Media Clips, and ATPeResources.com can be accessed by clicking on the appropriate button on the home screen. Clicking on the ATP web site button (www.go2atp.com) accesses information on related educational products. Unauthorized reproduction of the material on this CD-ROM is strictly prohibited.

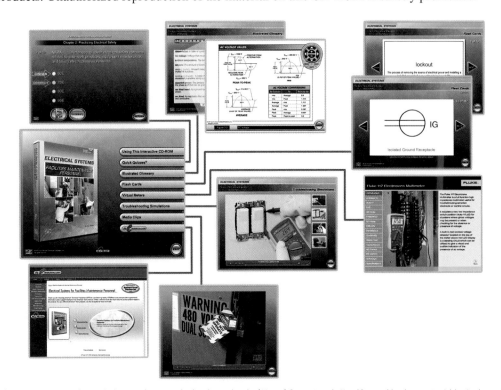

Microsoft, Windows, Windows Vista, PowerPoint, and Internet Explorer are either registered trademarks or trademarks of Microsoft Corporation in the United States and/or other countries. Adobe, Acrobat, and Reader are either registered trademarks of Adobe Systems Incorporated in the United States and/or other countries. Intel is a registered trademark of Intel Corporation in the United States and/or other countries. Firefox is a registered trademark of Mozilla Corporation in the United States and other countries. Apple, Macintosh, and Safari are registered trademarks of Apple, Inc. PowerPC is a registered trademark of International Business Machines Corporation. Quick Quiz, Quick Quizzes, and Master Math are either registered trademarks or trademarks of American Technical Publishers, Inc.